Isogeometric Analysis

for Engineers via MATLAB

Isogeometric Analysis

for Engineers via MATLAB

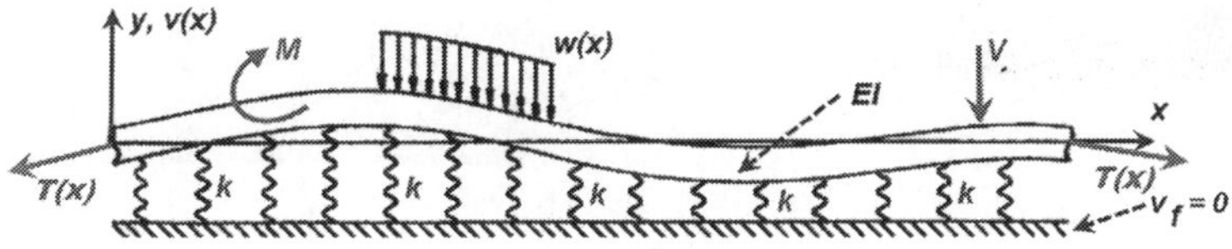

Ed Akin

Rice University, USA

World Scientific

NEW JERSEY • LONDON • SINGAPORE • BEIJING • SHANGHAI • HONG KONG • TAIPEI • CHENNAI

Published by

World Scientific Publishing Co. Pte. Ltd.

5 Toh Tuck Link, Singapore 596224

USA office: 27 Warren Street, Suite 401-402, Hackensack, NJ 07601

UK office: 57 Shelton Street, Covent Garden, London WC2H 9HE

Library of Congress Cataloging-in-Publication Data
Names: Akin, J. E., author.
Title: Isogeometric analysis for engineers via MATLAB / Ed Akin.
Description: New Jersey : World scientific, 2025. | Includes bibliographical references and index.
Identifiers: LCCN 2024002392 | ISBN 9789811288531 (hardcover) |
ISBN 9789811288548 (ebook) | ISBN 9789811288555 (ebook other)
Subjects: LCSH: Isogeometric analysis. | Engineering mathematics. | MATLAB.
Classification: LCC TA347.G46 A45 2025 | DDC 620.001/51--dc23/eng/20240907
LC record available at https://lccn.loc.gov/2024002392

British Library Cataloguing-in-Publication Data
A catalogue record for this book is available from the British Library.

For any available supplementary material, please visit
https://www.worldscientific.com/worldscibooks/10.1142/13737#t=suppl

Desk Editors: Balasubramanian Shanmugam/Steven Patt

Typeset by Stallion Press
Email: enquiries@stallionpress.com

Printed in Singapore

In Memory of
Professor J. Tinsley Oden
1936–2024
A friend and a scholar, second to none.

To

My wife, Tracy Akin, for her loving care and support during the long process of completing this work.

Preface

Isogeometric analysis (IGA) is one of the most popular new numerical analysis techniques being used worldwide by engineers and scientists. The method originated almost 25 years ago with the seminal works of Professor Thomas J.R. Hughes and his coworkers. Starting with Tom's excellent leadership role, and after many publications in finite element analysis (FEA), which applied his knowledge of advanced mathematics, B-splines, and computer-aided geometric design (CAGD) software, Tom questioned if the same polynomials forms used in CAGD to create exact engineering shapes could also be used in engineering analysis similar to the way in which interpolation polynomials are used in finite element analysis. If so, that would avoid the time-consuming mesh generation required by FEA and would yield improved accuracy over finite element analysis methods. Furthermore, success would allow the vast suite of FEA algorithms to be a sub-set of the new technology.

Professor Hughes' quest was indeed very successful and his team's seminal work on IGA (Cottrell *et al.*, 2007, 2009) has been applied to almost all of the engineering and science analysis fields where finite element analysis was the previous tool of choice. This is particularly true for fields that involve geometry changes since the prior time-consuming regeneration of finite element meshes has simply been replaced by modifications of the NURBS that defined the previous geometry.

Finite element studies usually require local refinements of the mesh in certain regions to provide the necessary accuracy and to reduce error estimates. The concept of local refinements was largely missing in the CAGD tools, and the IGA needs for such refinements helped lead to the use of T-splines which are now available as CAGD and IGA tools.

Isogeometric analysis relies heavily on the use of B-splines, NURBS (**N**on-**U**niform **R**ational **B**-**S**plines), and T-splines. The early developers of IGA had very strong backgrounds in applied mathematics and the mathematical evolution of NURBS. However, most BS-level engineers and scientists lack such a strong mathematics background. Thus, many readers of this book may start out thinking that the acronym **nurbs** has the alternate interpretation of **n**obody-**u**nderstands-**r**ational-**b**-**s**plines. Herein, the non-essential mathematical details about NURBS will be largely omitted but illustrated in figures with citations to published references and interactive web demonstrations of their most important properties. Also, matrix formulations more familiar to engineers will be used where practical.

Once the minimal foundational knowledge of the polynomial types used in IGA has been established, it will be applied to studies where FEA has proven to be so powerful. Examples include heat transfer, lubrication, stress analysis, structural mechanics, and vibrations. The IGA matrix notations to be developed are valid in all spatial dimensions (including space-time) and their applications will mainly be covered in one dimension and two dimensions.

The bibliography contains references that are primarily related to engineering aspects of IGA. Therein, I have short-changed Professor Hughes by omitting most of his numerous foundational papers on finite element analysis and isogeometric analysis. With modern search techniques, I thought engineers could easily find his important publications related to their fields of interest.

J. Ed Akin, Ph.D. P.E.
Rice University, Houston, Texas

About the Author

John Edward Akin is a Professor in the Department of Mechanical Engineering and a Professor in the Department of Computational and Applied Mathematics at Rice University in Houston, TX. At Rice, he has served as Chair of the M.E. Department and as Speaker of the Faculty. Dr. Akin received his B.S. and M.S. from Tennessee Technological University and his Ph.D. in Engineering Mechanics from Virginia Polytechnic Institute. Dr. Akin is a Fellow of the American Society of Mechanical Engineers (ASME) and a member of the American Society of Civil Engineers and the Society of Petroleum Engineers. He is a Registered Engineer and serves on the editorial board of the journal for *Multiscale and Multidisciplinary Modeling, Experiments and Design.* His areas of interest include numerical simulations and applications of computational mechanics in engineering.

Contents

IGA Notations

Acronyms

AMG	Automatic Mesh Generator
AP	Analysis Control Points
ASTS	Analysis Suitable T-Spline
BEM	Boundary Element Method
BRep	Computer-Aided Geometric Design
B-Spline	Basis function spline
BVP	Boundary Value Problem
CAD	Computer-Aided Design
CAE	Computer-Aided Engineering
CAGD	Computer-Aided Geometric Design
CAM	Computer-Aided Manufacturing
CN	Control Net (Control Mesh)
CP	Geometry Control Points
CSG	Constructive Solid Geometry
DOF	Degree Of Freedom
Element	Classic FEA Element Local Interpolation Domain
FDM	Finite Difference Method
FEA	Finite Element Analysis
FEM	Finite Element Method
F95	Fortran 1995 Standard
F23	Fortran 2023 Standard
GQ	Gaussian Quadrature
GUI	Graphical User Interface

HC	Homogeneous (Projective) Coordinates
iElement	IGA basis functions sub-region convertible to iGe for FEA, 'element'
IGA	Isogeometric Analysis
iGe	Extracted basis functions similar to an FEA element, IGA 'extract'
IVP	Initial Value Problem
LBB	The Ladyzhenskaya–Babuška–Brezzi inf-sup condition
LSQ	Least Squares
MWR	Methods of Weighted Residuals
NURBS	Non-Uniform Rational B-Splines
ODE	Ordinary Differential Equation
OOP	Object-Oriented Programming
OURBS	Open Uniform Rational B-Splines
PB	Point-Based spline
PDE	Partial Differential Equation
STL	Stereolithography
TSM	T-Spline Mesh (File type)

Roman letters

a, b, c	Algorithm natural coordinates, $-1 \leq a, b, c \leq 1$, in 1D, 2D, 3D
$\boldsymbol{B}^r$	Rectangular strain–displacement in a sub-region or 'iElement'
$B_{i,p}(u)$	The ith Bernstein basis function of degree p at point u, $1 \leq i \leq p+1$
C^0	Only the function value is continuous at the point
C^1	Function value and slope are continuous at the point
C^2	Function value, slope, and second derivative are continuous at the point
C^n	Continuity through the nth derivative, $\partial^n/\partial u^n$, of adjacent functions
$\boldsymbol{C}(u)$	B-spline curve
$\boldsymbol{c}$	System resultant source vector (column matrix)

$\boldsymbol{c}_s^{\boldsymbol{r}}$	Source vector for IGA sub-region, or 'iElement' (column matrix)
$[\boldsymbol{H}(u)]$	General row matrix of non-specific basis functions of degree p
$H_{i,p}(u)$	The ith Hermite polynomial of degree p at parametric point u
iff	If and only if
$L_{i,p}(u)$	The ith Lagrange polynomial of degree p at parametric point u
m	Multiplicity of a single knot
$N_{i,p}^{(d)}(d)$	The ith basis function of degree p in direction d, $d = r$ or s or t
$N_{i,p}(u)$	The ith B-spline basis function of degree p at point u, $1 \leq i \leq p+1$
$N_{i,j}^{p,q}(r,s)$	The ith, jth B-spline basis function product of degree p, q at point r, s
$N_{i,j,k}^{p,q,n}(r,s,t)$	The ith, jth, kth B-spline basis function of degree p, q, n at point r, s, t
n	The number of Bernstein basis functions and CP of a B-spline curve
n_{br}	Number of basis functions in parametric r-direction, n_br
n_{bs}	Number of basis functions in parametric s-direction, n_bs
n_{bt}	Number of basis functions in parametric t-direction, n_bt
n_{cr}	Number of analysis control points in parametric r-direction, n_br + 1
n_{cs}	Number of analysis control points in parametric s-direction, n_bs + 1
n_{ct}	Number of analysis control points in parametric t-direction, n_bt + 1
n_g	Number of generalized DOF per CP or node point, n_g
n_{kr}	Number of knot spans (iElement) in the r-parametric direction, n_kr
n_{ks}	Number of knot spans (iElement) in the s-parametric direction, n_ks

n_{kt}	Number of knot spans (iElement) in the t-parametric direction, n_kt
n_M	Multiplicity of repeated end knots n_M
n_{AP}	Number of analysis control points, n_AP
n_{CP}	Number of geometry control points $= n+1$, n_cp
n_{IP}	Number of interpolation points, n_ip
n_p	Number of parametric space dimensions, n_p
n_{pc}	Number of collocation points, n_pc
n_{qp}	Number of quadrature points, n_qp
n_s	Number of physical space dimensions, n_s
$\boldsymbol{P}$	Array of IGA control point values
p	Degree of a polynomial
$R_{i,p}(u)$	The ith NURBS basis function of degree p at point u, $1 \leq i \leq p+1$
$R_{i,j}^{p,q}(r,s)$	The ith, jth NURBS basis functions product of degree p, q at point r, s
$R_{i,j,k}^{p,q,n}(r,s,t)$	The ith, jth, kth NURBS basis function of degree p, q, n at point r, s, t
$\boldsymbol{S}$	System stiffness matrix
$\boldsymbol{S}(u,v)$	B-spline surface
$\boldsymbol{s}^e$	iElement stiffness, or conduction, square matrix
$\boldsymbol{s}^f$	Elastic foundation stiffness matrix or convection square matrix
$\boldsymbol{s}^r$	IGA sub-region 'iElement' or an iGe stiffness, or conduction, matrix
$\boldsymbol{T}$	Surface traction vector (force per unit area)
T	Temperature
r, s, t	Algorithm unit coordinates, $0 \leq r, s, t \leq 1$, in 1-, 2-, and 3-D
u	Unit coordinates space $0 \leq u \leq 1$ for CAGD, FEA, IGA in 1D
$\boldsymbol{V}(r,s,t)$	B-spline volume

Double struck letters

$\mathbb{A}^d$	Analysis knot vector in the parametric direction $d = 1$ to 4 (space-time)

$\mathbb{A}n$	Analysis numeric knot vector $\mathbb{A}n = \{n_1, n_2, \ldots, n_{m=n+p+1}\}$, $[n_i n_{i+1})$
$\mathbb{A}u$	Analysis unit knot vector $\mathbb{A}n \Rightarrow \mathbb{A}u = \{u_1 = 0, u_2, \ldots, u_{m=n+p+1} = 1\}$
$\mathbb{A}_L$	Analysis knot vector length, $m = n + p + 1$
$\mathbb{A}r$	Analysis knot vector in the parametric r-direction
$\mathbb{A}s$	Analysis knot vector in the parametric s-direction
$\mathbb{A}t$	Analysis knot vector in the parametric t-direction
$\mathbb{I}_k$	Integer control mesh segment k for knot vectors $\mathbb{A}$ or $\mathbb{K}$
$\mathbb{I}r$	Integer control space in the parametric r-direction
$\mathbb{I}s$	Integer control space in the parametric s-direction
$\mathbb{J}$	Jacobian matrix from an integer space to IGA space
$\mathbb{J}_{a:b}$	Jacobian matrix from space a into space b
$\mathbb{J}_{p:ks}$	Jacobian matrix from physical space into knot span space
$\mathbb{J}_{ks:gq}$	Jacobian matrix from knot span space into (Gaussian) quadrature space
$\mathbb{K}$	Geometric knot vector for $\mathbb{K}u = \{u_1, u_2, \ldots, u_{m=n+p+1}\}$, $[u_i u_{i+1})$
$\mathbb{K}_L$	Geometric knot vector length, $m = n + p + 1$
$\mathbb{K}n$	Geometric numeric knot vector $\mathbb{K}n = \{n_1, n_2, \ldots, n_{m=n+p+1}\}$, $[n_i n_{i+1})$
$\mathbb{K}r$	Geometric knot vector in the parametric r-direction
$\mathbb{K}s$	Geometric knot vector in the parametric s-direction
$\mathbb{K}t$	Geometric knot vector in the parametric t-direction
$\mathbb{R}$	Real numbers

Script letters

$\mathcal{B}$	A boundary differential operator in the IGA LSQ MWR
$\mathcal{L}$	A domain linear differential operator in the IGA LSQ MWR
$\mathcal{N}$	A domain nonlinear differential operator in the IGA LSQ MWR
$\mathcal{R}$	Total residual error in the IGA LSQ MWR, $\mathcal{R} = \mathcal{R}_\Omega + \mathcal{R}_\Gamma$
$\mathcal{R}_\Omega$	Residual error in the domain, Ω
$\mathcal{R}_\Gamma$	Residual error from the boundary, Γ

Greek letters

$\boldsymbol{\delta}$	All of the degrees of freedom in the system
$\boldsymbol{\delta}^R$	Degrees of freedom in an IGA sub-region, $\boldsymbol{\delta}^R \subset \boldsymbol{\delta}$
$\boldsymbol{\delta}^b$	DOF on a boundary of an IGA 'iElement' (or element), $\boldsymbol{\delta}^b \subset \boldsymbol{\delta}^e$
$\boldsymbol{\delta}^e$	Degrees of freedom in an IGA 'iElement' (or element), $\boldsymbol{\delta}^e \subset \boldsymbol{\delta}^r$
δ_k	The kth degree of freedom in the system vector, $\boldsymbol{\delta}$, or a sub-set
Δu_i	Span in a knot vector, $u_{i+1} - u_i \equiv \Delta u_i$
ε^r	Strain components in an IGA sub-region, 'iElement', or FEA element
ε_0	Initial strain components
Γ	Boundary of domain Ω, $\Gamma = \Gamma_0 \cup \Gamma_D \cup \Gamma_N \cup \Gamma_M$
Γ_D	Portion of Γ with Dirichlet boundary condition
Γ_N	Portion of Γ with Neumann boundary condition
Γ_M	Portion of Γ with mixed (Robin) boundary condition
Γ_0	Portion of Γ with natural (default) boundary condition
Ω	Domain of a partial differential equation
Ω^e	Domain of an IGA 'iGe', or FEA element

$\Omega^{(r)}$ Sub-region (r) of Ω defined in ⊞ or approximated in ⊡, $\Omega^e \subseteq \Omega^{(r)}$

$\hat{\Omega}$ Parametric region mapped to physical space Ω, requires a Jacobian

$\hat{\Omega}^{(r)}$ Parametric region mapped to $\Omega^{(r)}$, requires a Jacobian

$\Gamma^{(r)}$ Boundary of sub-region (r) in ⊞ or ⊡

ξ, η, ζ Natural coordinates, $-1 \leq \xi, \eta, \zeta \leq 1$, in 1-, 2-, 3-D in CAGD or IGA

Symbols

⊞ IGA region defined by B-splines, NURBS, or T-splines

⊡ Parametric IGA 'iElement' region, or FEA element sub-set ⊡ ⊂ ⊞

$\forall$ Mathematical shorthand meaning 'for all'

$\in$ Mathematical shorthand meaning 'is contained in'

$A \subset B$ A is a sub-set of a set of B

$A \subseteq B$ A is a sub-set of, or equals, set B

$A \cup B$ The union of set A with set B

[a,b] Closed region, $a \leq x \leq b$

[a,b) Half open region, $a \leq x \leq (b - \varepsilon)$, $\varepsilon \ll 1$

]a,b[Open region, $a < x < b$

$\langle \Delta n_j \rangle$ A knot interval list convertible to a knot vector, $\langle \Delta n_j \rangle \equiv \langle n_{j+1} - n_j \rangle$

$\forall$ Mathematical shorthand meaning 'for all'

$\oplus$ Numerical integration data in ⊞

$\odot$ Numerical integration data in ⊡

Δ Triangular or tetrahedra element sub-set, ⊞ ⊂ ⊡

Chapter 1

Comparing IGA and FEA Methods

1.1 Introduction

Both isogeometric analysis (IGA) and classical finite element analysis (FEA) approaches utilize piecewise polynomials of various types as 'basis functions' to describe the geometry and the primary unknowns. Both approaches typically utilize the Methods of Weighted Residuals or Variational Methods to establish integral forms that are equivalent to the partial differential equation(s) governing a particular field of application. By substituting their respective basis functions into the governing integral forms and numerically integrating the integral form over the domain, one can convert the original partial differential equation analysis into a large matrix analysis problem. To obtain the numerical solution, the matrix system must include in some way the essential (Dirichlet) and non-essential (Neumann and Robin) boundary conditions associated with the partial differential equation. Significant differences between IGA and FEA are highlighted in Table 1.1 and will be considered in detail in the following discussions, examples, and applications. (And no, NURBS does not mean nobody-understands-rational-b-splines.)

A significant difference between the classic FEA and IGA is highlighted in the second row with regard to continuity of an approximate solution across sub-patch interfaces and finite element interfaces. Physical items like displacements or temperatures typically vary continuously, except at singular points, with at least continuity

Table 1.1. Comparison of IGA and FEA basis functions and numerical procedures.

Basis Property	IGA Bernstein	IGA B-spline	IGA T-splines	IGA NURBS	FEA Lagrange	FEA 1D Hermite
Partition of unity	Yes	Yes	Yes	Yes	Yes	Yes
Boundary continuity	C^0	C^n, $n = p - 1 - k$	C^n, $n = p - 1 - k$	C^n, $n = p - 1 - k$	C^0	C^n, $n = 1, 2, 3 \ldots$
Positivity ≥ 0	Yes	Yes	Yes	Yes	No	No
Linearly independent	Yes	Yes	Yes	Yes	Yes	Yes
Variation diminishing	Yes	Yes	Yes	Yes	No	No
Isopar mapping	In sub-regions	In sub-regions	In sub-regions	In sub-regions	In elements	In elements
Unknowns at	CP	CP	CP	CP	Nodes	Nodes
Gradients at	CP	CP	CP	CP	QP to nodes	Nodes or QP
Dirichlet EBC	—	Special treatment	Special treatment	Special treatment	Nodes	Nodes
Basis active in	Full region	Knot spans	Knot spans	Knot spans	Element interior	Element interior
Basis overlap	No	Yes	Yes	Yes	No	No
Neumann NBC	Integral form	Integral form	Integral form	Integral form	Integral form	Integral form
Numeric integration	Collocation or GQ	Collocation or GQ	Collocation or GQ	Collocation or GQ	GQ	GQ

Notes: CP = Control Point, GQ = Gaussian quadratures, QP = Quadrature points, k = Number of repeated knots, n = Number of knots in knot vector, p = Polynomial degree.

of their gradient. In other words, physical items usually have at least continuity of their value, C^0, slope, C^1, and often their curvature, C^2. IGA solutions can easily match those desirable states in the analysis functions (basis functions). However, the column on Lagrangian interpolations, which are typically used in two- (2D) and three-dimensional (3D) FEA solutions, shows that they historically only provided continuity of the solution, C^0, across element boundaries. The last column on Hermite polynomials shows that higher levels of solution continuity are basically available only in one dimension where it is enforced at a point in 1D. It has proven very difficult to extend Hermite interpolation continuity to element interfaces in 2D or 3D FEA studies, but such continuity is natural with the basis functions of IGA.

Perhaps the biggest difference for beginners to understand and master when starting to study computational mechanics or converting from FEA to IGA is the row noting in which region a basis function (or interpolation function) is active. In FEA, the need to take the derivative of, or to integrate, an interpolation function is relatively easily satisfied because the integration takes place over the sum of the non-overlapping element domains:

$$\int_{\Omega} \text{Integral form } d\Omega = \sum_{e} \int_{\Omega^e} \text{Integral form } d\Omega,$$

since the element basis (interpolation) functions are considered to be identically zero outside of the element domain. That is, they do not overlap regions outside an element.

That is not the case for the piecewise basis functions in IGA studies because most of them overlap some sub-regions (knot spans) and some of them are identically zero in other sub-regions defined by the input knot vector (discussed later). Since the numerical integration can be expensive, it is desired to skip known sub-regions where the basis functions, and their derivatives, are identically zero. This means that the integral over the solution domain, Ω, must be summed over the IGA patch regions, Ω^r, but only integrating over the sub-regions, Ω^N, with their non-zero basis functions and their derivatives:

$$\int_{\Omega} \text{Integral form } d\Omega = \sum_{r} \left(\sum_{N} \int_{\Omega^N} [\text{Integral form } \neq 0] \ d\Omega \right)$$

When using numerical integration in FEA, it is necessary to map the element physical domain, Ω^e, into a non-dimensional quadrature domain, Ω^Q. That change of variables, $\Omega^e \leftrightarrow \Omega^Q$, requires only one Jacobian matrix to satisfy the rules of calculus. But, IGA usually requires two mappings and two Jacobian matrices, $\Omega^r \leftrightarrow \Omega^N \leftrightarrow \Omega^Q$. The additional IGA geometry mappings can affect the choice of the numerical integration procedure. FEA and IGA usually use Gaussian quadratures, but since the non-zero basis function regions can be a lot smaller than the Non-Uniform Rational B-Splines (NURBS) or T-spline region, using collocation on the governing differential equation has been shown to be efficient, as discussed later.

Another significant difference between FEA and IGA comes when enforcing non-zero essential (Dirichlet) boundary conditions before solving the governing matrix equations. Table 1.1 notes that essential boundary condition enforcement is done exactly at the nodes in FEA, but those essential boundary conditions are often only weakly approximated in an IGA study unless a control point happens to fall on the location of the boundary condition.

Another comparison of the other major differences between IGA and FEA is given in Table 1.2. The major differences in a typical stress analysis study are implied by the arrows in Fig. 1.1. The differences in other types of studies are seen by changing the interpretation of the primary unknown. For example, in a heat transfer study, deflections become temperatures and the post-processed stress components become heat flux components for other general field problems. Figure 1.2 highlights a major advantage of IGA over FEA in that it does not require the generation of a mesh before the analysis can begin. For practical problems, the FEA usually requires the use

Table 1.2. Comparison of IGA and FEA major differences.

NURBS and T-spline basis functions	Lagrange and Hermite basis functions
Exact geometry	Approximate geometry
Analysis control points (CPs)	Mesh nodes
Knot spans and knot vectors	Element mesh
Basis continuity is controllable	Basis continuity is mainly fixed
Basis functions do not interpolate CPs	Basis interpolate to nodes
No mesh generation	Requires mesh generation

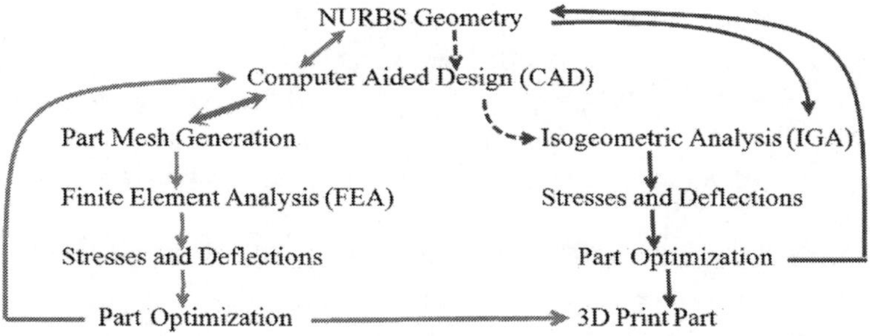

Figure 1.1. Primary analysis flows in FEA and IGA

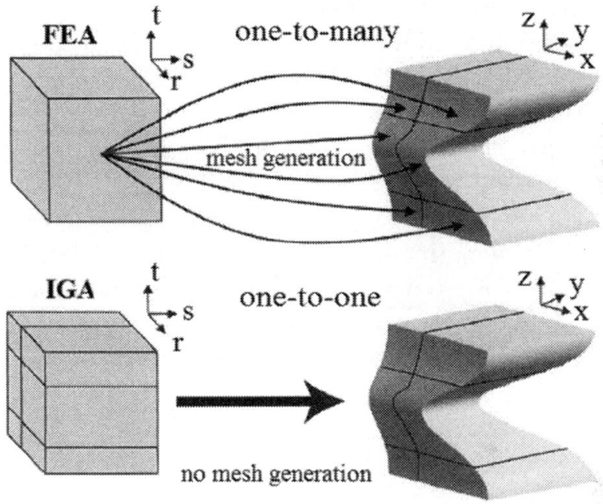

Figure 1.2. In FEA, a parametric element is mapped into many elements, while in IGA, parametric knot spans map to the exact geometry.

of a free-standing automatic mesh generation (AMG) program. Such programs are quite time consuming, require large amounts of memory, and often need to be rerun numerous times to refine a mesh or to correct undesirable features of the resulting mesh. Typically, the generation of an acceptable mesh often requires more than half of the total analysis time for complicated parts and assemblies.

AMG programs usually utilize the exact NURBS geometry output from some Computer Aided Geometric Design (CAGD) software. Often, localized details of the exact NURBS geometry produces so many small finite elements that the memory of the system is exceeded. This problem occurs so often that most AMG programs often include a 'de-feature' tool which suppresses or revises some features of the exact part geometry. Then, the de-featured NURBS model is saved to be used in the next attempt at using the automatic mesh generator.

In other words, an FEA analysis often starts with part geometry which is not exact and could miss important results, like stress concentrations. In addition, the Lagrangian basis functions in the elements of most FEA programs cannot exactly match the exact geometry of circles, cylinders, spheres, cones, etc., which are commonly part of any NURBS or T-spline model. Such element geometry errors become insignificant with highly refined meshes, but that greatly increases the number of nodes and run time for the analysis.

Some features of CAGD are tightly coupled with IGA and its carryover features from FEA. One of the goals of the present work is to outline several CAGD features without proof and to illustrate how such features eliminate some FEA steps and enhance other FEA steps to provide the engineer with an alternate, generally better, and faster analysis tool. As noted, the IGA process completely eliminates the need for the generation of a mesh. IGA uses a well-defined CAGD process for extracting sub-regions from a NURBS geometry, and the process has to be mastered by IGA users since those sub-regions are analogous to the concept of an element in FEA. Figure 1.3 illustrates the fact that, often, the IGA NURBS formulation requires fewer DOFs than an FEA solution.

Today, IGA has two main branches. Usually, it is necessary to read a publication about IGA to determine which of the two branches is being followed. Increasingly, it is the newer second extraction branch.

The first branch using the basis functions, which describe the exact geometry as analysis functions in each knot span (iElement). This approach leads to challenges, which have been mastered, in imposing both the Dirichlet (essential) boundary conditions and in creating graphical display of the analysis results. Those three tasks are easily accomplished in finite element methods. The second branch of IGA uses the exact geometry basis functions and 'extracts' continuous functions uses them as interpolation functions within an existing

FEA program, possibly with the addition of an optional interpolation sub-routine. Impressively, the extracted piecewise C^0 element interpolation functions recreate a segment of the original C^{p-1} IGA curve (or surface) such that the assembly of the full mesh of finite elements exactly reestablishes the full region of the C^{p-1} IGA analysis functions.

Both IGA and FEA utilize families of piecewise polynomials. The IGA polynomials are smoother and often have C^2 continuity across interfaces of analysis domains. This is in contrast to most FEA polynomials, which usually only have continuity of a value between adjacent elements. The next section will discuss different choices for selecting polynomials for analysis functions.

1.2 Piecewise polynomials

The customs of the computer graphics community are to count points and coefficients from 0 to n and use u to denote a non-dimensional unit coordinate space, $0 \leq u \leq 1$, which will be followed here when discussing geometric concepts. When discussing features that carry over to IGA applications, the notations of Hughes (2005) and the seminal book by Cottrell *et al.* (2009) will be followed. When discussing finite element concepts that preceded IGA, the notations of

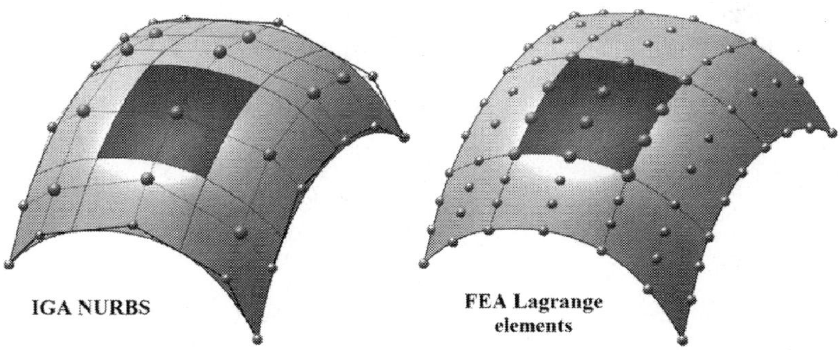

Figure 1.3. For the same number of sub-regions (iElements and elements), IGA requires fewer points for computing a solution and has higher continuity

Hughes (2020) will be followed. In both cases, such counts begin with 1 instead of 0. In IGA and FEA, the Greek letters, ξ, η, and ζ are used to denote unit coordinates in 1-, 2-, and 3-dimensional space, respectively. Later, when computer algorithms are listed and Greek letters are not available the notations, the Akin (2020) FEA book will be followed as additional IGA functions are presented. There, the unit coordinates set is usually listed as r, s, and t, respectively. Some common references utilize non-dimensional spaces over $-1 \leq a, b, c \leq 1$, herein called natural coordinates. It is easy to change from a natural coordinate system to a unit coordinate system to the next since $u = (a + 1)/2$ and $a = (2u - 1)$. Later, when dealing with knot locations, they are often provided in a non-dimensional parametric space of $c \leq u' \leq d$ that is often converted to $0 \leq u \leq 1$ to position the knot points along a parametric curve, $\mathbf{C}(u)$. The required linear transformation (mapping) is $u = (u' - c)/(d - c)$.

In the CAGD and IGA literature, the piecewise polynomial basis functions, and their derivatives, are defined mathematically by recurrence relations, which are listed in the Summary of Chapter 2. Those relations and alternate efficient algorithms for evaluating basis functions are detailed in *The NURBS Book* (Piegl and Tiller, 1997). Here, the discussion of piecewise polynomials will begin with their matrix forms.

In unit coordinates, a complete one-dimensional (1D) polynomial of degree p written in both scalar and matrix forms is

$$f(u) = c_0 + c_1 u + c_2 u^2 \cdots + c_p u^p = \begin{bmatrix} 1 & u & u^2 & \cdots & u^p \end{bmatrix} \begin{Bmatrix} c_0 \\ c_1 \\ c_2 \\ \vdots \\ c_p \end{Bmatrix} \equiv [\boldsymbol{F}(u)]\,\{\boldsymbol{c}\},$$

where $f(u)$ is a physical quantity, c_j are mathematical constants to be determined, and $[\boldsymbol{F}(u)]$ is a row matrix containing the complete polynomial of degree p. Usually, the mathematical constants are converted to physical items at CPs.

This change introduces a square $(p\,x\,p)$ transformation matrix, say $[\boldsymbol{M_p}]$, to the alternate matrix form of

$$P(u) = \left[1\,u\;u^2\ldots u^p\right][\boldsymbol{M_p}]\begin{Bmatrix} P_0 \\ P_1 \\ P_2 \\ \vdots \\ P_p \end{Bmatrix} = [\mathbf{F}(u)]\,[\mathbf{M_p}]\begin{Bmatrix} P_0 \\ P_1 \\ P_2 \\ \vdots \\ P_p \end{Bmatrix}$$

$$\equiv [\mathbf{H}(u)]\begin{Bmatrix} P_0 \\ P_1 \\ P_2 \\ \vdots \\ P_p \end{Bmatrix},$$

where here $[\mathbf{H}(u)]$ denotes any one of the many choices of basis functions that can act on the physical data, $[P_0\;P_1\;P_2\;\ldots\;P_p]'$, at the CPs for the polynomial curve.

In IGA, cubic and quintic polynomials are often used for approximating the solution of second- and fourth-order differential equations, respectively. In IGA, the common polynomial forms are smooth B-splines (Bernstein polynomial basis functions), NURBS, and T-splines, while FEA commonly uses piecewise smooth Lagrange or Hermite polynomials. Both FEA and IGA polynomial forms can be expressed in the above-mentioned matrix form. It can be shown that for the same polynomial degree, p, an FEA polynomial form can be converted to an IGA polynomial form, and vice versa.

> **Looking Ahead**: The ability to convert a non-interpolating IGA B-spline (segment) into an interpolating FEA Lagrange polynomial, or a non-interpolating Bernstein polynomial, is a key mathematical foundation for the extremely important 'Bezier Extraction Operator' and 'Lagrangian Extraction Operator', and is critical to having an IGA sub-set of basis functions that fit into existing FEA software with minimal changes to the FEA code.

Common finite element applications using Lagrange interpolation exactly interpolate information at equally spaced $(p+1)$ locations in the parametric space. For example, consider a quadratic Lagrangian polynomial ($p = 2$) to exactly interpolate ($p+1 = 3$) function and/or coordinate values:

$$P(u) = [1\ u\ u^2] \begin{Bmatrix} c_0 \\ c_1 \\ c_2 \end{Bmatrix} \equiv [\boldsymbol{F}(u)]\,\{\boldsymbol{c}\}$$

$$1 \times 1 = (1 \times 3)(3 \times 1).$$

Iff the polynomial set in the region $0 \le u \le 1$ and iff three equally spaced locations in the region are selected as $u_0 = 0$, $u_1 = 0.5$, $u_2 = 1$ and if the constraints are $P(u_0) = P_0$, $P(u_1) = P_1$ and $P(u_2) = P_2$, then writing these three identities in a matrix form gives

$$\begin{Bmatrix} P(u_0) \\ P(u_1) \\ P(u_2) \end{Bmatrix} \equiv \begin{Bmatrix} P_0 \\ P_1 \\ P_2 \end{Bmatrix} = \begin{bmatrix} 1 & 0 & 0^2 \\ 1 & 0.5 & 0.5^2 \\ 1 & 1 & 1^2 \end{bmatrix} \begin{Bmatrix} c_0 \\ c_1 \\ c_2 \end{Bmatrix}, \text{ or } \{\boldsymbol{P}\}^r = [\boldsymbol{M_p}]\{\boldsymbol{c}\},$$

where $\{\boldsymbol{P}\}^r$ denotes a column matrix containing the physical values of $P(u)$ in some sub-region Ω^r. Inverting the 3-by-3 matrix gives the mathematical constants in terms of the given physical data:

$$\begin{Bmatrix} c_0 \\ c_1 \\ c_2 \end{Bmatrix} = \begin{bmatrix} 1 & 0 & 0 \\ -3 & 4 & -1 \\ 2 & -4 & 2 \end{bmatrix} \begin{Bmatrix} P_0 \\ P_1 \\ P_2 \end{Bmatrix} = [\mathbf{M_p}]^{-1}\,\{\mathbf{P}\}^r.$$

Substituting the mathematical constants into the original polynomial gives the quadratic Lagrangian, $p = 2$, form of a basis function with equally spaced CPs over $0 \le u \le 1$ of

$$P(u) = [1\,u\,u^2] \begin{bmatrix} 1 & 0 & 0 \\ -3 & 4 & -1 \\ 2 & -4 & 2 \end{bmatrix} \begin{Bmatrix} P_0 \\ P_1 \\ P_2 \end{Bmatrix} = [\boldsymbol{F}(u)]\,[\boldsymbol{M_p}]^{-1}\,\{\boldsymbol{P}\}^r.$$

Conceptually, all parametric curves (or analysis solutions) can be thought of as

$$\begin{pmatrix} \text{Parametric} \\ \text{curves} \end{pmatrix} = \begin{pmatrix} \text{Complete} \\ \text{polynomial} \end{pmatrix} \begin{pmatrix} \text{Polynomial type} \\ \text{constraints} \end{pmatrix} \begin{pmatrix} \text{Control point} \\ \text{data} \end{pmatrix},$$

where the square matrix between the complete polynomial and the control data contains $(p+1)^2$ coefficients which can be constrained in different ways to make the resulting curve behave in a special way. Note that the nine coefficients in the quadratic Lagrange square matrix in this example simply come from requiring exact interpolation at specified parametric points. That constraint is significantly different from the constraints leading to the corresponding terms in the Bernstein polynomial basis functions (as shown in the following) where the square matrix coefficients are derived from different mathematical constraints on the polynomial. Later, spline matrices will be provided with no discussion of how such mathematical constraints were implemented.

In FEA studies, the first matrix product evaluated with the resulting row matrix known as the interpolation functions:

$$P(u) = [\boldsymbol{F}(u)]\,[\boldsymbol{M_p}]^{-1}\,\{\boldsymbol{P}\}^r \equiv [\boldsymbol{H}(u)]\,\{\boldsymbol{P}\}^r\,.$$

Because the FEA basis functions, $[\boldsymbol{H}(u)]$, were constrained to exactly interpolate the physical data at the CPs, it is recognized that the above-mentioned interpolation functions are the quadratic Lagrange basis functions:

$$P(u) = [\boldsymbol{H}(u)]\begin{Bmatrix} P_0 \\ P_1 \\ P_2 \end{Bmatrix} = [\boldsymbol{L_2}(u)]\begin{Bmatrix} P_0 \\ P_1 \\ P_2 \end{Bmatrix},$$

where $[\boldsymbol{L_2}(u)] = \left[(1-3u+2u^2)\;(4u-4u^2)\;(-u+2u^2)\right]$, and in scalar form the curve is defined as

$$P(u) = (1-3u+2u^2)P_0 + (4u-4u^2)P_1 + (2u^2-u)P_2.$$

With different positioning of the CPs, the nine coefficients of the geometry matrix, $[\boldsymbol{M_p}]$, are different. Note that at any parametric point, u, the sum of the basis functions is identically one:

$$\sum \boldsymbol{L_2}(u) = (1-3u+2u^2) + (4u-4u^2) + (-u+2u^2) \equiv 1.$$

In mathematical jargon, the basis functions represent a 'partition of unity', which is true for all valid FEA and IGA basis functions. An important distinction between IGA and FEA basis functions is that IGA basis functions are never negative for any value of the parametric

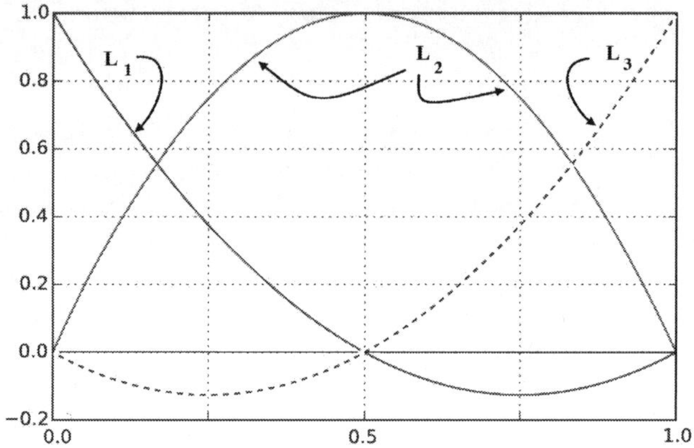

Figure 1.4. The quadratic Lagrange basis functions in unit coordinates

coordinate, but the FEA basis functions typically are negative over part of the parametric coordinates, as seen in Fig. 1.4.

The local parametric derivatives of IGA and FEA basis functions can also be written in matrix form. For example, a quadratic polynomial has

$$[\boldsymbol{F}(u)] = [1\,u\,u^2], \quad \frac{d[\boldsymbol{F}(u)]}{du} = [0\,1\,2u]\,, \quad \text{and} \quad \frac{d^2\,[\boldsymbol{F}(u)]}{du^2} = [0\,0\,2]\,.$$

The FEA element (sub-region) in physical space, Ω^e, where the data were obtained can also be defined by noting that $P(u)$ could be replaced with physical location $x(u)$ of the data points. Then, using

$$x(u) = (1 - 3u + 2u^2)x_0 + (4u - 4u^2)x_1 + (2u^2 - u)x_2 \equiv [\boldsymbol{H}(u)]\,\{\boldsymbol{x}^r\}$$

for a region with n_{cp} interpolation points makes it easy to plot the values of $P(u) = P(x)$ along the $x(u)$ axis. This concept of using a single (iso) type of parametric function for unknowns, coordinates, material properties, etc., is called the isoparametric approach in FEA and IGA studies. Since the Lagrangian forms have simple identities (constraints) to satisfy at each data point, the symbolic Matlab script listed in Script 1.1 can derive all of the above-mentioned algebra and derivatives in analytic form for uniform knot span lengths. The output from this script is given in Fig. 1.5. More complex algorithms are required for Bernstein polynomials.

```
Poly(u) = [1, u, u^2]
f(u)    = c3*u^2 + c2*u + c1
c_e     = f1, 4*f2 - 3*f1 - f3, 2*f1 - 4*f2 + 2*f3
H(u)    = [2*u^2 - 3*u + 1, - 4*u^2 + 4*u, 2*u^2 - u]
f(u)    = f1*(2*u^2-3*u+1)+f2*(-4*u^2+4*u)-f3*(-2*u^2+u)
df_du(u)= f1*(4*u - 3) + f3*(4*u - 1) - f2*(8*u - 4)

x(u)      = x1*(2*u^2-3*u+1)+x2*(-4*u^2+4*u)-x3*(-2*u^2+u)
dx_du(u)  = x1*(4*u-3)+x3*(4*u-1)-x2*(8*u-4)
df_dx(u)  = (f1*(4*u-3)+f3*(4*u-1)-f2*(8*u-4)) / ...
            (x1*(4*u-3)+x3*(4*u-1)-x2*(8*u-4))

x(u)      = x1*(2*u^2-3*u+1)+(x1/2+x3/2)*(-4*u^2+4*u) ...
            -x3*(-2*u^2+u)
J(u)      = x3-x1
df_dx(u)  = -(f1*(4*u-3)+f3*(4*u-1)-f2*(8*u-4))/(x1-x3)
```

Figure 1.5. Output of Lagrange symbolic interpolation

Once the script has defined the interpolation function $[\boldsymbol{H}(u)]$ to apply on the values to interpolate for $f(u)$, it applies that interpolation again to obtain the corresponding x-locations of the values to be interpolated, $x(u)$. If the interpolated point values, f_j^r, in a region are y-coordinates, $f(u) = y(u)$, and the $x(u)$ likewise combine to define a planar curve. The script uses the function '*inv*' to invert the small square matrix and the function '*diff*' to differentiate with respect to u, that is, $\partial y/\partial u$ and $\partial x/\partial u$. They in turn provide the slope of the curve

$$\frac{\partial y(u)}{\partial x} = \frac{\partial y/\partial u}{\partial x/\partial u},$$

even though the expression for $y(x)$ is not available. Also, in one dimension, the term $\partial x/\partial u \equiv [\boldsymbol{J}(u)]$ is the Jacobian from the parametric u-space to the physical x-space.

Script 1.1. Using Matlab to derive a Lagrange interpolation function

```
%                    Lagrange_interpolate.m
% Symbolic derivation of Lagrange p=2 interpolations
syms u u_n real                       % symbolic variables
syms c_e c1 c2 c3 real                     % math constants
syms f_e f1 f2 f3 real                % physical constants
u_n = [0, 1/2, 1]                   % equal point locations
```

```
c_e = [c1, c2, c3]'              % math constant vector
f_e = [f1, f2, f3]'                 % interpolated data
Poly (u) = [1, u, u^2]           % quadratic polynomial
f(u) = Poly(u) * c_e             % f at any 0 <= u <= 1
%
%        Identity u(r_n) = Poly (r_n) * c_n
for k = 1:3  ; % use interpolation identities ==> ==>
  G (k, :) = Poly (u_n(k))    ; % k-th row of identity
end     ; % for k-th identity at local pt <== <== <==
display (G)                       % local geometry matrix
%     iff the inverse exists, this is a valid
%  polynomial and local pt location combination
G_inv = inv (G)                       % attempt inversion
% G = [1,    0,    0]     G_inv = [ 1,  0,  0] % result
%     [1, 1/2, 1/4]               [-3,  4, -1] % result
%     [1,    1,    1]             [ 2, -4,  2] % result
c_e = G_inv * f_e              % replace math constants
%
H (u) = Poly (u) * G_inv    % element shape functions
%          f(u) = Poly (u) * G_inv * f_e
f(u) = H (u) * f_e             % solution interpolation
df_du = diff(f, u)       j     % parametric derivative
%
%       Done, now use physical x locations and
%     get df/dx for unequal and equal x intervals
syms x_e x1 x2 x3 real     % symbols for x-coordinates
x_e   = [x1, x2, x3]'                % unequal intervals
x(u)  = H(u) * x_e                    % interpolate x(u)
dx_du = diff (x, u)         % dx/du for unequal spaces
df_dx = df_du / dx_du       % df/dx for unequal spaces
%      use equal x intervals (constant Jacobian)
x_e=[x1 (x1+x3)/2 x3]'      % new linear mapping for x
x(u) = H(u) * x_e         % degenerates from p=2 to p=1
J(u) = diff (x, u) ;                      % J = new dx/du
J=simplify(J)          % special case constant Jacobian
df_dx= df_du / J               % df/dx for equal spaces
```

Using input values of both x- and y-coordinate values, and doing the interpolation twice, is not efficient. Recognizing that the control point inputs, $\{\boldsymbol{x}^r\}$, would often have multiple spatial coordinates and at least one degree of freedom, $\{\boldsymbol{\delta}^r\}$, at each control point, the input array should be upgraded to a rectangular array with one column for each input item so that the output results also becomes an array. Then, a Lagrangian interpolated quadratic planar curve would be

$$\begin{bmatrix} x(u) & y(u) & \delta(u) \end{bmatrix} = [u\ u^2] \begin{bmatrix} 1 & 0 & 0 \\ -3 & 4 & -1 \\ 2 & -4 & 2 \end{bmatrix} \begin{bmatrix} x_0 & y_0 & \delta_0 \\ x_1 & y_1 & \delta_1 \\ x_2 & y_2 & \delta_2 \end{bmatrix}.$$

$$(1 \times 3) = (1 \times 3)\ (3 \times 3)\ (3 \times 3)$$

Since the local gradient is known, two spatial coordinate vectors could be used to get the local gradient components and a third row to get the second partial derivative:

$$[xy\_d(u)] = \begin{bmatrix} x(u) & y(u) \\ \partial x(u)/\partial u & \partial y(u)/\partial u \\ \partial^2 x(u)/\partial u^2 & \partial^2 y(u)/\partial u^2 \end{bmatrix} = \begin{bmatrix} 1 & u & u^2 \\ 0 & 1 & 2u \\ 0 & 0 & 2 \end{bmatrix}$$

$$\times \begin{bmatrix} 1 & 0 & 0 \\ -3 & 4 & -1 \\ 2 & -4 & 2 \end{bmatrix} \begin{Bmatrix} x_0 & y_0 \\ x_1 & y_1 \\ x_2 & y_2 \end{Bmatrix}.$$

Clearly, this approach carries over to the different classes of polynomials that will be used in IGA.

Bezier curves are piecewise smooth interpolations of Bernstein polynomial, each of which is based on CPs that are generally not interpolation points. The Bezier curves can be written in the same matrix form used previously for the Lagrangian curves, but with different constraints determining the coefficients of the square matrix. The formulas for the coordinates of a single-span Bézier curve over the range $0 \leq u \leq 1$ are Bernstein polynomials. For example, the quadratic, $p = 2$, Bernstein polynomial is

$$B_2(u) = [1\,u\,u^2] \begin{bmatrix} 1 & 0 & 0 \\ -2 & 2 & 0 \\ 1 & -2 & 1 \end{bmatrix} \begin{Bmatrix} P_0 \\ P_1 \\ P_2 \end{Bmatrix} = [\boldsymbol{H}(u)]\,\{\boldsymbol{P}\},$$

or

$$B_2(u) = (1-u)^2 P_0 + 2u(1-u)P_1 + u^2 P_2.$$

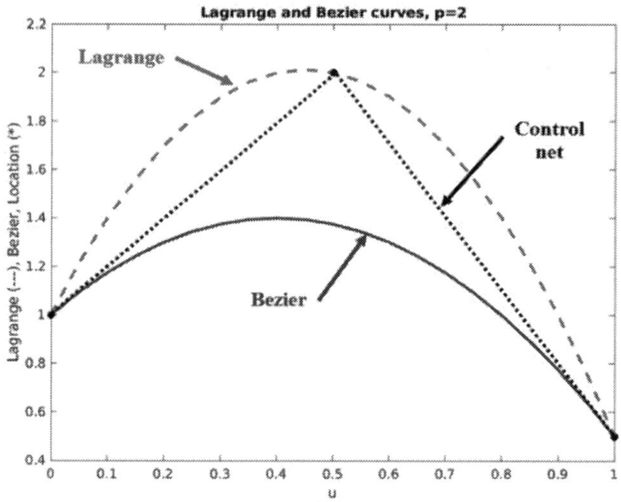

Figure 1.6. Quadratic Lagrange (top) and Bezier curves for the same three reference y-coordinates

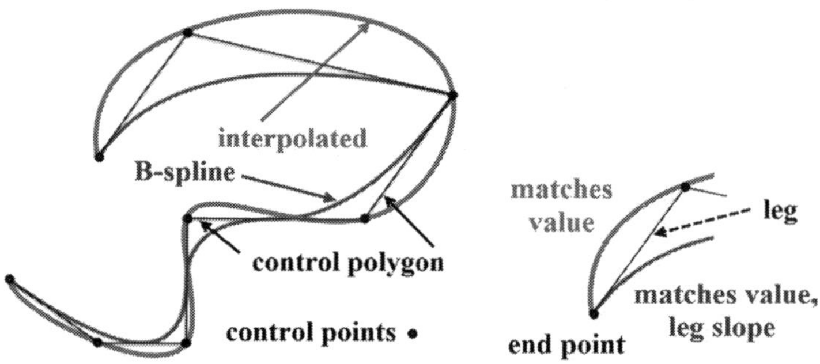

Figure 1.7. A multi-span interpolating curve versus a B-spline curve

The differences in these two quadratic curves for a single span using the same y-coordinates are illustrated in Fig. 1.6. The Lagrange form interpolates all three points. In contrast, the Bezier form interpolates the end points and is tangent to the control legs at each end. The differences between interpolating polynomials and B-spline planar curves over multi-span segments are shown in Fig. 1.7.

Note that since the position, u, is non-dimensional, both $B_2(u)$ and $L_2(u)$ have the same units as the control data.

Since $B_2(u)$ is a scalar it is equal to its transpose, $B_2(u) = B_2(u)^T$. Using the matrix algebra product transpose rule that the transpose of a matrix product is the product of the transpose of each term in reverse order, $(\boldsymbol{A} \cdot \boldsymbol{B} \cdot \boldsymbol{C})^T = \boldsymbol{C}^T \boldsymbol{B}^T \boldsymbol{A}^T$, this relation can also be written with the input data as the first row matrix.

The matrix symbol $\boldsymbol{H}(u)$ will be used generically to represent either the interpolation function matrix, which directly interpolates the function data $\boldsymbol{P}^r$, or the non-interpolator NURBS and B-splines used in CAGD and IGA, which are constructed from Bernstein interpolating polynomials. It is possible to extract a set of Bernstein polynomials from analysis B-splines, and the Bernstein polynomials can be used as interpolating polynomials in FEA.

Using the matrix form of Lagrange polynomials, Bernstein polynomials and B-spline curves are generally faster than computing their values with their original recursive definitions. In part, this is because the operations on small matrices are burned into GPU chips commonly found in many of today's computers.

The matrix form for a *single* cubic B-spline over $0 \le u \le 1$ is

$$B_3(u) = [1\ u\ u^2\ u^3] \begin{bmatrix} 1 & 0 & 0 & 0 \\ -3 & 3 & 0 & 0 \\ 3 & -6 & 3 & 0 \\ -1 & 3 & -3 & 1 \end{bmatrix} \begin{Bmatrix} P_0 \\ P_1 \\ P_2 \\ P_3 \end{Bmatrix}.$$

Similarly, the matrix form for a *single* quartic B-spline over $0 \le u \le 1$ is

$$B_4(u) = \begin{bmatrix} u & u^2 & u^3 & u^4 \end{bmatrix} \begin{bmatrix} 1 & 0 & 0 & 0 & 0 \\ -4 & 4 & 0 & 0 & 0 \\ 6 & -12 & 6 & 0 & 0 \\ -4 & 12 & -12 & 4 & 0 \\ 1 & -4 & 6 & -4 & 1 \end{bmatrix} \begin{Bmatrix} P_0 \\ P_1 \\ P_2 \\ P_3 \\ P_4 \end{Bmatrix}.$$

Note that the Bezier matrix forms are invertible triangular matrices. Also, the sum of all of the square matrix terms is unity (the measure of the u-space). The existence of alternate matrix forms allows the derivation of other mathematical properties and algorithms for improved efficiency.

The above-mentioned Lagrangian interpolation example shows the old FEA approach using data, $\boldsymbol{f}^e$, on the element region, Ω^e.

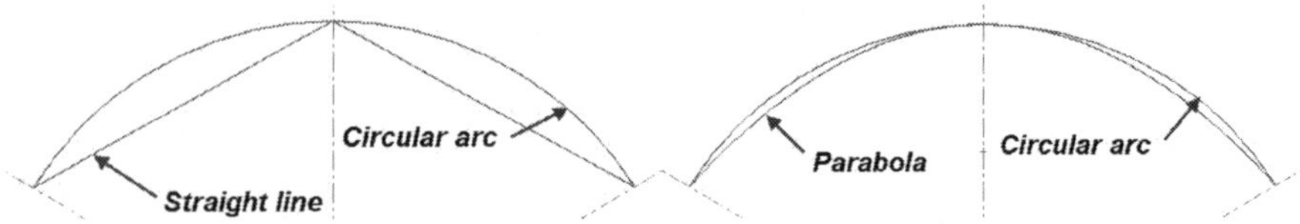

Figure 1.8. Linear and parabolic Lagrangian curves never match circular shapes

The extended concept in IGA is that the B-spline polynomials, $\boldsymbol{B}(u)$, or NURBS, $\boldsymbol{N}(u)$, no longer exactly interpolate FEA node data, but are defined by a larger set of data IGA CPs, $\boldsymbol{f}^r$. Also, each FEA element domain is generally contained within a much larger IGA domain, $\Omega^e \ll \Omega^r$.

Curve representations using just polynomials are simply not powerful enough to exactly match important engineering shapes like circles, ellipses, hyperbolas, and spheres. However, special ratios of two piecewise polynomials (a rational function) can exactly match those shapes. An example of the small 2-dimensional geometric boundary error due to polynomial approximations is seen in Figure 1.8 where linear and quadratic Lagrange polynomials pass through the same three points on a circular arc. The amount of geometrical error in approximating circular and elliptical shapes increases rapidly with the enclosed angle and decreases as the polynomial degree increases. Most FEA systems use $p = 2$ interpolation functions and are considered to give good approximations to a majority of engineering shapes, but good is not exact and IGA provides exact geometries through its use of NURBS.

1.3 Piecewise parametric polynomials

A piecewise polynomial is a polynomial defined by multiple sub-polynomials, where each sub-polynomial is non-zero in a different region of the divided domain. The piecewise qualifier is actually a different way of expressing the polynomial, rather than one that changes the characteristic of the polynomial.

In a 1-dimensional FEA study, the above-mentioned cubic Lagrangian polynomial is active in each piecewise part of the space (a finite element), but it is restricted to the full parametric range of

[0,1] in that space, while being only C^0 continuous at the ends of the element. It is not segmented over u. To achieve a C^1 function at the ends, a cubic function has to be changed to a different class of Hermite polynomials.

In other words, in FEA, it is only the domain that is piecewise, not the approximating polynomial. It is easy to extend the continuity requirements in one dimension, but it is quite difficult in higher dimensions. That was part of the motivation to consider the IGA approach. In IGA, both the domains and their internal polynomials are piecewise and the piecewise polynomials are automatically continuous to a selected degree as shown in Fig. 1.9. The IGA NURBS and T-splines are piecewise functions which are exact representations of the shape of the domain, but only approximate analysis results in a domain. The physical results in a domain are usually continuous, so a continuous IGA analysis approximation could be, and is, more accurate than FEA analysis approximations.

It is well known that interpolation using polynomials of a very high degree is often ill conditioned. A cubic polynomial ($p = 3$) is the lowest degree curve that can have an inflection point, that is, have a change in the sign of its slope. Therefore, it is the default curve in many CAGD systems. In FEA, the polynomials are discontinuous across element boundaries and typically range from linear to cubic. In IGA, the polynomials are continuous across sub-patch boundaries and typically range from cubic to fifth degree.

A *piecewise cubic* B-spline that is significantly different from the one considered earlier in that it is non-zero only over a segment of u and joins to the next segment of u with the same type of cubic B-spline with continuous values of the polynomial, its slope, and curvature has the matrix form

$$B_3(u) = \begin{bmatrix} 1 & u & u^2 & u^3 \end{bmatrix} \frac{1}{6} \begin{bmatrix} 1 & 4 & 1 & 0 \\ -3 & 0 & 3 & 0 \\ 3 & -6 & 3 & 0 \\ -1 & 3 & -3 & 1 \end{bmatrix} \begin{Bmatrix} P_0 \\ P_1 \\ P_2 \\ P_3 \end{Bmatrix}.$$

For this piecewise matrix form, the 16 terms in the square matrix were derived from 16 mathematical constraints that ensured the required continuities between the piecewise divisions of u, that the end conditions are met, that the polynomials are all positive, and

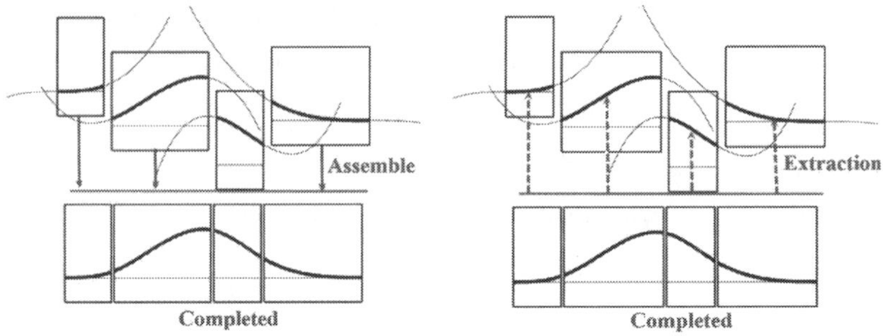

Figure 1.9. Exploded B-spline curve formed by Bernstein polynomial segments

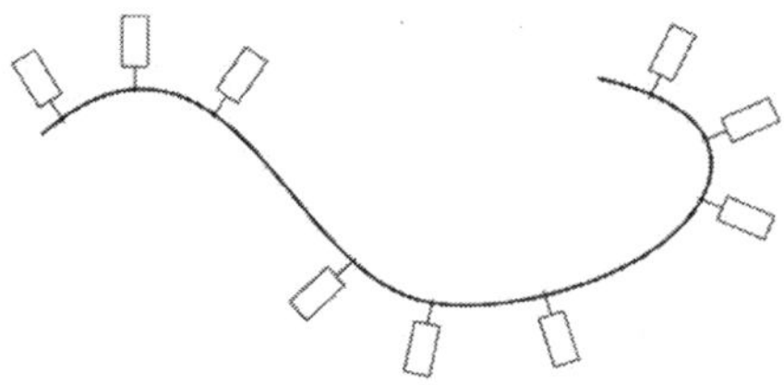

Figure 1.10. Physical beam spline with heavy ducks

that the resulting polynomials are a 'partition of unity' (they sum to unity at all u values). This and similar derivations require the presence of multiple knots at the two ends.

> **Looking Ahead**: As will be shown, 'extraction operators' can take piecewise C^2 IGA basis functions defining a curve segment over a knot span and create additional Lagrange (and other polynomial) CPs that fall on exactly the same piecewise C^2 curve. Those extractions create Lagrangian iGe regions that fit naturally into FEA codes like Lagrange elements, but they provide C^2 solution results, whereas the classic FEA version of Lagrange interpolation only provides C^0 continuity normal to the element boundaries. The same processes allow Bernstein polynomials to be extracted from B-splines to form similar iGe basis functions.

1.4 Splines overview

From the early days of building wooden sailing ships, metal automobiles, and aluminum aircraft, craftsmen needed to draw long curves passing through numerous points and having a continuous location, slope, and curvature. To do so, they used a longer thin flexible beam to bend smoothly through an arbitrary number of interpolation points held in place by heavy weights known as ducks, as shown in Fig. 1.10. The emphasis here is on the beam slope and curvature continuity, which today is denoted as C^2 continuity.

In late eighteenth century, mathematicians recognized such curves as special polynomials (Basis splines, or B-splines) and began studying them. When a polynomial exactly interpolates a set of data at the CPs, it is called an interpolatory polynomial; otherwise, it is called a non-interpolatory polynomial. An example of both is shown in Fig 1.7.

Along with the beginning of the age of digital computers, in the late 1950s, two French automotive designers, Paul de Casteljau at Citroen and Pierre E. Bézier, at Renault, simultaneously discovered what are called Bézier curves today. In the late 1970s, de Boor, Cox, and others developed a modern version of those splines and efficient computer algorithms for evaluating and displaying splines.

Since then, the use Bezier curves and their extension to NURBS have become commonplace in the geometric design of most complex engineering components. Recently, those geometric tools have been extended, and supplemented, for engineering analysis applications. Those extensions are now mature and this work will summarize the major engineering aspects of what has come to be known as IGA.

In 1967, Zienkiewicz introduced the seminal FEA book using interpolatory polynomials, mainly the C^0 families for 2D and 3D applications. Attempts to create 2D interpolations with C^1 or C^2 for FEA studies have been elusive and proved to be a gap in FEA analysis. The use of mathematical splines in IGA has proved to be an important extension of FEA concepts.

1.5 Overview of B-splines

The classic introduction to B-splines and NURBS for curves, surfaces, and volumes is *The NURBS Book* (Piegl and Tiller, 2012). Every reader should review that book as it provides many figures, examples, mathematical proofs, and efficient algorithms. The reader is encouraged to experience web-based interactive versions of those features at the Barton (2022) website, and to view the basic supporting theories and proofs given at the informative Shene (2016) site.

The equation for a B-spline curve in physical space is a piecewise linear combination of geometric Bernstein basis functions, $N_{i,p}(u)$, of degree p, and the spatial coordinate array of a corresponding set of CPs, $\boldsymbol{P}_i$, is a piecewise polynomial in u:

$$\boldsymbol{C}(u) = \sum_{i=1}^{i=(p+1)} N_{i,p}(u)\boldsymbol{P}_i \equiv [\boldsymbol{N_p}(u)][\boldsymbol{P}],\ 0 \leq u \leq \mathbf{1},$$

$$(1 \times n_s) = (1 \times (p+1))((p+1) \times n_s),$$

in size, and which have the 'partition of unity' and 'positivity' properties,

$$\sum_{i=1}^{i=p} N_{i,p}(u) \equiv 1, \quad \sum_{i=1}^{i=p} dN_{i,p}(u)/du \equiv 0, \quad \text{and } N_{i,p}(u) \geq 0,$$

and where the number of CPs is defined by the entries (knots) in the geometric knot vector, Kn, for that curve. The first property above is called a 'partition of unity', which is important in proving certain mathematical features. Both FEA and IGA use functions which are a partition of unity, but only the IGA functions have the benefit of being positive everywhere (the third property above).

The Cox-de Boor recursive definition of B-spline basis functions has been implemented as a symbolic script in Matlab in *symbolic_N_bases.m*, which is listed in Appendix 1. That script can be run to list the algebraic form of the various polynomial basis functions.

A symbolic Cox-de Boor recursive definition not only provides the analytic form of the non-zero $N_{i,p}(u)$ basis functions, and in which knot spans they occur, but such a script can also be extended to provide the local derivatives, $d^n N_{i,p}(u)/du^n$, in analytic form within those spans, as well as the analytic integral of the basis functions over the span. For example, a typical symbolic output for a quadratic two-span B-spline is repeated in Fig. 1.11.

Looking Ahead: In IGA, the mechanical strains at a CP depend on the first physical derivative of the CP's displacements using $(dN_{i,p}(u)/du)(du/dx(u))$. Likewise, the

```
Knot Vector [0, 0, 0, ½, 1, 1, 1], p = 2
Basis Functions, N(u)
Knot Span 1              Knot Span 2
[0. 0.5)                  [0.5, 1)
CP: 1-3                   2-4
[   (2*u - 1)^2,                    0]
[-2*u*(3*u - 2),   (2*u - 2)*(u - 1)]
[          2*u^2,  -6*u^2 + 8*u - 2]
[              0,        (2*u - 1)^2]
Sum:           1,                   1

Derivative of Basis Functions, dN(u)
[  8*u - 4,          0]
[4 - 12*u,     4*u - 4]
[      4*u,   8 - 12*u]
[        0,    8*u - 4]
Sum:     0,          0

2nd Derivative of Basis Functions, d2N(u)
[  8,   0]
[-12,   4]
[  4, -12]
[  0,   8]
```

Figure 1.11. Typical symbolic basis function results

post-processing integral of a bearing's lubrication pressure using $\int N_{i,p}(u)\,(dx(u)/du)\,du$ contributes to the total force being supported by a bearing.

The Cox-de Boor recursive definition is inefficient for numerical calculations needed in CAGD and IGA. Several efficient numerical algorithms for working with NURBS are given in *The NURBS Book* (Piegl and Tiller, 1997).

The above-mentioned four quadratic basis functions shown in Fig. 1.12 are actually six piecewise curves over two knot spans. Their combination over $0 \le u \le 1$ is shown in Fig. 1.13. Note that the basis functions have the same value and slope at their interface as show in Fig. 1.13. The quadratic basis functions are always non-negative, which contrasts with the Lagrange interpolation functions.

NURBS are rational polynomials, meaning they are the ratio of B-splines of the same order, but multiplied by different weights. Using the notation $N_{i,p}(u)$ to represent a non-zero B-spline of degree p acting on data at CP i, the weighted rational polynomial, $R_{i,p}$, defines a NURBS of the same degree at the same CP as

$$R_{i,p}(u) = \frac{N_{i,p}(u)\,w_i}{\sum_{j=1}^{i_{\max}} N_{j,p}\,(u)\,w_j}.$$

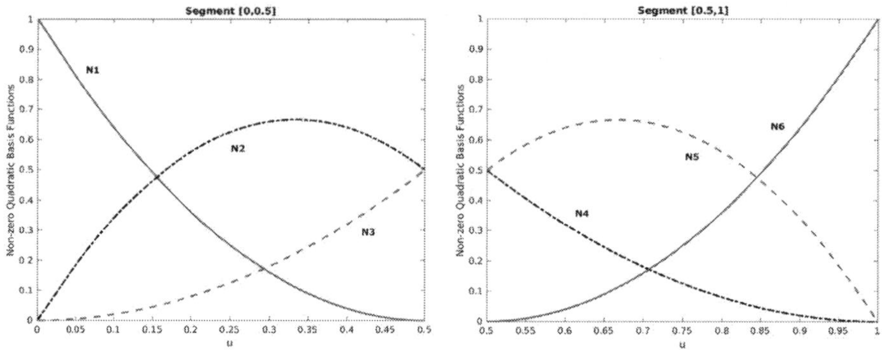

Figure 1.12. Basis function segments are C^{p-1} continuous at knot span interfaces

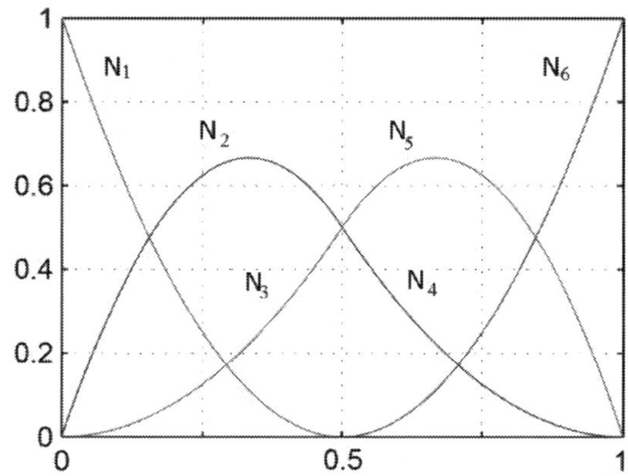

Figure 1.13. Completely positive $p = 2$ B-spline curve segments in contrast to Lagrange interpolation

NURBS are formed in a special way to preserve the desirable mathematical property of being a 'partition of unity' (as are FEA interpolations), which means that the sum of all of the NURBS on a knot span is algebraically equal to one, which is verified in the following:

$$\sum_{i=1}^{i_{\max}} R_{i,p}(u) = \sum_{i=1}^{i_{\max}} \left(\frac{N_{i,p}(u)\, w_i}{\sum_{j=1}^{i_{\max}} N_{j,p}(u)\, w_j} \right) = \left(\frac{\sum_{i=1}^{i_{\max}} N_{i,p}(u)\, w_i}{\sum_{j=1}^{i_{\max}} N_{j,p}(u)\, w_j} \right) \equiv 1,$$

where the $\mathrm{w_i}$ are CP weights selected by the user. This means that when weighted spatial NURBS curves are to be constructed, it now requires each CP to specify its spatial coordinates and its analysis weight value. Since the sum of the B-spline basis functions is always unity, if (as mentioned earlier) they are all multiplied by unity weights, $w_i \equiv 1$, it means the NURBS basis functions degenerate to the B-spline basis functions,

$$R_{i,p}(u) = \frac{N_{i,p}(u)\ w_i}{\sum_{j=1}^{i_{\max}} N_{j,p}\,(u)\,w_j}\ \forall\, w_j \equiv 1 \quad \text{is}\, N_{i,p}(u).$$

Many useful properties and geometric algorithms for NURBS curves and surfaces can be found in *The NURBS Book* by Piegl and Tiller (2012), especially in Chapters 5 and 6. That book also gives the recursive definition for higher-order NURBS.

Referring to the analytic form of the simple two-span quadratic basis functions in Fig. 1.11, the NURBS basis function denominator polynomial is always the weighted sum of the non-zero B-spline basis functions in a knot span. Here, for the above-mentioned four CPs controlling two knot spans, with the above-mentioned quadratic B-splines the denominator is shown by

$$\begin{aligned} R_{D,2}(u) = \sum_{i=1}^{i_{\max}} N_{i,p}\,(u)\,w_i &= N_{1,2}(u)w_1 + N_{2,2}(u)\,w_2 \\ &+ N_{3,2}(u)\,w_3 + N_{4,2}(u)\,w_4. \end{aligned}$$

In the first knot span,

$$\begin{aligned} R_{D1,2}(u) &= (1-2u)^2 w_1 + 2u(2-3u)w_2 \\ &+ (2u^2)w_3 + 0, \end{aligned}$$

and in the second knot span

$$\begin{aligned} R_{D2,2}(u) &= 0 + 2(1-u)^2 w_2. \\ &+ (-6u^2 + 8u - 2)w_3 + (1-2u)^2 w_4. \end{aligned}$$

Thus, the NURBS for the two knot spans are as follows:

Knot Span 1:

$$R_{1,2}(u) = \frac{N_{1,2}(u)w_1}{R_{D1,2}(u)} = \frac{(1-2u)^2 w_1}{R_{D1,2}(u)}.$$

$$R_{2,2}(u) = \frac{N_{2,2}(u)w_2}{R_{D1,2}(u)} = \frac{2u(2-3u)w_2}{R_{D1,2}(u)}.$$

$$R_{2,2}(u) = \frac{N_{3,2}(u)w_3}{R_{D1,2}(u)} = \frac{2u^2 w_3}{R_{D1,2}(u)}.$$

$$R_{2,2}(u) = 0.$$

Knot Span 2:

$$R_{1,2}(u) = 0.$$

$$R_{2,2}(u) = \frac{N_{2,2}(u)w_2}{R_{D2,2}(u)} = \frac{2(1-u)^2}{R_{D2,2}(u)}.$$

$$R_{2,2}(u) = \frac{N_{3,2}(u)w_2}{R_{D2,2}(u)} = \frac{(-6u^2+8u-2)w_3}{R_{D2,2}(u)}.$$

$$R_{2,2}(u) = \frac{N_{4,2}(u)w_2}{R_{D2,2}(u)} = \frac{(1-2u)^2 w_4}{R_{D2,2}(u)}.$$

The addition of weights to each control point gives added power to the user through the ability to assign weights, which default to unity. Even more important is the fact that NURBS can exactly create important engineering shapes like circles, cylinders, and cones, which B-splines cannot do.

To illustrate the geometric power of weights at CPs, consider Fig. 1.14. In it, the four CPs define a piecewise quadratic B-spline curve shown as the thick line. When the same B-spline basis functions are used to create a corresponding set of weighted NURBS, the user can assign more importance (gravity or weight) to one or more CPs. As noted, when all of the weights are unity, the thick B-spline curve and the NURBS curve are identical. But, if the third control point for the NURBS in Fig. 1.15 has its weight changed from unity, then it can define different curves (shown with thinner lines). When the third weight is greater than one, the NURBS curve is pulled closer to the third CP. But, if the third weight is less than one, the NURBS curve is not as close to that CP as the B-spline version is.

> **Looking Ahead**: Now, consider the IGA point of view. Assuming that an engineer has some basic understanding of the analysis problem and expects part of the solution domain near a CP to be of more importance, then the displacement control values, or temperature control values, near those regions can be weighted more heavily.

An important property of B-splines is that as the polynomial order rises, the curvature of a B-spline curve diminishes, as seen in

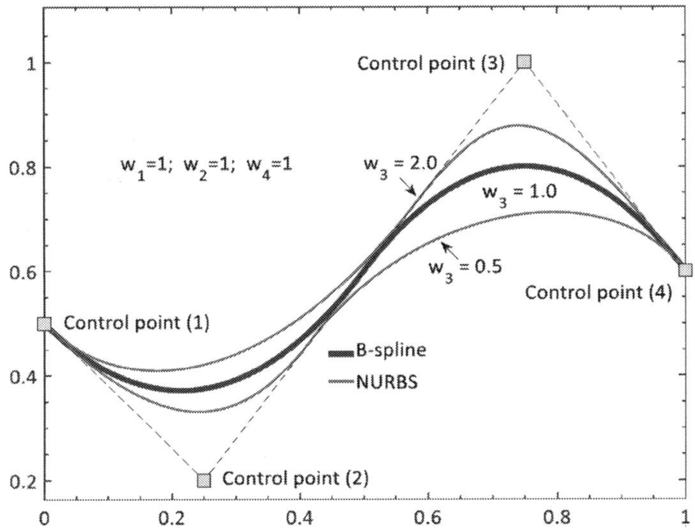

Figure 1.14. A NURBS curve with unity weights degenerates to a B-Spline

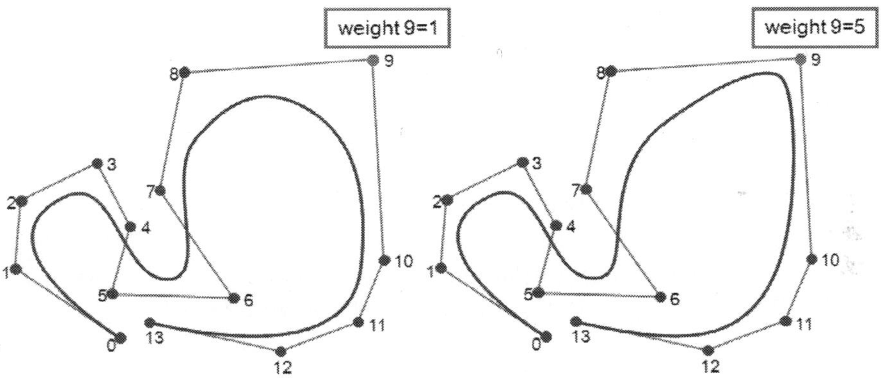

Figure 1.15. Effect of changing the weight at a control point on a B-spline

Fig. 1.16. This relative smoothness is often desirable for geometric features. However, for an IGA, raising the degree of the B-spline may smooth out local changes in the solution of an analytic model. Thus, IGA studies are usually less than degree 6. In FEA, the reverse condition occurs where raising the degree of a Lagrangian interpolation basis increases the curvature changes, so much so that they can

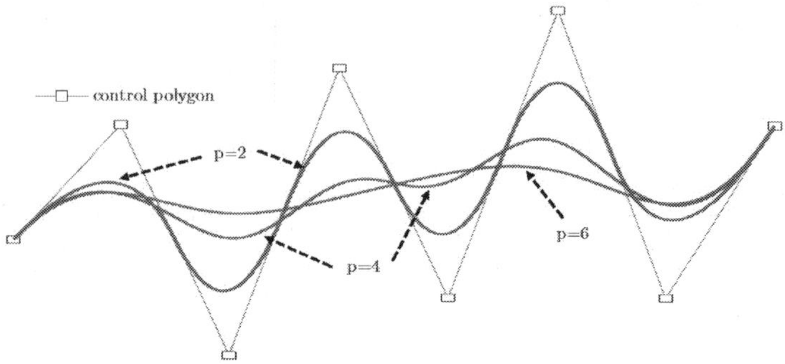

Figure 1.16. Raising the polynomial degree decreases the B-spline curve curvature

become ill conditioned at their ends. Thus, FEA interpolations usually do not go above degree 5, except for high-order (> 4) differential equations.

1.6 From CAGD with FEA to IGA

The use of non-interpolating Non-Uniform Rational B-Splines (NURBS) is central to IGA. The NURBS exact representation of model geometry is typically passed through independent pre-processing AMG software, which approximates the exact geometry and sub-divides it into repeating shapes (triangular, quadrilateral, tetrahedral, or hexahedral elements), each of which are connected to the adjacent shapes at node points. The NURBS in CAGD generally give the exact location, slope, and curvature of the surfaces of a model. The AMG just defines the spatial coordinates of every one of those node points, from the exact geometry, along with a connectivity list of nodes on each element. That pointwise approximate geometry data become input to an FEA study, which then uses a different set of polynomials with lower continuity to interpolate (approximate) the model geometry within each element shape.

Usually, an FEA study uses the same polynomials that interpolate the geometry coordinates to approximate the spatial shape of the unknown variables. This is known as the isoparametric formulation. Some FEA studies use an alternate super-parametric approach where there are more geometric nodes than analysis nodes. For example, the geometry could be interpolated element-wise with quadratic polynomials to reduce the geometric error while only linear polynomials are used to approximate the unknowns. It is less common to have a sub-parametric FEA study, which is the opposite of the super-parametric approach.

The use of Non-Uniform Rational B-Splines to exactly define the geometry of a model is a central theme of IGA, as is their extension for use as analysis functions. NURBS are an extension of Uniform Rational B-Splines (URBS), which use uniformly spaced interior knot points to be discussed later. By definition, any rational B-spline is a function that is the ratio of two B-splines (or Basis functions splines) acting on different data. To have insight into NURBS, analysts should have some understanding of B-spline curves, which are built from 'basis functions', or blending functions, which are known as Bernstein polynomials. The B-spline basis functions are piecewise polynomial smooth functions originally defined by a knot vector and the 'Cox-de Boor recursive definition', which is given in the summary, but more efficient numerical algorithms are available (Piegl and Tiller, 2012). Those polynomials can also be represented in matrix form, and can be transformed to other types of polynomials in matrix form.

Looking Ahead: The analytical forms and the implementation of Matlab functions for geometric basis functions used to define curves and their geometric properties will be used later to illustrate IGA and FEA example analysis solutions and for examples of numerical integration.

1.7 Calculus comparison

In IGA, the goal is to obtain numerical solutions to differential equations over a domain, exactly defined by NURBS and T-splines, by enhancing the geometric knot vectors, $\mathbb{K}n$, to become analysis knot vectors, $\mathbb{A}n$. The integer control mesh(s) is introduced in the next chapter to provide a data structure to locate sub-regions of the analysis domain space.

Looking ahead: FEA and IGA are both based on integral formulations. For example, in the mid-1750s, Euler identified the corresponding integral forms for almost all linear PDEs in physics and engineering. However, Euler's Variation Calculus requires a user to start with the integral and obtain the PDE and its boundary conditions from the integral. For example, Euler proved that solving the 2D integral

$$I = \int_{\Omega} \left[\frac{1}{2} \left(K \frac{\partial \varphi}{\partial x} \right)^2 + \frac{1}{2} \left(K \frac{\partial \varphi}{\partial y} \right)^2 - C\varphi \right] d\Omega + \int_{\Gamma} (q\varphi + a\varphi^2) \, d\Gamma = 0$$

is the same as solving Poisson's partial differential equation (PDE) for heat transfer, if both forms satisfy the same boundary conditions. In an IGA heat transfer study, the temperature, φ, will be approximated spatially by NURBS and/or T-splines. This integral form means that the derivatives of the NURBS and T-splines will be required as well as their integrals. An integer control mesh can identify where contributing NURBS and T-splines are non-zero so that integrations of zero can be avoided and one can increase the computational efficiency.

In contrast to Euler's variational calculus, an FEA or IGA user will want to start with a PDE and obtain the alternate governing integral form. Later, the Methods of Weighted Residuals will be used to accomplish it.

Looking Ahead: Let the symbol ⊞ denote any non-dimensional space over which an integral is to be evaluated. Then, the Gauss numerical integration rule is

$$I_{\boxplus} = \int_{\boxplus} F(u) \, d\boxplus \approx \sum_{q=1}^{n_q} F(u_q) \, w_q,$$

where u_q and w_q are integration points and integration weights, respectively, tabulated in ⊞ space for each n_q rule. If the function, $F(u)$, is a polynomial in u, then the approximately equal sign can be replaced with an equal sign if the number of summation points, n_q, is greater than or equal to $(p+1)/2$. Before this numerical integration can be used, the original physical integral

$$I = \int_{\Omega} \varphi(\boldsymbol{x})\, d\Omega$$

must have a change of variables from Ω to $\boxplus$, which requires a geometric Jacobian matric introduced in elementary calculus.

1.8 Other FEA to IGA comparisons

A background in FEA is not a prerequisite for the study of IGA, but a lot of the same algorithms are utilized in applying both types of studies. The main advantage of IGA over FEA is that the IGA basis functions are generally smoother (have higher continuity) than finite element interpolations. The main disadvantage of IGA compared to FEA is that the IGA basis functions generally have higher continuity than the common C^0 continuity of most finite element interpolations. This makes some essential (Dirichlet) boundary conditions very difficult to enforce in an IGA study, while they are trivial to enforce in an FEA study. Also, engineers are used to applying external forces, heat sources, etc., directly on the surface of a body, and that can rarely be done in an IGA study, but is relatively easy in an FEA study. Those challenges arise in IGA because, for many practical engineering geometries, the IGA CPs for the geometry/analysis are not physically on the body, and that is where the essential boundary conditions must be asserted. Still, there are operators that can convert a smooth IGA analysis sub-region into a C^0 IGA element that can be added into an existing FEA program. It takes some experience to know which of the approaches is best for a given application.

Unlike IGA, the FEA analysis basis (interpolation) functions were also used to approximate the geometrical shapes of components. Indeed, the FEA basis functions were often called 'shape functions' because they were used to define the geometric shape of each element. This is the opposite of IGA where the basis functions of the exact geometry are used to approximate the analysis solution in regions of the component.

Historically, FEA basis functions began with complete polynomials over triangles, then degenerated to line elements, extended to

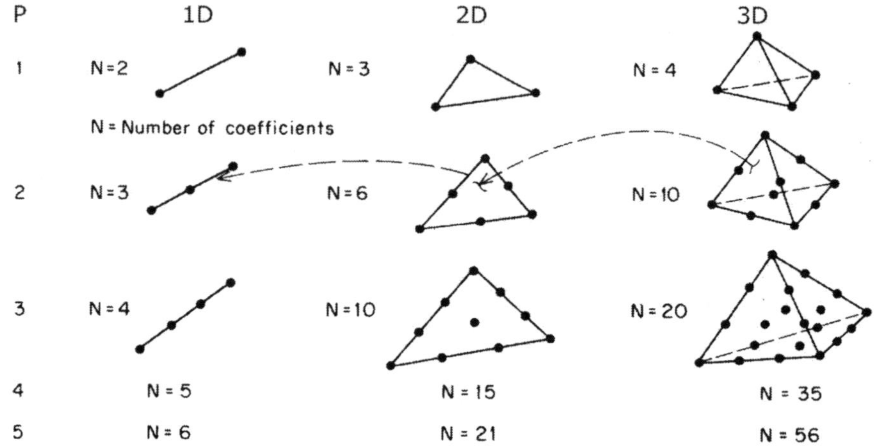

Figure 1.17. FEA nodes for simplex shapes with complete polynomials

complete polynomial tetrahedra, then incomplete polynomials over the interiors of quadrilaterals and hexahedra but with all having complete polynomials on their edges. Generally, those analysis interpolation functions were usually Lagrange polynomials. As shown in Fig. 1.17, simplex shapes with nodes (CPs) defining a complete polynomial are only C^0 compatible on the faces and/or edges with simplex elements of the same degree (P). IGA basis functions generally have higher levels of continuity between patches and result in more accurate analysis results (when boundary conditions are exactly satisfied) than FEA studies. Quadrilaterals for FEA studies only have complete polynomials on their edges, and their interior basis functions are incomplete. Their edges can be joined to simplex shapes of the same degree. Complete polynomials have an advantage in that they are unchanged by affine transformations, but quadrilaterals with incomplete polynomials on their interior give more accurate analysis results since they have a higher-degree polynomial on their interior.

Early computers were extremely memory limited so engineers viewed quadratic elements as the best trade-off between memory requirements and accuracy of the results. Meshes formed on the earliest digital computers were limited to one hundred or two hundred in number, were frequently made by hand, and were viewed with pin-and-ink plotters. Even then, cubic polynomials were available for analysis. Since a cubic polynomial is the lowest-order polynomial with the ability to have an inflection point (change in sign of the curvature), it tended to be used to define the boundaries of curved shapes, but those shapes were then sub-divided into quadratic or linear elements. The sub-division process was quickly automated by the Delaunay (1936) triangulation algorithm, which created straight-sided triangles, and later straight-edged tetrahedra. Straight edged triangles and tetrahedra have constant geometric Jacobians, which significantly reduces the cost of numerically integrating element matrices. To retain the constant Jacobian feature while upgrading the interpolations to second degree, the three edge nodes of quadratic elements were placed at the midpoints of the straight triangle sides. In other words, while quadratic interpolation was usually used for the analysis solution (and mass properties), geometry interpolation is often degenerated to a linear polynomial.

Before Delaunay tessellation was common, the mathematical concept of conformal mapping was sometimes used to create FEA meshes. That process is similar to an important IGA process which begins with a set of integer cells (or cubes) defining an integer space. A conformal map of a rectangular grid into any other shape is a function that locally preserves the interior corner angles, but not necessarily the length of a cell edge between two corners. Figure 1.18 illustrates how a mapping of integer cells can approximate shapes and allow for refinements. By selecting the shortest diagonal of a mapped cell, the resulting quadrilateral can be split into two triangular elements. Also, combining cells into sub-groups allows the creation of higher-degree elements.

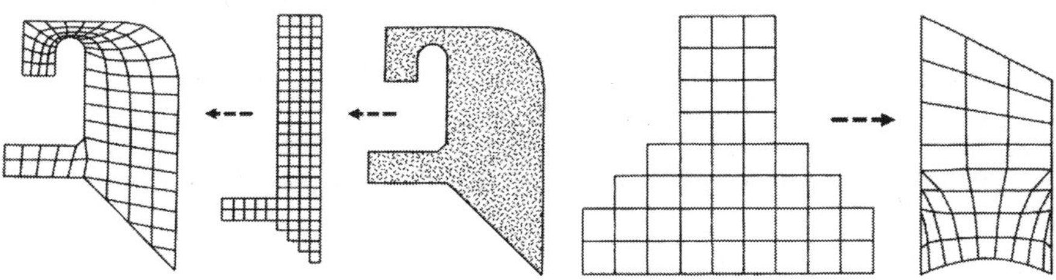

Figure 1.18. Approximate conformal mappings for FEA mesh and IGA NURBS and refinements

When cathode ray tubes became available, meshes could be created interactively, but with limitations. When Graphical User Interfaces (GUIs) became available, there was pressure to develop much more user-friendly tools which led to the parallel field of Computer-Aided Geometric Design (CAGD) for defining exact engineering shapes. CAGD, in turn, led to NURBS, which are now also used in IGA analysis. The IGA need for an analysis refinement capability led to the development of T-splines, which are also now quite popular for geometry creation in CADG.

> **Looking Ahead:** In IGA studies of surfaces (or solids), a rectangular grid of cells based on integer knot vectors is mapped with NURBS and/or T-splines to form exact geometry patches which are also used to represent the analysis solution within the patches. Only the T-splines can be refined.

1.9 Continuity considerations

A distinct advantage of IGA studies is that the analysis basis functions easily have smoother piecewise solution results (higher continuity) than FEA results. But, IGA accomplishes it by using only solution values at the CPs as system degrees of freedom. Differential equations higher than second order can require one to specify the solution slope as an essential (Dirichlet) boundary condition. Since IGA has only solution values as CP degrees of freedom, such essential conditions must be approximated. Some FEA formulations can satisfy those essential conditions exactly. The difficulty of IGA in satisfying some Dirichlet conditions in higher dimensions is discussed in detail in later sections.

It is common in both FEA and IGA studies to have a constant number of degrees of freedom at each control point. This is not a requirement, but it makes the assignments of system degree of freedom numbers in the final system matrices almost trivial.

1.10 Non-dimensional parametric coordinates

Most of the dimensionless parametric coordinate systems used in the literature for FEA analysis geometry are also found in the IGA literature and are summarized here. The most common parametric

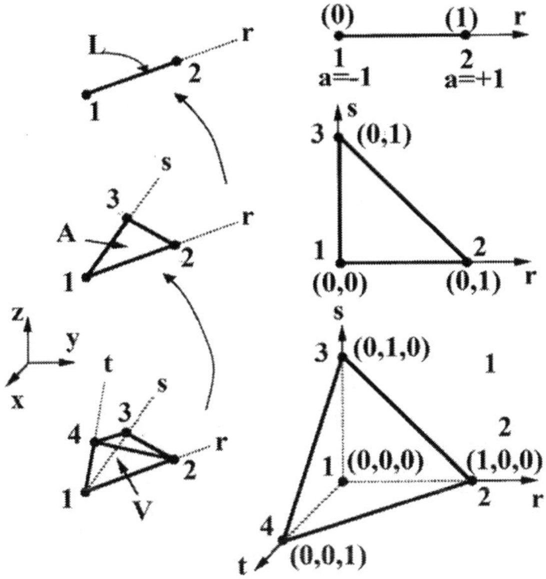

Figure 1.19. The non-dimensional unit coordinate systems in FEA and IGA

coordinates for lines, low-order triangles, and low-order tetrahedra are the 'unit coordinates' $0 \leq r, s, t \leq 1$ shown in Fig. 1.19. Common parametric coordinates for lines, quadrilaterals, and hexahedra are the 'natural coordinates' $-1 \leq a, b, c \leq 1$ as well as the unit coordinates. As noted in the top right of Fig. 1.19, these two coordinates are related by $r = (a + 1)/2$, etc.

The concept of forming basis functions for quadrilateral and hexahedra using tensor products of the 1D basis functions is used as illustrated in Fig. 1.18. For quadrilateral and hexahedra elements, using natural coordinates for the tensor products allows a very compact definition of the base functions for quadrilaterals and hexahedra (bricks). But, for low polynomial degrees, the quadrilaterals and hexahedra are also easy to write in terms of unit coordinates, as illustrated in Fig. 1.20.

Looking Ahead: A vast majority of IGA basis functions for geometry and/or analysis are expressed in unit coordinates.

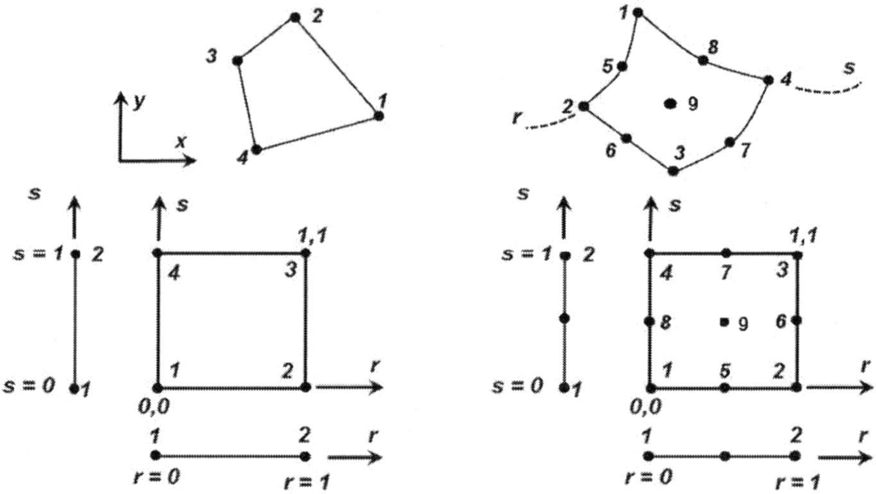

Figure 1.20. Lagrange quadrilaterals are tensor products of 1D functions

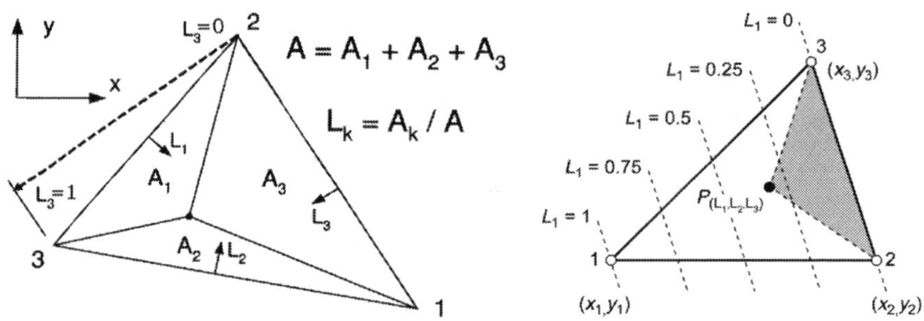

Figure 1.21. Area (Barycentric) coordinates

For lines, triangles, and tetrahedra, barycentric coordinates are popular because they simplify writing the basis functions (but with the additional expense of complicating their derivatives). For simplex shapes, there are as many barycentric coordinates as element vertices. Some alternate names are 'area coordinates' and 'volume coordinates'. The area coordinates range over $0 \leq L_k \leq 1$ and are defined as the ratio of the area between an edge and an interior point over the total area of the triangle. As illustrated in Fig. 1.21, the sum of the barycentric coordinates are unity. Herein, unit coordinates

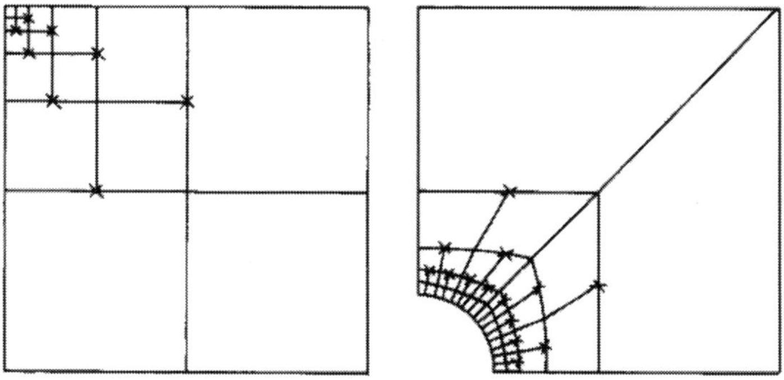

Figure 1.22. Adding a compatible fifth node (Q5s) allows the refinement into two adjacent quadrilaterals

are used for all simplex shapes and barycentric coordinates are included just for completeness (along with their quadrature data later).

1.11 Element interpolation functions

Newcomers to IGA may or may not be familiar with the basis function (interpolation function) technology of the well-established FEA. Also, some IGA results will be compared to FEA solutions of the same problems for validations or to note difficulties in each approach.

In the early days of memory-limited computers, Serendipity basis functions were developed to avoid having interior CPs (nodes) in quadrilaterals and bricks and to significantly reduce their numerical integration costs. Their basis functions are easily written in condensed forms using natural coordinates, and the linear, quadratic, cubic, and quartic quadrilateral basis functions are given in the Summary along linear, quadratic, and cubic bricks.

> **Looking Ahead**: There was an early need in FEA for the h-refinement option to improve analysis accuracy. That need led to the development of the five-node transition element (Gregory and Whiteman, 1974). That transition element, shown in Fig. 1.22, is quite analogous to the very important T-splines in IGA, which were also developed to provide refinements in T-meshes as well as reduce the total number of control point degrees of freedom.

1.12 Summary

1.12.1 *Concepts*

Isogeometric Analysis (IGA): Polynomial analysis basis functions use geometric basis functions, which exactly capture the geometry and usually have at least C^2 continuity. The analysis basis functions approximate the unknown analysis solution. The analysis basis functions can be refined (via T-splines) without modifying the exact geometry.

Finite Element Analysis (FEA): Polynomial analysis basis functions approximate the unknown solution on a piecewise basis and usually have just C^0 continuity. The analysis basis functions are also used as geometry basis functions, which rarely match the exact geometry except for flat surfaces. FEA geometry basis functions can never match important engineering shapes like circles, cones, and other conic sections. The analysis basis functions can be refined (via hanging nodes), but retain approximate geometry.

Polynomial Basis Functions: Polynomial basis functions are complete polynomials that take different names for special forms and have special capabilities. In IGA, Bernstein polynomials, B-splines, NURBS, and T-splines are most common. In FEA, Lagrange polynomials are common, along with Hermite polynomials in 1D. Polynomials of one form can be converted to, or extracted from, another form having the same polynomial degree.

1.12.2 *Terminology*

Herein, the FEA word 'node' is only used to denote an FEA point, and IGA discussions will refer to an 'index point', 'knot point', or 'control point' depending on whether the point is in the index, parametric, or physical space. The FEA word 'element' in IGA discussions is replaced with 'sub-region' or 'iElement' when the region's smooth basis functions are B-splines, NURBS, or T-splines. An iElement is a rectangular region in which all blending functions are internally C^∞ (without repeated knots) and have $C^{(p-1)}$ continuity across adjacent iElement boundaries. The new term iGe (iso-Geometric-element) is introduced when a Bézier or Lagrange extraction is applied to an iElement, forming element-like basis functions that can be inserted into a classic FEA program.

1.12.3 *Cox-de Boor recursion formula (for B-spline basis functions in unit coordinates)*

Given an analysis numeric knot vector (discussed next) with non-decreasing knots $n = \{u_1, u_2, \ldots, u_n\}$, the polynomial basis functions of degree p for the i-th knot span $[u_i, u_{i+1})$ are defined by recursively starting with the zeroth-order basis function ($p = 0$) as follows:

$$N_{i,p=0}(u) = \begin{cases} 1 & \text{if } u_i \leq u < u_{i+1} \\ 0 & \text{otherwise} \end{cases},$$

and for $p \geq 1$

$$N_{i,p}(u) = \frac{u - u_i}{u_{i+p} - u_i} N_{i,p-1}(u) + \frac{u_{i+p+1} - u}{u_{i+p+1} - u_{i+1}} N_{i+1,p-1}(u),$$

all of which satisfy $N_{i,p}(u) \geq 0$ and $\sum_{i=1}^{i=n} N_{i,p}(u) = 1$, and are linearly independent. From these, the parametric derivative is

$$\frac{d}{du} N_{i,p}(u) = \frac{p}{u_{i+p} - u_i} N_{i,p-1}(u) - \frac{p}{u_{i+p+1} - u_{i+1}} N_{i+1,p-1}(u).$$

1.12.4 *Lagrange basis functions*

$$L_{i,p}(u) = \prod_{\substack{m \neq i}}^{1<m\leq(p+1)} \frac{u - u_m}{u_i - u_m}, \text{ with } \sum_{i=1}^{i=p+1} L_{i,p}(u) = 1.$$

For example,

$$L_{1,p}(u) = \prod_{\substack{m \neq 1}}^{1<m\leq(p+1)} \frac{u - u_m}{u_i - u_m}$$
$$= \left(\frac{u - u_2}{u_1 - u_2}\right)\left(\frac{u - u_3}{u_1 - u_3}\right)\cdots\left(\frac{u - u_{p+1}}{u_i - u_{p+1}}\right)$$

1.12.5 *B-spline curve*

A B-spline curve in D-dimensional space, $\mathbf{C}(u)$, is constructed as a piecewise linear combination of B-spline basis functions (Bernstein polynomials) with their n control point values, $\boldsymbol{B}_i$, as coefficients

$$\boldsymbol{C}(u) = \sum_{i=1}^{n} N_{i,p}(u)\boldsymbol{B}_i.$$

In CAGD, the ith row of $\boldsymbol{B}$ contains the d-spatial coordinates, and often their weight. In IGA, that row is extended to contain associated degree-of-freedom (DOF) values, and their weight.

A B-spline curve is defined with a series of polynomials. The number of polynomials needed depends on the number of CPs and the order of the curve. A 'knot vector' of length $n+p+1$ determines where in the parameter range these polynomials start and stop. They are called 'knots' because the values tie the polynomials together. The gap between two consecutive knots is called the 'knot span'.

1.12.6 *B-spline surface*

A parametric surface is defined by two generally non-orthogonal parametric coordinates, say u and v. A B-spline surface is created by a rectangular domain formed by the tensor product of two B-spline curves each of which is associated with a knot vector. The first set of basis functions, $N_{i,p}(u)$, is of degree p and its defining knot vector has $n + p + 1$ knots. The second set of basis functions, say $M_{j,m}(v)$, is of degree q and its defining knot vector has $m + q + 1$ knots.

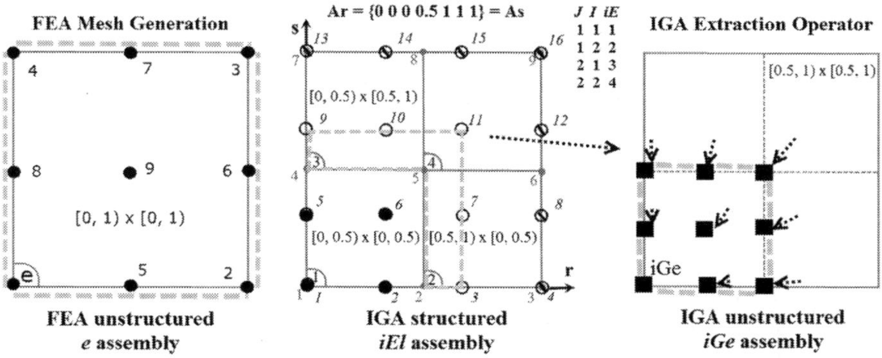

Figure 1.23. Parametric space for an element, an iElement, and an iGe extraction

The tensor product B-spline surface is defined as

$$\boldsymbol{S}(u, v) = \sum_{i=1}^{n} \sum_{j=1}^{m} N_{i,p}(u)\, M_{j,m}(v) \boldsymbol{B}_{i,j},$$

where the basis function tensor product is also non-negative and a partition of unity.

1.13 Three analysis entities

Essentially, this book is about extending the concept of an 'element', in the FEA sense, to two alternate analysis domains developed by Prof. Tom Hughes and his coworkers. Unlike most IGA literature, herein, the three different mathematical approaches have been assigned different names so as to emphasize the approach being discussed. To partially justify that approach, a conceptual generalization of their differences is sketched in Fig. 1.23. A major difference between the classic FEA study and the earliest IGA studies is that the FEA nodes always lie on, and inside, the boundary of an element, while the structured iElement has CPs located outside the physical size of the knot span defining the iElement. That is, a finite element has no outside support (dashed lines in Fig. 1.23), while an iElement has an inverted 'L-shaped' or 'ring-shaped' region of support that is outside the knot span sizes.

After a lengthy discussion of iElements in the forms of B-splines, NURBS, or T-splines, the very important subject of using mathematical matrix extraction techniques to form an IGA extract (iElement $\rightarrow$ iGe) that has a data structure that fits common FEA programs (with minor changes) and still retains the higher continuity and accuracy of the iElements will be addressed. Being able to have the option of replacing a classic C^0 finite element approximation with an IGA extract with higher continuity greatly enhances the large investment expended to develop current finite element systems for various fields of application.

1.14 Comments

Industrial experience attributes about 80% of the total analysis time to converting a complex geometric model into an FEA mesh (not needed in IGA) vs. 20% to the FEA itself.

FEA shape functions are given directly as bivariate functions, but the IGA basis functions are defined for each direction separately.

FEA geometry is simply given by x-, y-, z-coordinates of nodes and the element connection list. IGA geometry is defined by knot vectors and x-, y-, z-coordinates of CPs.

IGA writes displacements and reactions at the CPs, in contrast to FEA writing them at the nodes.

In the iElement version of IGA, continuous basis functions are created and assembled over loops in each parametric direction, whereas FEA (and iGe extracts) uses a single loop over elements.

On a per-degree-of-freedom basis, in the field of computational mechanics, IGA is shown to be superior to FEA in that IGA exhibits increased accuracy and robustness.

IGA evaluates analysis matrices by looping over NURBS patches, then looping over iElements, and then quadrature spaces. FEA begins with a mesh (hopefully) and loops over elements, and then quadrature spaces.

For approximately the same degree, p, and the same number of global degrees of freedom, IGA shows strain energy that is closer to the exact value than does FEA.

The IGA geometric representation (e.g., NURBS and T-splines) is typically continuous and smooth, whereas the FEA representation is typically continuous but not smooth.

IGA utilizes a single (iso) class of basis functions for both geometry and engineering analysis.

An IGA parameter space is partitioned into sub-regions (iElements) by a knot vector in each direction.

An IGA iElement exists between two non-repeated knots; thus, a knot vector defines the iElement mesh in its direction.

IGA offers smoother analysis results for smooth data than FEA does.

Numerical integration using Gauss quadratures is not ideal for the IGA continuous basis functions as it is for finite element polynomials.

Control points, CPs, serve as the degrees of freedom (dofs) in IGA.

All conic sections can be exactly represented by rational quadratic polynomial curves.

Knot spans define iElement, and iGe, domains.

An FEA database is easily enriched by additional connection data and extraction data needed by iGe extracts.

Each 1D polynomial basis function of degree p has $(p+1)$ knots which support it.

Increasing a basis function degree, p, gives smoother curves (and analysis results).

A Bézier curve is defined as an interpolation of points, but a B-spline curve is an interpolation of piecewise curves.

Having high continuity across the 'iElement' boundaries gives a continuously defined flux, or stress, in a homogeneous material (vs. FEA discontinuous fluxes and stresses).

B-spline curves, and iElements, require three things: Knot Vectors, Basis Functions, and CPs.

To construct a B-spline curve, a set of physical CPs, $[\boldsymbol{P}] = [\boldsymbol{x}; \boldsymbol{y}; \boldsymbol{z}]$, is required.

A B-Spline works *because* of the concept of a knot vector (next chapter).

Tensor product basis functions do not allow local mesh refinement needed in analysis.

An affine transformation of a B-spline curve is obtained by applying the transformation to its CPs.

NURBS basis functions with all weights equal to one become B-spline basis functions.

Bézier extraction (decomposition) is used to compute a set of Bernstein polynomials from NURBS for use in a classic FEA program.

Lagrange extraction (decomposition) is used to compute a set of Lagrange polynomials from Bézier extraction for use in a classic FEA program.

An IGA 2D analysis refinement 'T-junction' is analogous to an FEA 'hanging node'.

Natural coordinates, $-1 \leq a \leq 1$, $a = (2u - 1)$, and unit coordinates, $0 \leq u \leq 1$, $u = (a + 1)/2$, are equally valid for forming basis functions and quadrature rules.

Chapter 2

Knot Vectors, Knots, and Control Points

2.1 Introduction

One of the confusing but very important topics for newcomers to Isogeometric Analysis (IGA) is the use of knot vectors which contain a sequence on non-diminishing real numbers, say r_k, called knots. The knots are probably called that because they are locations that 'tie together' the support regions (knot spans) of the set of piecewise polynomials which define a curve governed by the knot vector. Knots also are points that lie on a curve defined by a knot vector. The shape of the curve is defined by the user's choice of control point (CP) spatial coordinates. Since knot vectors are so well known in the CAGD community, the famous and very useful *The NURBS Book* (Piegl and Tiller, 1997) only devotes a few lines discussing them, but its index lists many algorithms for using them.

In CAGD, the knot vector is just the knot vector even though it can be presented in a few different ways. In geometry literature, the knots are numbered from 0, but in engineering literature, they are numbered from 1. The IGA concept provides additional important ways to view a knot vector. Herein, the geometric and analytic views of a knot vector are given different symbols, though that is neither common nor necessary.

2.2 Knot vectors

An open knot vector provides several pieces of information. The number of repeated end values is the same as the number of coefficients $(p+1)$ needed to describe a complete polynomial basis function of degree p. The knot vector length, m, less the number of coefficients, gives the number of user-defined control points (CPs) that govern a curve shape. Each entry in a knot vector becomes a real number parametric coordinate, say r_k, with an integer subscript. The integer subscripts create an integer space to help visualize the support region of the basis functions. The real value, r_k, locates the knot in the r-direction parametric space. That parametric location, r_k, is then mapped into a physical space knot point (x_k, y_k). The sequential knot integer subscripts are never repeated; but, the parametric coordinates can be repeated, giving $r_k = r_{k+1}$. Straight lines of integers in index space become real number lines in the parametric space and curves physical space. When knots are repeated, their parametric lines and physical curves will fall on top of each other. In other words, the three types of grid representations (considered in detail later) can look different but help understand the support regions where basis functions are non-zero.

A 'uniform knot vector' has equally spaced intervals; otherwise, it is called a 'non-uniform knot vector'. A non-uniform knot vector tends to bias a curve toward the smaller intervals as shown in Fig. 2.1, and in IGA, it focuses the analysis basis functions toward the smaller intervals.

It is common to create a geometry knot vector where the knots are a sequence of integers, fractions, or real numbers and where repeated

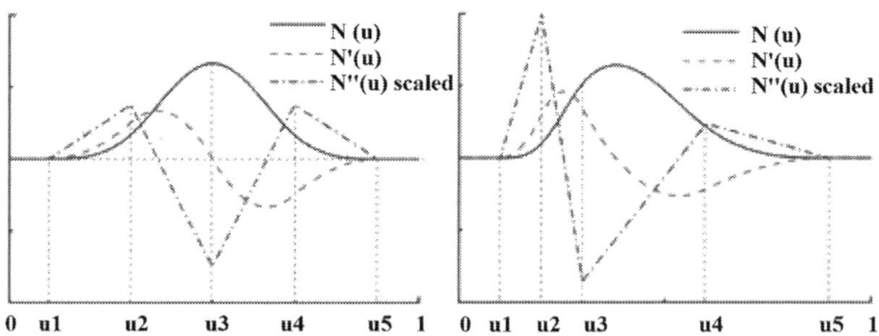

Figure 2.1. Cubic B-spline curve and its scaled derivatives for a uniform (left) and non-uniform knot vector

entries at each end define the polynomial degree, p, of the curve governed, in part, by the knot vector. A knot vector is said to be open if its first and last knot values are repeated $(p+1)$ times. Most IGA studies use Open Non-uniform Rational B-Splines (ONURBS), but ONURBS are still commonly referred to and pronounced as NURBS.

An IGA study with a curve usually begins with the exact NURBS, B-spline, or T-spline geometrical shape of a domain defined by the geometric knot vector

$$\mathbb{K}\boldsymbol{n} = \{n_1, n_2, \ldots, n_{m=n+p+1}\}, \text{ having knot spans } [n_i,\ n_{i+1}),$$

where the knot span notation $[a, b)$ is a half open region, $a \leq x \leq (b - \varepsilon),\ \varepsilon \ll 1$.

Historically, knot spans just represented a segment of the geometry. In IGA, they become an analysis entity and they define the range over which the IGA matrices for that entity must be integrated. The first type of IGA analysis region, introduced in the seminal paper by Hughes *et al.* (2005) was the knot span with its geometric NURBS or B-splines also used to approximate the analysis variable. That analysis domain is clearly analogous to the FEA element domain, and is often referred to as an 'element' in the IGA literature. The IGA domain is distinctly different from the FEA element, and herein it is denoted as an 'iElement' to emphasize that fact (even though it is still pronounced 'element'). The spelling implies that the domain is ***i***mproved (by higher continuity) and ***E***xpanded (by having a larger support region that an FEA element occupying the same volume).

2.3 Typical knot vectors

Probably, the most common style for a geometric cubic B-spline knot vector would include sequential integers like $\mathbb{K}\boldsymbol{n} = \{0, 0, 0, 0, 1, 2, 3, 4, 4, 4, 4\}$ to define the curve and its representation in the 'index space'. An equivalent integer knot vector $\mathbb{K}\boldsymbol{n} = \{1, 1, 1, 1, 2, 3, 4, 5, 5, 5, 5\}$, etc. The number of repeated knots at the beginning and end of an open knot vector is the number of coefficients needed, $p + 1$, to define the complete polynomials of degree p it represents. By inspection of the above-mentioned knot vectors, that number is $4 = p + 1$, and the above-mentioned example knot vectors will define piecewise cubic, $p = 3$, basis functions. The length of a knot vector, $m = 11 = n + (p + 1)$, defines how many user-selected CPs, n, are needed to uniquely define the set of piecewise cubic polynomials over the physical domain. In geometry and CAGD, the $n = 11 - 4 = 7$ spatial coordinates, and their weights, at the CPs define the exact physical geometry with which the engineering analysis of that region begins.

Note that all open knot vectors have repeated numbers as knots, but there is a clear sequence of the unique increasing knot values. The unique entries in each knot vector are important as beginning points for IGA data structures. The first three example geometric knot vectors listed earlier are (unique $\mathbb{K}\boldsymbol{n}$) $= (0, 1, 2, 3, 4)$, (unique $\mathbb{K}\boldsymbol{n}$) $= (1, 2, 3, 5)$, and (unique $\mathbb{K}\boldsymbol{u}$) $= (0, 0.25, 0.5, 0.75, 1)$. The Matlab function *unique(vector)* returns the three unique lists noted earlier.

Importantly, define the number of iElements (knot spans) in the r-direction as

$$n_{kr} = \text{length}(\text{unique}(\mathbb{K}\boldsymbol{r})) - 1.$$

In a 3D study, the additional number of iElements in the parametric s- and t-directions would be

$$n_{ks} = \text{length}(\text{unique}(\mathbb{K}\boldsymbol{s})) - 1, \quad n_{kt} = \text{length}(\text{unique}(\mathbb{K}\boldsymbol{t})) - 1.$$

Clearly, their product defines the number of iElements active in an analysis, but not the total number of degrees of freedom (dofs) at the analysis control points. For that number, the number of polynomial coefficients in each parametric direction is also needed. Let the polynomial degrees in the three parametric dimensions be p, q, and c. Then, the number of analysis control points (degrees of freedom) in each parametric direction is

$$n_{cr} = n_{kr}(p + 1), \quad n_{cs} = n_{ks}(q + 1), \quad n_{ct} = n_{kt}(c + 1).$$

Leads to the total number of CP in the 'mesh' in 1D, 2D, or 3D are $n_m = n_{cr}$, or $n_m = n_{cr}n_{cs}$, or $n_m = n_{cr}n_{cs}n_{ct}$, respectively. Finally, the number of system analysis dofs $n_d = n_m n_g$, where $n_g \geq 1$ denotes the number of generalized unknown dofs at each CP. The sequence of unique pairs in a knot vector is the number of knot spans with *non-zero* basis functions. Each such knot span in IGA studies becomes an analysis region (the iElement). Having that number is the first step in building a database for automating IGA studies.

There are an infinite number of ways to write the knot vector for a single unique curve. They can all be reduced to a single unique unitary one. Herein, for calculating the analysis basis functions, the unique unitary knot vector is usually used. For the first example, a knot vector subtracting 1 from every entry and then dividing by 4 gives its unique form: $\mathbb{K}\boldsymbol{u} = \{0, 0, 0, 0, 0.25, 0.5, 0.75, 1, 1, 1, 1\}$. For the unique $\mathbb{K}\boldsymbol{u}$, the ranges of those analysis regions are $[0, 0.25)$, $[0.25, 0.5)$, $[0.5, 0.75)$ and $[0.75, 1)$. Those are essential data because they identify the intervals over which their respective non-zero basis functions must be integrated to build any IGA analysis matrix. Therefore, they are stored for each parametric direction for each knot span. For the current example, where the number of knot spans (iElements) in the r-direction is $n_{kr} = 4$, the array would be stored for later integration use as

$$[\text{range\_r}] = \begin{bmatrix} 0 & 0.25 \\ 0.25 & 0.5 \\ 0.5 & 0.75 \\ 0.75 & 1 \end{bmatrix}.$$

For a non-unitary form like $\mathbb{K}\boldsymbol{n} = \{0, 0, 0, 0, 1, 2, 3, 4, 4, 4, 4\}$, the ranges are

$$[\text{range\_n}] = \begin{bmatrix} 0 & 1 \\ 1 & 2 \\ 2 & 3 \\ 3 & 4 \end{bmatrix}.$$

When numerical integration is used, the parametric ith span interval, $\langle \Delta r_i \rangle = \langle r_{i+1} - r_i \rangle$, also sets the constant Jacobian needed to map into the non-dimensional quadrature space (as shown later).

2.4 Knot vector constants

Together, in 2D, the number of knot spans represents a grid of $n_{kr} \times n_{ks}$ rectangular cells in parametric space (and in index space) that become curvilinear quadrilateral analysis regions (iElements) in 2D or 3D physical space.

An IGA study mainly utilizes open knot vectors with the breakdown shown in Fig. 2.2.

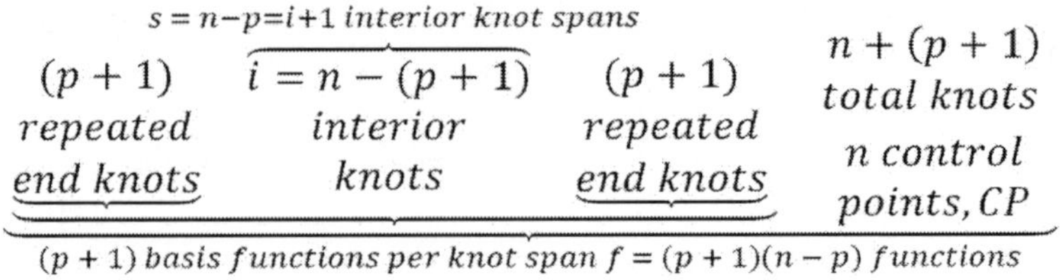

Figure 2.2. Breakdown of an open knot vector

For example, a geometric knot vector,

$$\mathbb{K}\boldsymbol{u} = \left\{0, 0, 0, 0, \frac{1}{3}, \frac{2}{3}, 1, 1, 1, 1\right\},$$

$$\text{Unique } \mathbb{K}\boldsymbol{u} = \left(0, \frac{1}{3}, \frac{2}{3}, 1\right),$$

$$(p+1) = 4, \quad p = 3, \quad m = 10 = n + (p+1), \; n = 6$$

$$\text{Control Points,}$$

$$i = n - (p+1) = 2 \quad \text{Interior knots,}$$

$$s = i + 1 = n - p = 6 - 3 = 3 \quad \text{Interior knot spans,}$$

$$f = (p+1)s = (4)(3) = 12 \quad \text{Total piecewise basis functions,}$$

$$\text{Knot Span Ranges : } \left[0, \frac{1}{3}\right), \left[\frac{1}{3}, \frac{2}{3}\right), \left[\frac{2}{3}, 1\right)$$

$$\text{with CP } 1 - 4, 2 - 5, 3 - 6,$$

and those 12 basis functions are listed in the Appendix A1 where the knot span is their column number. Likewise, an open uniform cubic ($p = 3$) B-spline with 7 CPs can have a numeric knot vector

$$\begin{aligned} \mathbb{K}\boldsymbol{u} &= \{0, 0, 0, 0, 1, 2, 3, 4, 4, 4, 4\} \\ &\rightarrow \left\{0, 0, 0, 0, \frac{1}{4}, \frac{2}{4}, \frac{3}{4}, \frac{4}{4}, \frac{4}{4}, \frac{4}{4}, \frac{4}{4}\right\}, \\ \text{Unique } \mathbb{K}\boldsymbol{u} &= \left(0, \frac{1}{4}, \frac{2}{4}, \frac{3}{4}, \frac{4}{4}\right), \end{aligned}$$

$$\begin{aligned}
(p+1) = 4, \quad p = 3, \quad & m = 11 = n + (p+1),\ n = 7 \\
& \text{control points,} \\
i &= n - (p+1) = 3 \quad \text{Interior knots,} \\
s &= 7 - 3 = 4 \quad \text{Interior knot spans,} \\
f &= (4)(4) = 16 \quad \text{Total piecewise basis functions,} \\
& \text{Knot Spans}: \left[0, \frac{1}{4}\right), \left[\frac{1}{4}, \frac{2}{4}\right), \left[\frac{2}{4}, \frac{3}{4}\right), \left[\frac{3}{4}, \frac{4}{4}\right) \\
& \text{with CP } 1-4, 2-5, 3-6, 4-7.
\end{aligned}$$

In the IGA literature, a knot vector is usually represented with a capital Greek letter, like Xi Ξ, and the individual entries are the lower case Greek letter, ξ, with integer subscripts such as

$$\Xi = \{\xi_1, \xi_2, \ldots, \xi_{m=n+p+1}\}.$$

But, advanced standardized programming languages like Fortran 2023, C++, python, and the Matlab environment do not yet include Greek letters in their character sets. To present algorithms related to CAGD and IGA, the literature often spells out the names of the Greek letters in the source code. To avoid that, the letters r, s, and t will be used to represent parametric spatial directions and x, y, and z will represent physical space directions. This allows later Matlab scripts to use the same characters as those appearing in equations in the text.

Only the normalized relative difference between the knot values is importance in defining the shape of a curve. Adding a constant to all knot values will generate the same curve. Likewise, by multiplying all knots by a constant factor leaves the curve unchanged. The tensor product of two such knot vectors defines a structured NURBS surface; three tensor products define a structured NURBS solid shape. In time-dependent IGA studies, it is possible to use four tensor products to create a (non-obvious) NURBS space–time domain (e.g., Liu *et al.*, 2022).

> **Looking Ahead**: A geometry of structured (tensor product) cells does not allow for a computationally efficient refinement needed in most engineering studies. Thus, several topics need to be mastered before introducing T-spline refinements which are extracted from NURBS by using 'local analysis vectors'.

2.5 Analysis knot vectors

The IGA concept began with the use of the exact geometry CP to define the spatial distribution of the solution of an analysis. That is, the geometry CP is usually thought of in IGA literature as also referring to additional analysis quantities like temperatures or displacements, and their weights. That means a CP would contain geometry weights and analysis weights. This is one reason why, herein, the concept of a separate analysis CP, AP, is introduced to contain any information about study results and their weights. The exact geometry knot vector can be a sub-set of an analysis knot vector, $\mathbb{K}\boldsymbol{n} \subseteq \mathbb{A}\boldsymbol{n}$.

In an IGA study, analysis accuracy almost always requires local refinement by introducing additional knots and thus analysis CPs, AP, into the analysis knot vector, $\mathbb{A}\boldsymbol{n}$, beyond those in the original geometric knot vector, $\mathbb{K}\boldsymbol{n}$, to increase the number of sub-regions, without changing the original exact geometric shape of the curve (or domain). The analysis numeric knot vector in the 1D r-parametric direction is denoted as the knot sequence

$$\mathbb{A}\boldsymbol{r} = \{r_1, r_2, \ldots, r_{mr=nr+(pr+1)}\}, \quad \text{having knot spans } [r_i, r_{i+1})$$

or its unitary form

$$\mathbb{A}\boldsymbol{u} = \{0, 0, \ldots, 1, 1_{m=n+(p+1)}\}, \quad \text{having knot spans } [u_i, u_{i+1}),$$

Where, for analysis basis functions of polynomial degree p, there still are $p+1$ leading zeros and $p+1$ trailing ones in the knot vector of length m.

The key points in the conceptual change in notation from $\mathbb{K}\boldsymbol{u}$ to $\mathbb{A}\boldsymbol{u}$ are that an analysis knot vector can retain exactly the same geometry created in a CAGD system but can have a larger or smaller number of analysis control points (APs) which have expanded data which add the analysis dofs and their weights in addition to the new CP coordinates and weights originally in the geometric vector, $\mathbb{K}\boldsymbol{n}$.

> **Looking Ahead**: Adding more analysis CPs, APs, to a geometric knot vector (called knot insertion) does not change the exact geometry and allows a segment of the geometry to be extracted and utilized in a standard FEA code as an element-like entity (herein called an iGe or an extract) with enhanced IGA abilities.

Also note that the entries in any knot vector, like r_k, are considered to be real numbers with sequential integer subscripts. The integer subscripts never repeat but the sequential real numbers can, and frequently do, repeat. The repetitive real values of some knots make it difficult to visualize them in a mesh. To aid in visualizing each basis function's support regions, it is important to learn (shortly) about using the subscripts, k shown as uniformly distributed in an integer index space; the parametric values, r_k, shown as non-uniformly distributed in an parametric space; and the geometric shapes and the analysis results, shown as non-uniformly distributed in physical space.

The first and last CPs always fall at the ends of the parametric coordinates. The interior CP locations are assumed to be equally spaced in parametric space, but in physical space the CP locations are selected by the user. There are special physical interior CP locations that can be selected to force the Jacobian mapping from parametric space to physical space to be a constant. This allows some analytical integrals to be illustrated here. But, such location choices are usually not made in a practical analysis.

The Cox-de Boor recursion formula for B-spline basis functions of degree p are

$$p = 0{:}\quad N_{j,0}(u) = \begin{cases} 1 & \text{if } u_j \leq u \leq u_{j+1} \\ 0 & \text{otherwise} \end{cases},$$

$$p = 1, 2, 3, \ldots \qquad N_{j,p} = \frac{u - u_j}{u_{j+p} - u_j} N_{j,p-1}(u)$$

$$+ \frac{u_{j+1} - u}{u_{j+p+1} - u_{j+1}} N_{j+1,p-1}(u).$$

Both the numeric and symbolic algorithms for evaluating the basis functions can use a general numeric knot vector. Note from the recursion formula for B-spline basis functions that it is always just the relative spans, $\Delta n_i = n_{i+p} - n_i$ or $\Delta u_i = u_{i+p} - u_i$, that determine the curve shape or the spatial distribution of an analysis variable along that curve. In Appendix 1, several example basis functions (from symbolic Matlab) are listed. In it, the unique unitary knot vector descriptions are listed first, followed by some integer forms.

However, most of the examples herein utilize the normalized unit coordinate form of the analysis knot vector:

$$\mathbb{A}\boldsymbol{u} = \{u_1 = 0, u_2, \ldots, u_{m=n+(p+1)} = 1\}, \text{ having knot spans } [u_i, u_{i+1})$$

and where the parametric coordinate domain is $0 \leq u \leq 1$. The scaling for this change is simple; subtract the first number from every knot value and divide the knot vector by the new last entry. For example, the conversion of the above-mentioned cubic curve knot vector gives

$$\mathbb{A}\boldsymbol{n} = \{1, 1, 1, 1, 2, 3, 5, 5, 5, 5\} \Rightarrow$$
$$\mathbb{A}\boldsymbol{u} = \{0, 0, 0, 0, 0.25, 0.5, 0.75, 1, 1, 1, 1\}.$$

For example, what looks like a single planar B-spline cubic curve (Piegl and Tiller, 2012) shown in Fig. 2.3(b) is built using the geometric knot vector $\mathbb{K}\boldsymbol{u} = \{1, 1, 1, 1, 2, 3, 5, 5, 5, 5\}$ and consists of seven basis functions with four cubic polynomials in each of the four interior knot spans [Piegl and Tiller, 1997). The shape of the interpolated curve is defined by the x- and y-coordinates at the CPs (and their weights). The curve has $C^{p-1} = C^2$ continuity, so at the junction of two knot spans, the position, slope, and curvature are continuous, and have C^∞ between the knots.

In a practical IGA study, the analysis knot vector begins as the geometric knot vector, which captures the exact geometry to be used. Then, the analysis knot vector inserts another 10, 40, or 100 uniform knots to enhance the analysis results without changing the exact geometry. In such cases, it is usually not practical to create their analytic form, as done here for simple examples, and the basis functions

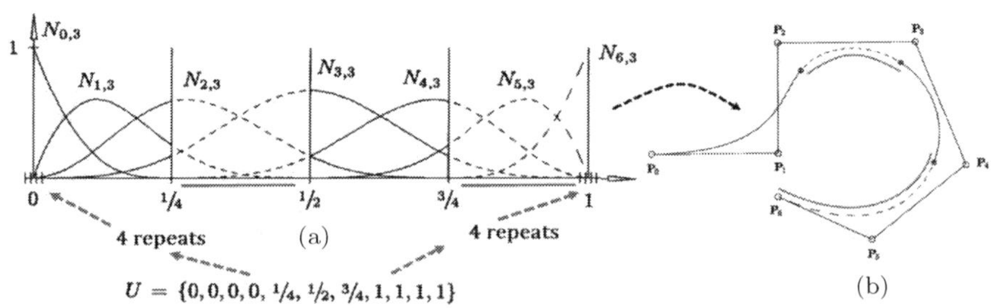

Figure 2.3. **A geometric knot vector and B-splines for a planar cubic curve with seven CPs (Piegl and Tiller, 2012)**

and their derivatives are evaluated numerically at the quadrature points. Some quadratic basis functions are evaluated by hand in Example 2.1.

Example 2.1. Quadratic basis functions via recursive rules
Given: The open analysis knot vector is given as

$$\{u_1 = 0, u_2, \ldots, u_{m=n+p+1} = 1\} = \left\{0, 0, 0, \frac{1}{2}, 1, 1, 1\right\}.$$

The three repeated end knots give $3 = p+1$ meaning that the final basis functions are to be polynomial degree $p = 2$ and its length of $m = 7 = n + p + 1$ gives that there are $n = 7 - 3 = 4$ final basis functions: $N_{1,2}(u)$, $N_{2,2}(u)$, $N_{3,2}(u)$, and $N_{4,2}(u)$. Determine their analytic forms over the two knot spans of $[0, 1/2)$ and $[1/2, 1)$.

Solution: An analysis unitary knot vector with non-decreasing knots for polynomial basis functions of degree p : $\mathbb{A}\boldsymbol{u} = \{u_1 = 0, u_2, \ldots, u_{m=n+p+1} = 1\}$, then the functions for the ith knot span $[u_i, u_{i+1})$ are defined by recursively starting with the zeroth-order basis function $(p = 0)$ as follows:

$$N_{i,p=0}(u) = \begin{cases} 1 & \text{if } u_i \leq u < u_{i+1} \\ 0 & \text{otherwise} \end{cases},$$

and for $p \geq 1$,

$$N_{i,p}(u) = \frac{u - u_i}{u_{i+p} - u_i} N_{i,p-1}(u) + \frac{u_{i+p+1} - u}{u_{i+p+1} - u_{i+1}} N_{i+1,p-1}(u).$$

Beginning with $p = 0$, and $i = 1$, there is no $u_1 \leq u < u_2$ so $N_{1,0}(u) \equiv 0$. Likewise, there is no $u_2 \leq u < u_3$, so $N_{2,0}(u) \equiv 0$. Next,

$$N_{3,0}(u) = \begin{cases} 1 & \text{if } 0 \leq u < 1/2 \\ 0 & \text{otherwise} \end{cases} \quad \text{and}$$

$$N_{4,0}(u) = \begin{cases} 1 & \text{if } 1/2 \leq u < 1 \\ 0 & \text{otherwise} \end{cases}.$$

Similarly, since there is no $u_5 \le u < u_6$, etc., $N_{5,0}(u) \equiv 0 \equiv N_{6,0}(u) \equiv 0$ and they terminate the number of $p = 0$ basis functions because $n = 7-1-0 = 6$. Going on to the five $p = 1$ basis functions,

$$N_{1,1}(u) = \frac{u - u_1}{u_2 - u_1} N_{1,0}(u) + \frac{u_3 - u}{u_3 - u_2} N_{2,0}(u)$$
$$= \frac{u-0}{0-0}0 + \frac{0-u}{0-0}0 = 0,$$

$$N_{2,1}(u) = \frac{u - u_2}{u_3 - u_2} N_{2,0}(u) + \frac{u_4 - u}{u_4 - u_3} N_{3,0}(u)$$
$$= \frac{u-0}{0-0}0 + \frac{1/2-u}{1/2-0} N_{3,0}(u),$$

$$N_{2,1}(u) = 0 + (1-2u)\begin{cases} 1 & \text{if } 1/2 \le u < 1 \\ 0 & \text{otherwise} \end{cases},$$

$$N_{3,1}(u) = \frac{u - u_3}{u_4 - u_3} N_{3,0}(u) + \frac{u_5 - u}{u_5 - u_4} N_{4,0}(u)$$
$$= \frac{u-0}{1/2-0} N_{3,0}(u) + \frac{1-u}{1-1/2} N_{4,0}(u),$$

$$N_{3,1}(u) = 2u\begin{cases} 1 & \text{if } 0 \le u < 1/2 \\ 0 & \text{otherwise} \end{cases} + 2(1-u)\begin{cases} 1 & \text{if } 1/2 \le u < 1 \\ 0 & \text{otherwise} \end{cases}.$$

$$N_{3,1}(u) = \begin{cases} 2u & \text{if } 0 \le u < 1/2 \\ 2(1-u) & \text{if } 1/2 \le u < 1 \\ 0 & \text{otherwise,} \end{cases}$$

$$N_{4,1}(u) = \frac{u - u_4}{u_5 - u_4} N_{4,0}(u) + \frac{u_6 - u}{u_6 - u_5} N_{5,0}(u)$$
$$= \frac{u-1/2}{1-1/2} N_{4,0}(u) + \frac{1-u}{1-1/2},$$

$$N_{4,1}(u) = (2u-1)\begin{cases} 1 & \text{if } 1/2 \le u < 1 \\ 0 & \text{otherwise} \end{cases} \quad \text{and } N_{5,1}(u) = 0 + 0 = 0.$$

The final four quadratic basis functions become

$$N_{1,2}(u) = \frac{u - u_1}{u_3 - u_1} N_{1,1}(u) + \frac{u_4 - u}{u_4 - u_2} N_{2,1}(u),$$

$$N_{1,2}(u) = \frac{u-0}{0-0}N_{1,1}(u) + \frac{1/2-u}{1/2-0}N_{2,1}(u)$$
$$= 0 + (1-2u)^2 \begin{cases} 1 & \text{if } 1/2 \le u < 1 \\ 0 & \text{otherwise} \end{cases},$$
$$N_{2,2}(u) = \frac{u-u_2}{u_4-u_2}N_{2,1}(u) + \frac{u_5-u}{u_5-u_3}N_{3,1}(u)$$
$$= 2u\ N_{2,1}(u) + (1-u)N_{3,1}(u),$$
$$N_{2,2}(u) = 2u(1-2u)\begin{cases} 1 & \text{if } 1/2 \le u < 1 \\ 0 & \text{otherwise} \end{cases}$$
$$+(1-u)\begin{cases} 2u & \text{if } 0 \le u < 1/2 \\ 2(1-u) & \text{if } 1/2 \le u < 1 \\ 0 & \text{otherwise} \end{cases},$$
$$N_{2,2}(u) = \begin{cases} 2u(2-3u) & \text{if } 0 \le u < 1/2 \\ (2-2u)2(1-u) & \text{if } 1/2 \le u < 1 \end{cases},$$
$$N_{3,2}(u) = \frac{u-u_3}{u_5-u_3}N_{3,1}(u) + \frac{u_6-u}{u_6-u_4}N_{4,1}(u),$$
$$N_{3,2}(u) = \frac{u-0}{1-0}N_{3,1}(u) + \frac{1-u}{1-1/2}N_{4,1}(u)$$
$$= uN_{3,1}(u) + 2(1-u)N_{4,1}(u),$$
$$N_{3,2}(u) = u\begin{cases} 2u & \text{if } 0 \le u < 1/2 \\ 2(1-u) & \text{if } 1/2 \le u < 1 \\ 0 & \text{otherwise} \end{cases}$$
$$+\,2(1-u)\begin{cases} (2u-1) & \text{if } 1/2 \le u < 1 \\ 0 & \text{otherwise} \end{cases},$$
$$N_{3,2}(u) = \begin{cases} 2u^2 & \text{if } 0 \le u < 1/2 \\ 2(1-u)(3u-1) & \text{if } 1/2 \le u < 1 \\ 0 & \text{otherwise} \end{cases},$$
$$N_{4,2}(u) = \frac{u-u_4}{u_6-u_4}N_{i,p-1}(u) + \frac{u_7-u}{u_7-u_5}N_{5,1}(u),$$

$$N_{4,2}(u) = \frac{u - 1/2}{1 - 1/2} N_{4,1}(u) + \frac{1 - u}{1 - 1/2} N_{5,1}(u)$$
$$= (2u - 1)N_{4,1}(u) + 2(1 - u)N_{5,1}(u),$$
$$N_{4,2}(u) = (2u - 1)\begin{cases} (2u - 1) & \text{if } 1/2 \le u < 1 \\ 0 & \text{otherwise} \end{cases}.$$

2.6 Properties of basis functions

B-spline, NURBS, and T-spline basis functions will be considered in detail later for various IGA study applications. They all have the mathematical property seen in Fig. 2.4 in that all such basis functions are non-negative. They also have the important mathematical property called a 'partition of unity', which means that at any parametric point the sum of all of the basis functions is one. The partition-of-unity property also means that within any knot span the sum of all non-zero basis functions in the knot span (iElement) is zero. This property is illustrated in Fig. 2.4 where three basis functions in a knot span sum to the unit black line (right image).

The concept of combining piecewise curves for a cubic spline having four knot spans is sketched in Fig. 2.5. The thin lines outside each of the four spans are local cubic polynomials governed by the CP coordinates for that knot span. IGA only uses the thick part of each local curve which falls inside the knot span. The thin segments

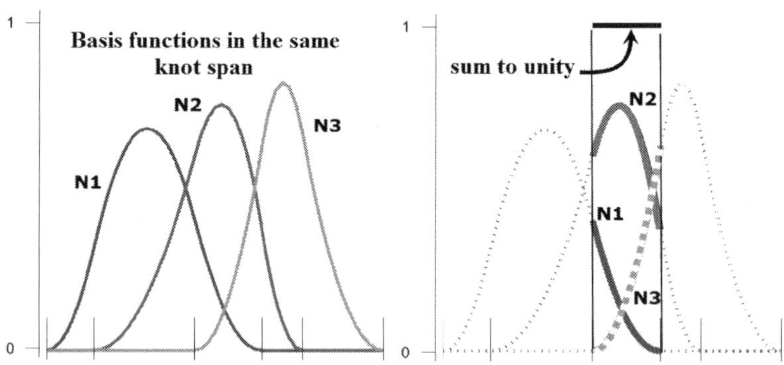

Figure 2.4. Basis functions in a knot span (iElement) sum to one

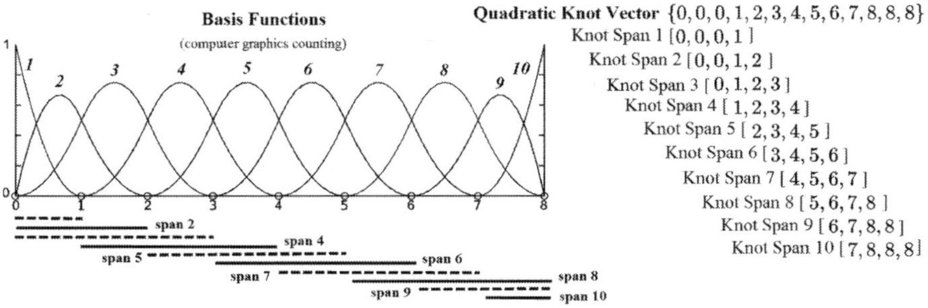

Figure 2.5. The ten knot spans controlling nine uniform quadratic basis functions

outside a knot span are shown here to illustrate that at their end knots they match the position, slope, and curvature of the cubic of the thin line for the next span. That is, they have $C^{p-1} = C^2$ continuity at the shared knots. Again, between their two end knots, each segment has C^∞ curvature.

> **Looking Ahead: The IGA** basis functions, for iElements introduced in 2007, are not directly compatible with FEA data structures. The use of 'extraction operators', introduced in 2011, created a powerful sub-set (iGe $\subset$ iElement) which maintains the C^{p-1} basis function continuity and only requires easy modifications to FEA codes to create the improved IGA analysis features.

A knot vector defines the sub-regions (knot spans, or iElements), and the level of continuity between the two knot spans connected by a knot point, $u_k \leq u_{k+1}$, defaults to C^{p-1}. Thus, the above-mentioned cubic, $p = 3$, will have $C^{p-1} = C^2$ continuity between each of the piecewise cubic segments along a curve. In other words, the curve segments' position, slope, and curvature will be continuous as the default. The difference between two adjacent interior knots, $u_{k+1} - u_k \equiv \Delta u_k$, is a 'knot span' for the basis functions in that interval. If two adjacent knots are repeated, $u_i \equiv u_{i+1}$, it can be considered as a geometric curve of zero span, and thus zero support, which reduces the degree of continuity of the adjacent piecewise curves by one for each consecutive repeat occurrence at that knot.

> **Looking Ahead:** Differential equations being solved with IGA studies often have coefficients and/or source terms with spatial discontinuities. An experienced analyst in a field will recognize

that such discontinuities will probably require one to reduce the continuity at some knots to enhance the accuracy of such a study.

The typical support regions associated with each knot span are shown for the quadratic, $p = 2$, knot vector $\mathbb{K}\boldsymbol{u} = \{0,0,0,1,2,3,4,5,6,7,8,8,8\}$ in Fig. 2.5. Its unique entries are $(0,1,2,3,4,5,6,7,8)$, yielding eight knot spans of $[0,1)$, $[1,2)$, $[2,3)$, $[3,4)$, $[4,5)$, $[5,6)$, $[6,7)$, and $[7,8)$. The knot vector of length $m = 13$ requires $n = 13 - 3 = 10$ CP for the curve geometry with three controlling each knot span. The shape of the curve in physical space depends on the spatial coordinates assigned by the user to the 10 geometric CPs.

When the basis functions, $N_{i,p}(u)$, defined by the above-mentioned recurrence relations, are multiplied by the CP coordinates x_i and y_i, they define a planar curve

$$\boldsymbol{C}(u) = \begin{Bmatrix} x(u) \\ y(u) \end{Bmatrix} = \sum_{i=1}^{CP} N_{i,p}(u) \begin{Bmatrix} x_i \\ y_i \end{Bmatrix}^r = [\mathbf{N}(u)]\{\boldsymbol{x}\_\boldsymbol{y}\}^r$$

like the one in Fig. 2.6 where the user-defined CP x- and y-coordinates are connected by straight thin lines to form the 'control net' for the B-spline curve (thick lines). The physical curve arc lengths between knot points are NOT directly proportional to the knot span size (because the geometric Jacobian between parametric and physical spaces is usually NOT a constant).

Geometric CPs, and analysis CPs, can have weights associated with them. To give a feel for how geometric knot vectors, CPs, and control weights influence a geometric curve (and later analysis spatial distributions), a few still images will be displayed. The

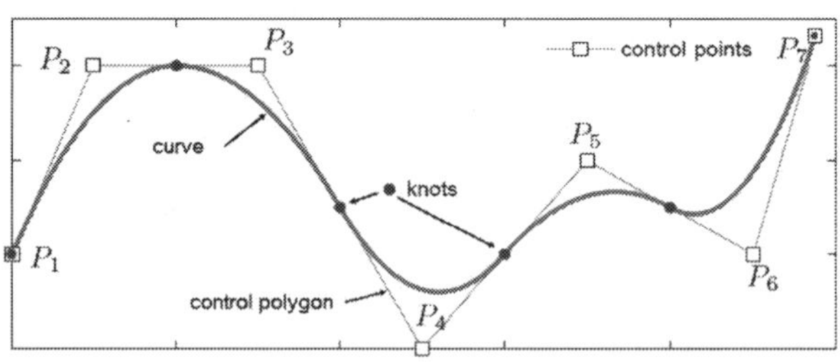

Figure 2.6. A B-spline curve (thick line) defined by CP coordinates

reader is encouraged to experience interactive versions of those features at the Barton (2022) website, and view the basic supporting theories and proofs given at the informative Shene (2016) site.

Another important property of B-splines is that when the polynomial order rises, the curvature of a B-spline curve diminishes (flattens the shape). The relative smoothness is often desirable for geometric features. However, for an IGA, raising the degree of the B-spline may smooth out local changes in the solution of an analytic model. Thus, IGA studies are usually less than degree 6. In FEA, the reverse condition occurs, where raising the degree of a Lagrangian interpolation basis increases the curvature changes, so much so that they can become ill conditioned at their ends. Thus, FEA interpolations usually do not go above degree 5, except for high-order (>4) ODE. For reasons to be discussed later, it is easier to use odd-degree basis functions for 2D and 3D analyses. This behavior of B-splines and NURBS is sketched in Fig. 2.6.

2.7 Index space

It has become useful in IGA to employ a sequential integer control mesh of knot vectors as a regular integer space of unit cells, $I = \{I_1, I_2, \ldots, I_{m=n+p+1}\}$, to represent a curve. For example, a 1-dimensional list of knot coordinates could be sketched as

1	2	3	4	5	6

$$\begin{aligned} \mathbb{I} &= \{1 \quad 2 \quad 3 \quad 4 \quad 5 \quad 6\} \\ \text{or } \mathbb{K}\boldsymbol{n} &= \{n_1 \quad n_2 \quad n_3 \quad n_4 \quad n_5 \quad n_6\}, \quad \mathbb{K}_L = 6, \end{aligned}$$

where the integer control mesh can extract the corresponding knot values and can be used to draw lines, or cells, with lengths representing each knot span direction, and place the span number (or iElement number) at the lower-left corner (LLC) of each non-empty cell.

In two and three dimensions, this type of integer control mesh sketch becomes a grid of unit square cells and unit cubes, respectively, which help visualize the very important non-zero NURBS contributions in portions of overlapping analysis regions of curves, surfaces, and volumes.

The most important feature of the integer index space and its relation to the parametric space and physical space is that the index space helps identify (visualize) the non-zero support regions of all basis functions associated with the knot points. When there are no repeated knots, the difference in the unique knot vector entries indicates the presence of an iElement in that index direction.

Consider the tensor product of two knot vectors and their index space, parametric space, physical space, and the numerical integration space. Since linear basis functions are the same in IGA studies and FEA studies, introducing the new concepts with two knot vectors to define quadratic forms with one interior knot:

$$\mathbb{A}\boldsymbol{r} = \{r_1, r_2, r_3, r_4, r_5, r_6, r_7\} = \{0, 0, 0, 0.5, 1, 1, 1\},$$

$$\mathbb{A}\boldsymbol{s} = \{s_1, s_2, s_3, s_4, s_5, s_6, s_7\} = \{0, 0, 0, 0.5, 1, 1, 1\}.$$

The tensor product is shown in Fig. 2.7 in index space (left) and parametric space. The index space shows nine knot points and four NURBS iElements in the knot spans. The sequential numbers at the LLC of each cell define an integer array that tracks the (I, J) coordinates of each NURBS iElement:

The parametric space viewpoint for the first time recognizes the analysis CPs needed to construct the basis functions for each of the

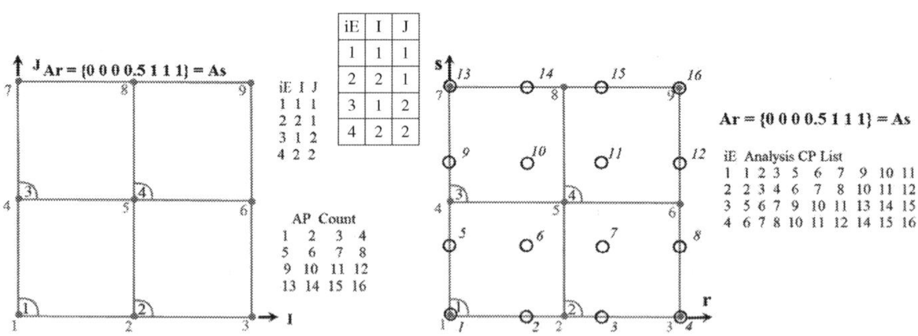

Figure 2.7. Index space and parametric space for bi-quadratic knot vectors

four iElements. For an individual basis function of degree p in the r-direction, $p+1$, analysis solution values are required. The analytic forms of the three (non-zero) quadratic basis functions in the r-direction knot vector $\mathbb{A}\boldsymbol{r} = \{0, 0, 0, 0.5, 1, 1, 1\}$ for each of the two r-knot spans are listed in Appendix 1 as follows:

Count,	**r-Basis Functions,**	**Nr(r), per knot span**
1, 0	$[(1-2r)^2,$	*0*]
2, 4	$[2r(2-3r),$	$2(1-r)^2]$
3, 5	$[2r^2,$	$8r-2-6r^2]$
0, 6	[0,	$(1-2r)^2]$
AP	**1–3**	**2–4**
Span	**1**	**2**

Note that iElements 1 and 3 fall in the first r-knot span, while numbers 2 and 4 are in the second span. The s-direction knot vector has been chosen in this example to be the same when r is replaced with an s.

A two-span quadratic requires four analysis control points (APs) in each parametric direction, giving a total of 16 APs. The index space does not show the analysis control points. The parametric space does show the 16 CPs in their default equally spaced positions. The 9 knot point numbers, 4 iElement numbers, and 16 APs are assigned left to right and then bottom to top in their respective space. In physical space, the CPs are shown at the actual positions assigned by the CP input coordinates. The index space and the parametric space (right image) both show the four NURBS regions. The above-mentioned table defines where the corner of a cell is located for each of those regions. For example, NURBS region number 3 is associated with the index coordinates of (1, 2) and the parametric coordinates of (r_1, s_2). This information becomes very important later when an analyst needs to refine that region locally by converting the NURBS basis functions at (r_1, s_2) with T-spline basis functions 'anchored' to that knot span.

> **Looking Ahead**: B-spline, NURBS, and T-spline 'anchors' will be defined later for odd–odd degree knot vectors, even–even knot vectors, and mixed odd–even knot vectors. The odd–odd combination is by far easier to understand and will be used in most examples herein because it has the basic functions evaluated at each knot.

The 16 AP in the parametric space are counted sequentially in the r-direction and then in the s-direction. For the IGA database, it is necessary to have an integer array that lists which of the 16 APs controls each of the four iElements. To see those associations, the above-mentioned parametric space will be repeated four times so as to highlight which of the nine of 16 APs control each of the four iElements. Remembering that a 1D quadratic basis function requires three CPs in each parametric direction, in each knot span, Fig. 2.8 highlights the $3 \times 3 = 9$ APs which control the first and second iElements. The figure also lists the analysis control points connected to each iElement. The same data for iElements 3 and 4 follow in Fig. 2.9. In the last four figures, the 4×9 array named AP_list is

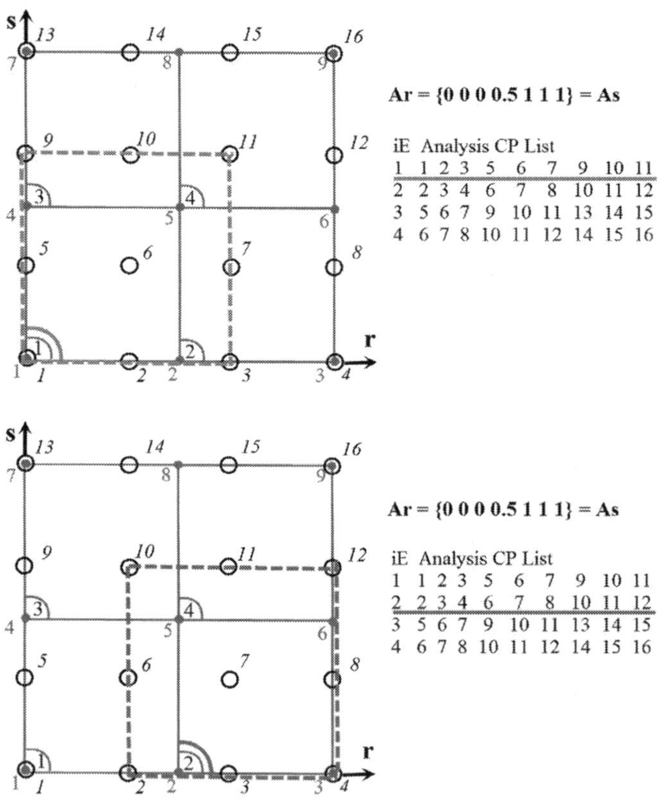

Figure 2.8. Identifying the nine APs controlling iElements 1 and 2 (underlined)

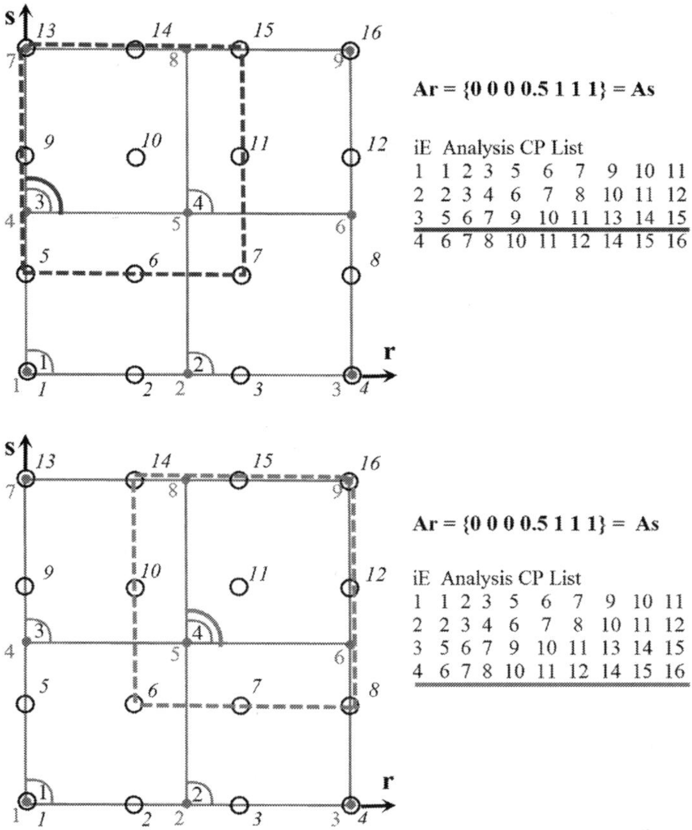

Figure 2.9. Identifying the nine APs controlling iElements 3 and 4 (underlined)

a **very important** connection list for each of the four iElements, giving 9 out of the 16 APs values that scatter into the system matrices (assuming that the number of generalized unknowns per AP is one, $n_g \equiv 1$).

The lines in index space are also shown in parametric space, but there, the lines can fall on top of each other when only zero-valued basis functions lie between the lines. In physical space, the nine knot points, 16 CPs, mesh lines, and sometimes the iElement numbers are shown in the physical space simply as visual aids. Of course, the four square iElements in index space and parametric space become curvilinear quadrilaterals as implied in Fig. 2.10.

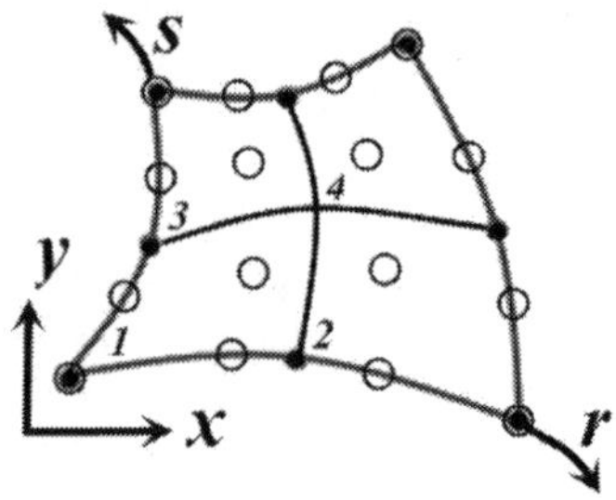

Figure 2.10. Parametric domains are generally curved in physical space

The 16 analysis control point locations are only seen correctly in physical space since their coordinates are set by the user. In parametric space, only the number of APs in each direction is known and they are shown uniformly spaced by default.

What changes in these graphical aids when there are repeated entries in the knot vector? Repeated entries represent parametric cells with no area, and thus identically zero basis functions. However, each entry is counted in the index space, even when repeated in parametric space. Let prior r-direction analysis knot vector be changed to have a repeated knot, $r_4 = r_5$: $\mathbb{A}\boldsymbol{r} = \{r_1, r_2, r_3, r_4, r_5, r_6, r_7, r_8\} = \{0, 0, 0, 0.5, 0.5, 1, 1, 1\}$. In other words, the second unique value $(0, 0.5, 1)$ being repeated flags a knot span with zero basis functions, as shown in the index space of Fig. 2.11. Those two null cells appear in index space, but they are not numbered, and the following non-null knot spans (iElements) have the LLC indices of (3, 1) for the second iElement and (3, 2) for the fourth one. In the parametric space, there are only three vertical lines versus four in the index space because the two interior vertical lines fall on top of each other in the parametric space and physical space. The physical lines will again normally appear as in Fig. 2.12.

To review the 2D concepts of index space and parametric space, herein the index space is always plotted with axes which are integers ranging from 1 to the highest number in the knot vectors in each direction. There is no uniform standard for the concept of an integer space, and some IGA studies include the first and last repeated integers (which define the polynomial degree) as extra border lines around the integer plots. In some of the original literature the implied concept of an index count appeared to allow for refinements in the

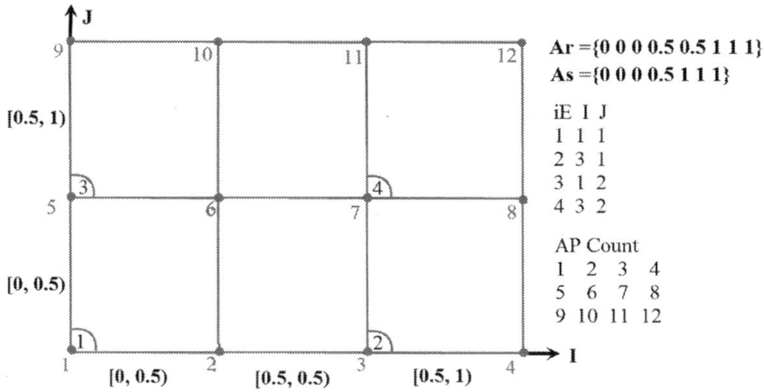

Figure 2.11. Null cells are seen in the index space

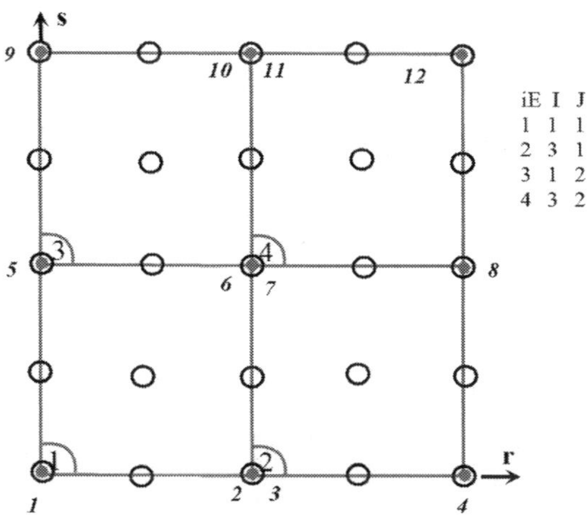

Figure 2.12. Null cells are not seen in parametric and physical spaces

analysis by introducing, gaps in the parametric coordinates at I and J values. The index space allows for visualizing iElement locations via their lower-left corner coordinates. Figure 2.13 shows that an integer space plot, for the two knot vectors shown there, was created by simply looping over rows and columns. The same script formed the parametric space view in Fig. 2.14. Note the important difference that the axes of Fig. 2.14 are decimal numbers that are the digital r- and s-value entries in the knot vectors at the I and J locations,

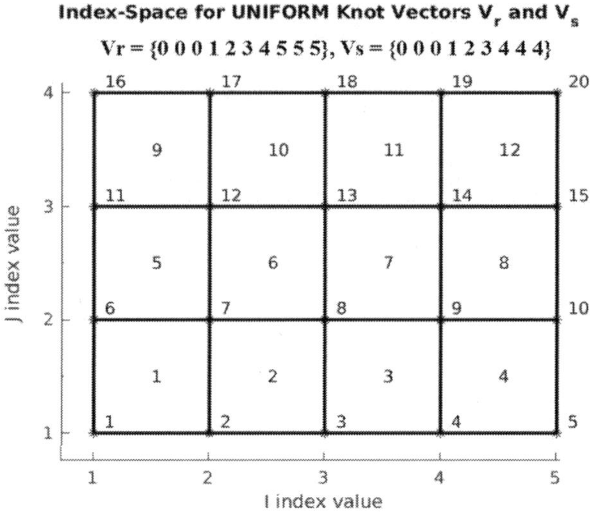

Figure 2.13. Index space requires integer valued axes

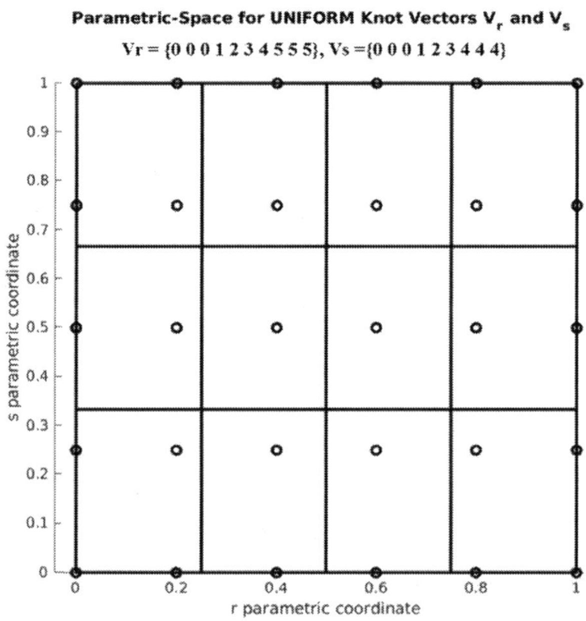

Figure 2.14. The parametric space axes are real numbers

respectively, after they are converted to lie in the range $0 \leq r \leq 1$, etc., for s. Also, Fig. 2.14 also contains open circles denoting the control point locations in parametric space. Actually, the numerical algorithm for evaluating the basis functions accepts any increasing sequential real numbers for what correspond to the axes in these two figures.

An additional feature of a parametric space sketch is that it is the first time that the number of analysis control points in each direction is known and their relative locations, assuming equal spacing, can be visualized. The actual locations of the analysis control points and the knots cannot be displayed accurately until the user, or CAGD software exchange file, specifies their physical locations.

2.8 T-Meshes

The tensor product of B-Splines and NURBS over rectangular or hexahedral integer spaces is not present in the T-Splines used for analysis (and geometric) refinements. Thus, the calculations of the T-Spline basis functions are no longer based on global knot vectors. Indeed, the knot vectors for a domain containing T-Splines are derived from the T-mesh using a concept called 'anchors' to be discussed later, where analysis refinements will be addressed. The presence of a T-mesh is made clear by a cell plot where at least one cell has a T-shaped junction (five vertices) on its perimeter. They occur twice in Fig. 2.15.

The original publications on T-splines described their basis functions in terms of knot 'intervals', $\langle \Delta n_j \rangle \equiv \langle n_{j+1} - n_j \rangle$. A list of knot

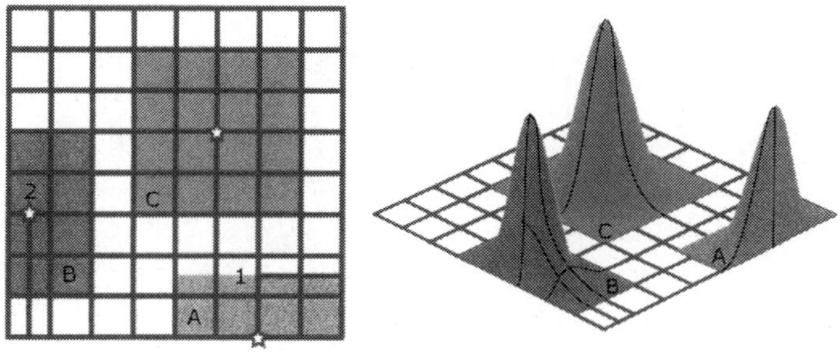

Figure 2.15. Non-zero basis regions of two bi-cubic T-splines (Thomas *et al.*, 2015)

intervals, say $\langle 1,3,2\rangle$, can be converted to a knot vector by making the first entry arbitrary and summing the intervals sequentially as

$$\langle 1,3,2\rangle \leftrightarrow \{a, a+1, a+1+3, a+1+3+2\} = \{a, a+1, a+4, a+6\}.$$

For $a=0$: $\langle 1,3,2\rangle \leftrightarrow \{0,1,4,6\}$ or $\langle \Delta n_j\rangle \leftrightarrow \mathbb{K}\boldsymbol{n}$. The current literature on T-splines, which are very important for analysis refinement, is split between the knot vector approach used herein and the knot interval approach.

2.9 Comments

In an IGA study, the dofs are at the control (analysis) points, not the knots.

The IGA parameter space is partitioned into iElements by a knot vector in each direction.

The iElements are also known as knot spans, because they span between knot values.

In IGA, the geometry (analysis result) is defined by knot vectors and CP coordinates (analysis point values).

The length of any open knot vector is equal to the number of CPs, n_{CP}, plus the order of the curve, $(p+1)$.

The magnitude of the knots in a knot vector does not matter. The knot vectors $\{0\ 0\ 0\ 1\ 2\ 2\ 2\}$, and $\{100\ 100\ 100\ 200\ 300\ 300\ 300\}$ can be converted to the knot vector $\{0\ 0\ 0\ 0.5\ 1\ 1\ 1\}$, and all produce the same curve, but with different coefficients.

Only the relative parametric difference between the knots is important in defining the B-spline (or iElement). Adding a constant to all knot values will generate the same B-spline. Likewise, multiplying all knots with a constant factor leaves the B-spline unchanged.

No vertex in the index space corresponds to any particular CP.

Except at the knots, a B-spline, and iElement, will have C^{∞} continuity. At the knots, the spline has C^{p-m}continuity, where m is the multiplicity of the knot in the knot vector.

The images of the knot lines under NURBS mapping create the boundaries of the physical iElements.

B-splines are defined by 'global knot vectors' while T-splines, require 'local knot vectors' centered about an 'anchor point'.

Normalization of the knot vector is irrelevant for the shape of the basis functions, which do not depend on absolute knot values, but on the relative distances between knot values.

A knot vector can be converted to a knot interval for representing T-splines, $\mathbb{K}\boldsymbol{n} \leftrightarrow \langle \Delta n_j \rangle$. For example, $\mathbb{K}\boldsymbol{n} = \{0, 1, 4, 6\} \leftrightarrow \langle 1, 3, 2 \rangle = \langle \Delta n_j \rangle$.

Knots may be inserted into a knot vector without changing the geometric shape or the properties of a B-spline curve.

The number of basis functions only equals the number of knots when the polynomials in each direction are odd. Otherwise, other basis function 'anchors' are required.

2.10 Summary

Recursive relation for B-splines of degree p:

$$p = 0: \quad N_{j,0}(u) = \begin{cases} 1 & \textit{if } u_j \leq u \leq u_{j+1} \\ 0 & \text{otherwise} \end{cases},$$

$$p = 1, 2, 3, \ldots \quad N_{j,p} = \frac{u - u_j}{u_{j+p} - u_j} N_{j,p-1}(u)$$

$$+ \frac{u_{j+1} - u}{u_{j+p+1} - u_{j+1}} N_{j+1,p-1}(u),$$

$x(u)/\partial u = [\boldsymbol{N}(u)/\partial u]\{\boldsymbol{x}_{CP}\}$, $\sum_j dN_j(u)/du = 0$. Function: *symbolic_N_bases.m*, and in *num_bases_line.m*.

Recursive relation for derivatives of B-splines of degree p:

$$N'_{j,p} = \frac{p}{u_{j+p} - u_j} N_{j,p-1}(u) + \frac{p}{u_{j+p+1} - u_{j+1}} N_{j+1,p-1}(u).$$

Both are implemented in Matlab algorithm *symbolic_N_bases.m*, and in *num_bases_line.m*.

Chapter 3

Parametric and Physical Derivatives

3.1 Introduction

In FEA, physical derivatives are related to only one parametric coordinate system. In Isogeometric Analysis (IGA), two parametric coordinate systems are used sequentially to define physical derivatives. When relating one coordinate system to another coordinate system, it requires one to establish the Jacobian matrix for the transformation. The determinant of the Jacobian matrix must be evaluated to relate a differential measure in one system to a differential measure in the other system. The inverse Jacobian matrix is required to evaluate the physical derivatives in terms of known parametric derivatives.

From the 'chain rule' of calculus, the relationship between the partial derivatives of a function $f(x, y, z) = f(r, s, t)$ in a parametric coordinate system, say $(r,\ s,\ t)$, and a 'physical' coordinate system, say (x, y, z), are

$$\begin{aligned}
\frac{\partial f}{\partial r} &= \frac{\partial f}{\partial x}\frac{\partial x}{\partial r} + \frac{\partial f}{\partial y}\frac{\partial y}{\partial r} + \frac{\partial f}{\partial z}\frac{\partial z}{\partial r}, \\
\frac{\partial f}{\partial s} &= \frac{\partial f}{\partial x}\frac{\partial x}{\partial s} + \frac{\partial f}{\partial y}\frac{\partial y}{\partial s} + \frac{\partial f}{\partial z}\frac{\partial z}{\partial s}, \\
\frac{\partial f}{\partial t} &= \frac{\partial f}{\partial x}\frac{\partial x}{\partial t} + \frac{\partial f}{\partial y}\frac{\partial y}{\partial t} + \frac{\partial f}{\partial z}\frac{\partial z}{\partial t},
\end{aligned} \tag{3.1}$$

or in a matrix form this defines the Jacobian matrix, $[\boldsymbol{J}(r,s,t)]$,

$$\begin{Bmatrix} \frac{\partial f}{\partial r} \\ \frac{\partial f}{\partial s} \\ \frac{\partial f}{\partial t} \end{Bmatrix} = \begin{bmatrix} \frac{\partial x}{\partial r} & \frac{\partial y}{\partial r} & \frac{\partial z}{\partial r} \\ \frac{\partial x}{\partial s} & \frac{\partial y}{\partial s} & \frac{\partial z}{\partial s} \\ \frac{\partial x}{\partial t} & \frac{\partial y}{\partial t} & \frac{\partial z}{\partial t} \end{bmatrix} \begin{Bmatrix} \frac{\partial f}{\partial x} \\ \frac{\partial f}{\partial y} \\ \frac{\partial f}{\partial z} \end{Bmatrix} \equiv [\boldsymbol{J}(r,s,t)] \begin{Bmatrix} \frac{\partial f}{\partial x} \\ \frac{\partial f}{\partial y} \\ \frac{\partial f}{\partial z} \end{Bmatrix}.$$

The Jacobian matrix is not always square; it has as many rows as parametric coordinates and as many columns as 'physical' coordinates. The relevance of different partitions of the full 3D Jacobian matrix is denoted in the left of Fig. 3.1, and its various rectangular sub-sections are illustrated on the right.

As illustrated in Fig. 3.1, for a planar curve, the parametric Jacobian contributions appear as a row matrix and its components relate the physical and parametric lengths:

$$\left(\frac{dl}{du}\right)^2 = \left(\frac{dx}{du}\right)^2 + \left(\frac{dy}{du}\right)^2 \rightarrow \frac{dl}{du} = \sqrt{\left(\frac{dx}{du}\right)^2 + \left(\frac{dy}{du}\right)^2},$$

which usually have to be evaluated numerically.

In practice, a full 2D or 3D Jacobian matrix, of a specific finite region $\boldsymbol{J}^r(\boxdot)$, is usually evaluated numerically as the product of the local derivatives of the non-zero B-spline basis functions, $\boldsymbol{N}$, supported by the iElement control points. At a specific local point, the

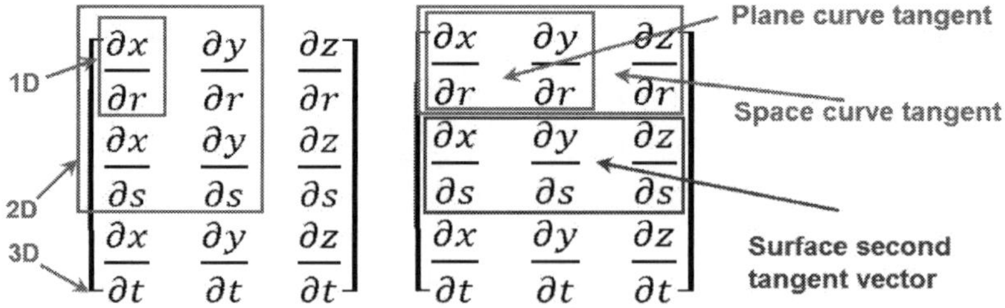

Figure 3.1. **Partitions of a coordinate transformation Jacobian matrix [Akin, 2022]**

Jacobian matrix is either square or rectangular:

$$[\boldsymbol{J}^r(\boxdot)] = \frac{\boldsymbol{\partial N}(\boxdot)}{\boldsymbol{\partial}\boxdot}\boldsymbol{x}^r$$

of size

$$(n_p \times n_2) = (n_p \times n_{CP})(n_{CP} \times n_2),$$

where the first (parametric) space having n_p parametric dimensions is mapped into a second space which has either $n_2 = n_s$ physical space dimensions or is mapped into a second parametric space (in IGA), say $n_2 = n_I$. Usually, a second parametric space is one containing the non-dimensional numerical integration quadrature data.

The determinant of the Jacobian of a coordinate transformation relates the 'physical' differential volume to the differential parametric volume:

$$d\Omega = |\boldsymbol{J}(\boxdot)|\, d\,\boxdot.$$

Since the IGA approach requires element-wise integration of weak form integrals in physical space (except for the strong collocation version), that relation is always needed. For example, computing the physical length of a single parametric curve requires

$$L^r = \int_0^{L^r} dl = \int_{u=0}^{u=1} |\boldsymbol{J}(u)| du = \int_{u=0}^{u=1} \sqrt{\left(\frac{dx}{du}\right)^2 + \left(\frac{dy}{du}\right)^2}\, du,$$

where the physical coordinates are defined by the B-spline polynomials acting on the coordinates of the control points. In practice, such integrals must be evaluated by numerical integration, which approximates an integral by a summation of weighted values of the integrand at specially tabulated weights and locations called quadrature points:

$$L^r \approx \sum\nolimits_{q=1}^{n_q} \sqrt{\left(\frac{dx(u_q)}{du}\right)^2 + \left(\frac{dy(u_q)}{du}\right)^2}\, w_q.$$

For straight lines and rectangular IGA patches with multiple knot spans (iElements) in each direction, the second Jacobian is a square matrix. When a curve has multiple iElements (knot spans), each contribution to the physical length must be evaluated by integration over the knot span and summed. Then,

$$L^r = \sum_{e=1}^{e=n_e-1} \int_{u^e}^{u^{e+1}} |\boldsymbol{J}(u)| du \approx \sum_{e=1}^{e=n_e-1} (u^{e+1} - u^e)$$

$$\times \sum_{q=1}^{q=n_q} |\boldsymbol{J}(u_q)| w_q \text{ with } u^e \leq u_q < u^{(e+1)},$$

$$L^r \approx \sum_{e=1}^{e=n_e-1} (u^{e+1} - u^e) \sum_{q=1}^{q=n_q} \sqrt{\left(\frac{dx(u_q)}{du}\right)^2 + \left(\frac{dy(u_q)}{du}\right)^2} w_q.$$

3.2 Summary

Jacobian and determinant:	$[\boldsymbol{J^r}\ (\boxdot)] = \frac{\boldsymbol{\partial N()}}{\boldsymbol{\partial}} \boldsymbol{x^r},$ $\lvert\boldsymbol{J}(\boxdot)\rvert = \det([\boldsymbol{J^r}\ (\boxdot)])$
Differential volumes:	$dx = \lvert J(r)\rvert\, dr = \left\lvert \frac{\partial x(r)}{\partial r} \right\rvert dr$ $dA = dx\, dy = \lvert J\,(r,\ s)\rvert\ dr\, ds$ $dV = dx\, dy\, dz = \lvert J(r,\ s,\ t)\rvert\ dr\, ds\, dt$
Physical derivatives from inverse Jacobian:	$\frac{\partial()}{\partial \mathrm{x^r}} = [J^r\ (\boxdot)]^{-1} \frac{\partial()}{\partial \boxdot}$
Physical integrals:	$L = \int_0^L dl = \int_{r=0}^{r=1} \lvert\boldsymbol{J}(r)\rvert\ dr,$ etc.

Chapter 4

Review of PDEs and Boundary Conditions

4.1 Introduction and terminology

The goal of isogeometric analysis (IGA) is to extend the use of the exact geometry construction functions of computer-aided geometric design (CAGD), which describe the shapes of modern parts, to also solve the partial differential equations (PDEs) within those parts and assemblies. Such studies are required for the engineering analysis and design features of the geometric shapes. PDEs are of fundamental important in engineering and applied mathematics. Thus, a brief review of PDEs will be covered here before summarizing the CAGD functions to be extended to the solution of PDEs. Also, in this chapter, some of the advantages and disadvantages of how IGA conducts its numerical approximations of the PDEs and their boundary conditions (BCs) will be discussed.

Most of the PDEs and ordinary differential equations (ODEs) important in engineering applications are even order, in that their highest derivative is of integer order $n = 2m$. Such equations will be the focus of this introduction to IGA. Frequently, a PDE is derived by applying a balance or equilibrium concept to a differential volume, where 'differential' is used in the calculus concept, meaning infinitesimally small. Over time, it has been discovered that many physical relations and laws can be expressed in the form of PDEs. The qualifier 'partial' means that the function or process represented by a PDE is dependent upon partial derivatives taken with respect to (wrt) two or more independent variables, usually spatial coordinates. When only one independent variable is present, the corresponding equation is called an ordinary differential equation (ODE). The order of the highest derivative is called the order of the PDE.

The function, say u, which satisfies a PDE inside an application domain, will be referred to as the solution. For example, a very common second-order PDE in engineering analysis is the Poisson's Equation in two spatial dimensions

$$\kappa_x(x, y)\frac{\partial^2 u(x, y)}{\partial x^2} + \kappa_y(x, y)\frac{\partial^2 u(x, y)}{\partial y^2} = Q(x, y). \tag{4.1}$$

In the case of heat conduction, the solution $u(xy)$ is the temperature within a domain, Ω, and along the boundary, Γ, of that domain. In Equation (4.1), κ_x and κ_y are the measured thermal conductivities of the material in the domain. The internal rate of heat generation in the domain is a 'source term' denoted by Q. It is frequently zero (which defines Laplace's Equation), but can arise from electrical resistance or chemical reactions.

A PDE usually has values of the required solution, u, and/or ODEs of its normal derivative, $\partial u/\partial n$, specified at non-overlapping segments of the domain boundary. Those conditions are called 'boundary conditions' (BCs). If the PDE contains time as one of the variables, so that $u(x, y) \to u(x, y, t)$, then the values at the starting time (usually $t = 0$) are called 'initial conditions' (ICs). An example of a PDE having ICs is the 1-dimensional wave equation:

$$c(x)\frac{\partial^2 u(x, t)}{\partial x^2} = \frac{\partial^2 u(x, t)}{\partial t^2}, \tag{4.2}$$

where c denotes the speed of the wave in the material.

A 'strong solution' of a PDE is a function that has all the partial derivatives in the equation continuous in the domain and satisfies the PDE everywhere in the domain. Later, the FEA and IGA solutions will be based on an equivalent governing integral 'weak solution' where the highest derivative in the PDE is not present in the integral. These two definitions refer to the mathematical continuity, C^n, of a function, which means that the function and its nth derivative are continuous at a point, where C^0 denotes the function value, C^1 denotes a continuous gradient (slope) of the function,

C^2 denotes a continuous curvature (second derivatives), and so on to C^∞. Likewise, C^{-1} denotes a discontinuous function. Both IGA and FEA solution functions are C^∞ within their sub-patch and element, respectively, but FEA solution functions are usually only C^0 across element boundaries, while IGA solution functions are usually $C^{n>0}$ across sub-patch boundaries, which is a significant advantage for IGA.

Another common type of PDE analysis is where a coefficient in the governing differential equation, say λ, is an unknown global constant to be determined, called an eigenvalue. There is a group of solutions to such a differential equation. This class of analysis is called an 'eigen-problem' from the German word *eigen* meaning belonging distinctly to a group. Common engineering applications include the vibration of electrical and mechanical systems and electromagnetic waveguides for microwave ovens and other electrical components. Likewise, the surface elevations of shallow bodies of water, like harbors, are subjected to periodic excitations by the moon. Tidal studies are called Seiche motion studies and are governed by eigen-problem solutions. These problems, and many others, are described by PDEs like the scalar Helmholtz differential equation:

$$k\,\nabla^2 u(x,\,y,\,t) + \lambda\, u(x,y,t) = 0, \tag{4.3}$$

which is solved for a group for multiple eigenvalues, λ_n, and a corresponding set of solution eigen-vectors to define the group of solutions. The magnitudes of the eigenvalues are important, but only the spatial shape of the eigen-vectors matters and their magnitudes are usually scaled to range over -1 to $+1$.

Eigen-problem solutions require much more computation effort than solving a linear system of equations. In theory, a continuous system has an infinite number of eigenvalues, and an FEA or IGA approximation of a system has as many eigenvalues as there are degrees of freedom (after enforcing the EBC). Most engineering applications require just the first few or the last few eigenvalues and eigen-vectors. There are efficient algorithms for both approaches.

For waveguides and other applications, the eigenvalue can be a complex number, but for mechanical vibrations, it is a non-negative number. For mechanical and structural vibrations, a positive eigenvalue corresponds to the square of the natural frequency of vibration,

$\lambda = \omega^2$, and a zero value, $\lambda = 0$, corresponds to a rigid body (non-vibrating) motion. There are at most six rigid body motions of a solid (three translations and three rotations). Theoretically, a mechanical vibration analysis should not yield a negative eigenvalue, $\lambda < 0$, but numerical errors can produce them as an approximation of a rigid body motion.

4.2 Post-processing

Once a numerical or analytical solution has been obtained for a PDE and its BCs, it is almost always necessary to post-process that solution to obtain at least its integral and/or gradient vector. Usually, combinations of the gradient vector are utilized to define physical quantities related to the original problem. For example, Fourier's Law defines the heat flux vector per unit area as the negative of the thermal conductivity times the temperature gradient vector:

$$\vec{q} = -\left(k_x \frac{\partial u}{\partial x}\right)\vec{\boldsymbol{i}} - \left(k_y \frac{\partial u}{\partial y}\right)\vec{\boldsymbol{j}},$$

or in matrix notation for an orthotropic material

$$\vec{q} = \begin{Bmatrix} q_x \\ q_y \end{Bmatrix} = \boldsymbol{q} = -\left[\boldsymbol{\kappa}\right]\boldsymbol{\nabla}^{\boldsymbol{T}} u = -\begin{bmatrix} k_{xx} & 0 \\ 0 & k_{yy} \end{bmatrix} \begin{Bmatrix} \partial u/\partial x \\ \partial u/\partial y \end{Bmatrix}, \tag{4.4}$$

where in general $[\boldsymbol{\kappa}]$ contains the directionally dependent thermal conductivities of an anisotropic material:

$$[\boldsymbol{\kappa}] = \begin{bmatrix} k_{xx} & k_{xy} \\ k_{xy} & k_{yy} \end{bmatrix}. \tag{4.5}$$

A second example of post-processing components of the solution gradient vector involves the normal and engineering shear strain components

$$\varepsilon_{xx} = \frac{\partial u}{\partial x} \quad \text{and} \quad \gamma_{xy} = \frac{\partial u}{\partial y} + \frac{\partial v}{\partial x},$$

respectively, where u and v are components of the displacement vector.

Less commonly, the solution result of interest is the integral over the domain of the solution of the PDE. One example is water flow in an open channel. There, the PDE governs the velocity of the water flowing normal to the cross section of the channel. After the velocity distribution has been obtained, the integral of that solution over the channel cross section gives the volume rate of flow in the channel.

For both FEA and IGA, the numerical operations required to form the matrices in the equivalent form are same ones needed in post-processing. Since writing and reading binary files are often the second-fastest operations on a digital computer, many of the numerical arrays used in evaluating the integral forms are written to binary storage and later recovered to speed the required post-processing steps.

4.3 Essential and non-essential boundary conditions

Essential (Dirichlet) boundary conditions (EBCs) specify the values of the primary unknowns (dependent variables) on a portion of the boundary, Γ_D, of the domain. BCs other than the essential type are usually written in terms of an ODE written wrt the solution gradient normal to the boundary, as summarized in Tables 4.1 and 4.2. Often, the normal flux entering the body is known and is specified on a portion of the boundary, say Γ_N, as one of the Neumann or non-essential boundary conditions (NBCs). It indirectly specifies the normal gradient of the solution at a point. A more complicated non-essential condition is the mixed condition (MBC), also known as a Cauchy or Robin condition, on a portion of the boundary, say Γ_M, that couples the normal gradient of the solution to the unknown value of the solution on that boundary segment. That mixed boundary condition usually occurs as a convection boundary condition (CBC).

If a region of the boundary, $\Gamma \equiv \Gamma_0$, has no BCs imposed, then the default boundary condition on the boundary is that the normal gradient of the solution and the normal flux is zero. This is called the natural boundary condition (NatBC), and it does not require any input data in an IGA study. The entire boundary is formed by the union of those non-overlapping boundary condition regions, $\Gamma = \Gamma_0 \cup \Gamma_D \cup \Gamma_N \cup \Gamma_M$.

Table 4.1. Boundary conditions for even-order PDEs

Partial Differential Equation (PDE)	$\frac{\partial^{2m}u}{\partial x^{2m}}+\cdots+\frac{\partial u}{\partial x}+g(x)u+q(x)=0$
Essential Boundary Conditions (EBCs)	$u,\ \frac{\partial u}{\partial n},\ldots,\frac{\partial^{(m-1)}u}{\partial n^{(m-1)}}$
Non-essential Boundary Conditions (NBCs)	$\frac{\partial^{(m)}u}{\partial n^{(m)}},\frac{\partial^{(m+1)}u}{\partial n^{(m+1)}},\ldots,\frac{\partial^{(2m-1)}u}{\partial n^{(2m-1)}}$

Note: n denotes normal to the boundary.

Table 4.2. Example of even-order ODE boundary conditions

1D Application	**ODE**	$n=\mathbf{2}m$	**EBC [0 to** $m-$**1]**	**NBC [**m **to 2**$m-$**1]**
Axial conduction ODE	$k\,\frac{\partial^2 u}{\partial x^2}=p$	2	u	$k\,\frac{\partial u}{\partial n}=q$
Beam bending ODE	$EI\,\frac{\partial^4 u}{\partial x^4}=p$	4	u and $\frac{\partial u}{\partial n}$	$EI\frac{\partial^2 u}{\partial n^2}=M$ and $EI\,\frac{\partial^3 u}{\partial n^3}=V$

Let the a PDE or ODE have an even order as $n=2m$ for $m\geq 1$, then the EBCs involve specifying the value of u and its $(m-1)$ derivatives. The zeroth derivative is the function value. For even-order differential equations, the NBCs involve specifying the $(m\text{th})$ to $(2m-1)$ order derivatives. Those common rules are summarized in Table 4.1. Two common even-order ODE applications are illustrated in Table 4.2.

4.4 Manufactured solutions

When validating an IGA program and its application to a new specific shape and/or differential equation, the question of how to assess the accuracy of the solution arises. This is particularly true when the PDE is nonlinear as it is claimed that there is no known solution and the accuracy must be asserted by continuously refining the model (increasing the number of degrees of freedom) or by comparing to another approximate solution. The assertion of the lack of a known solution is an oversight. When an exact solution is not available to validate a numerical procedure, one can be manufactured.

For any PDE, the 'exact' solution $u(x, y)$ over any shape can be, for example, to consider the Poisson Equation (4.3) in the introduction

$$\kappa_x(x,\, y)\frac{\partial^2 u(x,\, y)}{\partial x^2} + \kappa_y(x,\, y)\frac{\partial^2 u(x,\, y)}{\partial y^2} = Q(x,\, y).$$

To manufacture the solution and its BCs, assume a spatial solution $a(x, y) \to u(x, y)$. To establish a varying essential (Dirichlet) boundary condition (Section 2.5) on a portion of the boundary, just substitute the coordinates of that region, $(x,\, y) \in \Gamma_D$, into the assumed spatial form there to define that boundary condition, $a_{\text{EBC}}(x,\, y) \in \Gamma_D$. Likewise, where there is an ODE normal to the boundary defining a Neumann condition on Γ_N or a mixed condition on Γ_M, manufacture those conditions as well. It will be shown later that those non-essential conditions appear in the integral form and, regardless of their complexity, they can be numerically integrated in an IGA solution.

Likewise, all of the partial derivatives appearing in the PDE are applied to the assumed solution. In this example, all partial derivatives, $\partial a(x, y)/\partial x$ through $\partial^2 a(x,\, y)/\partial y^2$, are known (as are time derivatives if present in the PDE). Substituting the present derivatives and the known coefficients, like $\kappa_x(x,\, y)$ and $\kappa_x(x,\, y)$, into the PDE defines the manufactured source term, $Q_{\text{DEFINED}}(x, y)$:

$$\kappa_x(x,\, y)\frac{\partial^2 a(x,\, y)}{\partial x^2} + \kappa_y(x,\, y)\frac{\partial^2 a(x,\, y)}{\partial y^2} \Longrightarrow Q_{\text{DEFINED}}(x, y).$$

Again, any spatial source terms will appear in the integral form and, regardless of their complexity, they can be numerically integrated in an FEA or IGA solution. In summary, manufacturing a known solution to be compared to an FEA or IGA solution to assess their accuracy is possible. The accuracy of a manufactured solution just depends on any error in the numerical integration of the source term and/or the NBCs. Such errors are easily controlled by increasing the number of terms in the numerical integration summation. Today, applying symbolic software processing can provide the required partial derivatives and solutions for other messy steps in manufacturing a solution.

4.5 Equivalent integral forms

Both IGA and FEA always start with the strong governing PDE converted to an equivalent weak governing integral over the solution domain and its boundary. Then, a spatial approximation of the solution, with unknown constants, is substituted into the integral form to convert it to a governing matrix form that can be solved numerically to find the constants that best satisfy the integral form and thus the original PDE.

Usually, the governing integral form is one that can be proven to give the same solution as the PDE when the integral form satisfies the same BCs. The equivalent integral forms for most linear even-order PDEs and ODEs in engineering and physics have been known since about the 1750s. For most of the IGA applications to be considered herein, the integral forms are stated without proof. Also, weak equivalent integral forms can be obtained using various spatial approximations of the residual in the methods of weighted residuals (MWRs).

For linear elliptical PDEs, there are two types of integral forms that have the same BCs and whose solutions can be shown to be an exact solution of the PDE. Those two forms are Euler's historical Theorem of Variational Calculus (circa 1750) and the Galerkin Weighted Residual Method (circa 1915). Those two forms yield identical definitions for the IGA and FEA matrices. The Galerkin method, mainly used herein, can be used to derive an integral form for nonlinear problems, like fluid flow. Another MWRs, which can be used for IGA and FEA, is the Least Squares Method, but it has continuity requirements that are difficult to satisfy in 2D and 3D FEA studies.

4.6 Discontinuity

When considering the continuity of approximate solutions of PDEs and ODEs, it is necessary to consider how representing a 'point source' impacts an IGA analysis polynomial solution. In general, it is known that an analytical solution for a point source introduces infinite spatial derivatives (a singularity) at such a point. In finite element analysis, this is addressed by one of three ways. In the p-method FEA, where the element size (h) is held constant while the polynomial degree (p) is increased, point sources are simply not allowed as input and must be distributed over a small region.

In the more common h-method FEA, where the polynomial degree is the same in all elements, it is usually required that the point source occurs at a vertex node on an element interface so that the C^0 elements automatically have a discontinuous gradient normal to the element boundaries at that vertex. Less commonly, in the h-method FEA, a point singularity is specified at a point interior to a single element and then distributed to all of that element's nodes via a Dirac distribution. This smears the point source over the whole element.

A point singularity has an impact on analysis knot vector definitions as well. As an example, consider the 4th-order ODE for the Euler thin beam theory description of the transverse deflection, $v(x)$:

$$\frac{d^2}{dx^2}\left[EI\frac{d^2v}{dx^2}\right] - f(x) = 0.$$

At a point along a beam, the following quantities are considered in engineering studies:

$EI\, v''''^{(x)} = f(x)$	load per unit length,	Differential equation,
$EI\, v'''(x) = V(x)$	transverse shear force,	Nonessential boundary condition,
$EI\, v''(x) = M(x)$	bending moment,	Nonessential boundary condition,
$v'(x) = \theta(x)$	slope,	Essential boundary condition,
$v(x) =$	deflection,	Essential boundary condition.

In an analytic engineering static analysis of structures, it is common to represent a very localized transverse load near x_j as a point transverse shear force, $V(x_j)$, and/or to represent a very localized bending moment (couple) near x_k as $M(x_k)$ acting normal to the plane of the deflections, $v(x)$. In other words, it is common in structural mechanics to have analytic beam solutions with jumps in the second and or third derivatives.

If the ODE was homogeneous, $f(x) \equiv 0$, then repeated integrations would yield an analytic solution that is a cubic polynomial. Typically, the transverse force per unit length is input as a polynomial, usually of degree 2. Those very common approximations result in a solution that is a degree 5 polynomial. Therefore, it would be logical in an IGA study to use degree 3 or degree 5 B-spline polynomials as the analysis functions over the length of a beam. To allow for a jump in the second or third derivative, it will be necessary to introduce a required number of repeated knots at such points while at least letting the first and second derivatives be continuous across said knots.

Similarly, it is common in heat conduction for point heat sources to occur in the domain of the governing second-order ODE, thus requiring a discontinuous slope at such points and corresponding repeated interior knots in the IGA knot vector.

4.7 Summary

Boundary conditions for even-order PDEs

Partial Differential Equation (PDE)	$\frac{\partial^{2m} u}{\partial x^{2m}} + \cdots + \frac{\partial u}{\partial x} + g(x)u + q(x) = 0$
Essential Boundary Conditions (EBCs)	$u, \frac{\partial u}{\partial n}, \ldots, \frac{\partial^{(m-1)} u}{\partial n^{(m-1)}}$
Non-essential Boundary Conditions (NBCs)	$\frac{\partial^{(m)} u}{\partial n^{(m)}}, \frac{\partial^{(m+1)} u}{\partial n^{(m+1)}}, \ldots, \frac{\partial^{(2m-1)} u}{\partial n^{(2m-1)}}$

Note: n denotes normal to the boundary $\mathbf{y}$.

Example of even-order ODE boundary conditions

1D Application	**ODE**	$n = 2m$	**EBC [0 to** $m-1$**]**	**NBC [**m **to** $2m-1$**]**
Axial Conduction ODE	$k\,\dfrac{\partial^2 u}{\partial x^2} = p$	2	u	$k\,\dfrac{\partial u}{\partial n} = q$
Beam Bending ODE	$EI\,\dfrac{\partial^4 u}{\partial x^4} = p$	4	u and $\dfrac{\partial u}{\partial n}$	$EI\dfrac{\partial^2 u}{\partial n^2} = M$ and $EI\,\dfrac{\partial^3 u}{\partial n^3} = V$

Dirac distribution is defined as $\delta(x - x_k) = \begin{cases} 0\ if\ x \neq x_k \\ 1\ if\ x = x_k \end{cases}$ with the integral property $\int_{-\infty}^{\infty} f(x, y, z)\,\delta(x - x_k)\,dx \equiv f(x_k, \mathrm{y_k}, \mathrm{z_k})$.

Method of weighted residuals:
Assumed solution: $u^r(x) = \boldsymbol{N}^r(x)\boldsymbol{\delta^r}$ in Ω^r.
Residual error: $R(x) \equiv L_\Omega\, u(x) - f(x) \neq 0$.
Weighted residual:

$$I_{MWR} = < w(x),\, R(x) > \equiv \int_\Omega w(x)\, R(x)\, d\Omega = 0.$$

Weighting functions:
Galerkin: $w(x) = u(x)$,
Collocation: $w(x) = \delta(x - x_k), \quad k = 1, 2, \ldots, \text{size}(\boldsymbol{\delta})$,
Least squares: $w(x) = \partial R(x)/\partial\boldsymbol{\delta}$.

Chapter 5

Methods of Weighted Residuals

5.1 Introduction

Several differential equations related to engineering applications have been known since the mid-1700s. Finding an analytic solution to a differential equation over a domain of arbitrary shape and satisfying the boundary conditions on the boundary of that domain have always been very difficult. Thus, engineers and applied mathematicians have long sought approximate solutions based on numerical techniques. Several methods for converting differential equations to solve a matrix system were published beginning in the early 1900s. They were known by various names such as 'trial solutions with undetermined parameters' and 'error distribution principles'. They are all based on attacking the error residual that develops when an approximate solution is substituted into a differential equation. Crandall's book (1956) entitled *Engineering Analysis* was the first to gather the most practical residual error approaches under the name of the 'method of weighted residuals' (MWRs). Crandall demonstrated various MWR options applied to obtain matric based solutions to initial value (transient) equations, differential equations, and eigenvalue problems. Crandall described the MWR categories as the Least Squares Method (1908), Galerkin Method (1915), Sub-Domain Method (1930), Method of Moments (1932), and Collocations (1937). Those methods are stable for self-adjunct differential equations. For non-self-adjunct applications, the Petrov–Galerkin Method (1940) or 'up-wind method' is needed. The Petrov–Galerkin method (originally in Russian) was applied to finite elements by Mitchell and Wait [1977] and later recognized as the Petrov approach.

The governing differential equation(s) and the boundary conditions of many physical problems can be written symbolically as

$$\begin{aligned} L_\Omega\, u^* - f^* = 0, &\quad \text{in } \Omega, \text{ with u} = u_E \text{ on } \Gamma_E, \\ L_N\, u^* - g^* = 0, &\quad \text{on } \Gamma_N, \end{aligned}$$

where $u^*(x, y, z)$ is the solution to be found, L_Ω is the governing differential operator in the domain Ω that is limited by its boundary, Γ, which consists of a portion with essential boundary conditions, Γ_E, combined with a non-overlapping portion with any non-essential boundary conditions, Γ_N,

$$\Gamma = \Gamma_E \cup \Gamma_N,$$

where L_N is the differential operator that defines the non-essential boundary condition (if any) on Γ_N, f^* is a known function of space inside the domain, Ω, and g^* is a known function of position on the boundary portion Γ_N.

If an approximate solution is assumed as $u(x, y, z) \neq u^*(x,\, y,\, z)$ and/or, in the data, is approximated by $f(x, y, z) \neq f^*(x, y, z)$ or $(s) \neq g^*(s)$, then the above-mentioned problems will not be zero, but will define non-zero spatially dependent residual errors:

$$\begin{aligned} R_\Omega(x,\, y,\, z) = L_\Omega\, u - f \neq 0, &\quad \text{in}\, \Omega, \\ R_N(s) = L_N\, u - g \neq 0, &\quad \text{on } \Gamma_N. \end{aligned}$$

Note that if essential boundary conditions are applied everywhere on the boundary, then operator L_N and operator R_N are not present. Before finite elements were invented, it was usually required that the assumed solution satisfy, in advance, any non-essential boundary operator. When that was done, there was no residual boundary error, $R_N \equiv 0$.

The residual error(s) will not vanish everywhere except in the unlikely case where the assume approximation happens to be the exact solution. Thus, these methods do the next best thing and require that weighted integral(s) of the residual(s) vanish:

$$I_{kj} = \int\limits_{\Omega} R_{\Omega}(x, y, z)\, w_k(x, y, z)\, d\Omega$$
$$+ \alpha \int\limits_{\Gamma_N} R_N(s)\, W_j(s)\, d\Gamma = 0, \quad 1 \leq k \leq n, \quad 1 \leq j \leq m$$

or

$$\boldsymbol{I} = \int_{\Omega} R_{\Omega}(x, y, z)\boldsymbol{w}(x, y, z)\, d\Omega + \alpha \int_{\Gamma_N} R_N(s)\, \boldsymbol{W}(s)\, d\Gamma = \mathbf{0},$$

where the total number of unknown constants in the approximate solution is $M = n + m$, and where α is an arbitrary scalar constant that scales the relatively importance of any non-essential boundary conditions. α is also used to achieve dimensional homogeneity. (Any valid equation must have the same dimensions in every term.) Most analysts simply pick $\alpha = 1$, with the appropriate units. Others suggest that α be some ratio of some mathematical norm of the domain integral over the norm of the boundary integral, which has the proper units.

It should be noted that the boundary operator is not needed $(m = 0)$ when the domain integral can be simplified through integration by parts. That is because integration by parts automatically introduces the boundary operator into the residual R_{Ω}. This situation occurs in the Galerkin form of weighted residuals (defined in the following), which is the main residual method utilized in this class.

5.2 Common methods

The most common residual error methods used in simulations are the Galerkin method and the Least Squares method. Two other common methods are the Collocation Method (which avoids the integrals) and the Method of Moments. To simplify this introduction, only the domain weights will be discussed (since R_N is often not present). The approximate solution is often taken as the sum of spatial functions and constants, D_k, to be determined:

$$u(x, y, z) = \sum_{k=1}^{M} D_k h_k(x, y, x) = \boldsymbol{h}(x, y, z)\, \boldsymbol{D} = \boldsymbol{D}^{\boldsymbol{T}} \boldsymbol{h}^{\boldsymbol{T}}(x, y, z).$$

Then, the Galerkin method defines the domain weights as

$$w_k(x, y, x) = \frac{\partial u}{\partial D_k} = h_k(x, y, x) \quad \text{or} \quad \mathbf{w} = \frac{\partial u}{\partial \boldsymbol{D}} = \boldsymbol{h}(x, y, z).$$

The Least Squares method defines the domain weights as

$$w_k(x, y, x) = \frac{\partial R_\Omega(x, y, z)}{\partial D_k} \text{ or } \boldsymbol{w}(x, y, x) = \frac{\partial R_\Omega(x, y, z)}{\partial \boldsymbol{D}}.$$

The Method of Moments defines its weights as

$$w_k(x, y, x) = x^{k-1}\, y^{k-1}\, z^{k-1}, \quad 1 \leq k \leq n.$$

The Collocation Method avoids the integration by defining its weights as the Dirac Distribution. Effectively, that means that the residual error is set to zero at k arbitrary points since

$$w_k(x, y, x) = \delta(x - x_k)\, \delta(y - y_k)\, \delta(z - z_k),$$

where the distribution is defined as

$$\delta(x - x_k) = \begin{cases} 0 & \text{if } x \neq x_k \\ 1 & \text{if } x = x_k \end{cases},$$

so its integral is simply the integrand evaluated at the point

$$\int_{-\infty}^{\infty} f(x, y, z)\, \delta(x - x_k)\, dx = f(x_k, y, z).$$

This property yields the simple relation wherein the differential equation is exactly satisfied at the arbitrary points: $R_\Omega(x_k, y_k, z_k) \equiv 0$. While that is the fastest, cheapest way to generate a system of equations to solve for the unknown constants, there is no mathematical guidance as to how to pick the arbitrary locations so as to give some 'best' solution. By way of comparison, the Galerkin method is proven to yield the 'best' solution in a certain mathematical norm.

5.3 One-dimensional examples

Here, the concept of approximating the solution of a differential equation by the MWRs is introduced as it was originally used on a global basis, before the development of finite element analysis. Originally, the approach required that we guess the solution over the entire domain and that our guess exactly satisfy both the essential and non-essential boundary conditions. Later, we will introduce the simple but important change that the finite element approach brings to the MWR process. (The finite element approach to the MWR will remove the important pre-condition of satisfying the boundary conditions first and will thereby greatly increase the power of the MWR when applied to domains of practical shapes.) Guessing a spatial solution that satisfies the boundary conditions is very difficult in 2D and 3D space, but it is relatively easy in one dimension.

5.4 Global MWR (pre-IGA methods)

To illustrate an original global (or single element solution) MWR solution, consider the following 1D ordinary differential equation:

$$L_\Omega(u^*(x), Q^*(x)) = 0, \ x \in]0, 1[,$$

where L is an operator that includes derivatives of the exact solution u^* and a spatially dependent source term $Q^*(x)$ and where the differential equation defines the problem in the domain between $x = 0$ and $x = 1$, but not at those end boundary points since the essential or non-essential boundary conditions must be applied there. For example, the operator might represent a second-order ordinary differential equation like

$$L_\Omega(u^*) + Q^*(x) = 0, \ x \in]0, 1[,$$

with the linear operator being $L_\Omega(u^*) \equiv \frac{d^2u^*}{dx^2} + u^*(x)$.

If an approximate solution $u(x) \neq u^*(x)$ and/or approximate spatial source term with $Q(x) \neq Q^*(x)$, then the operator will not identically equal zero, but will define a *local residual error* $R(x)$:

$$L_\Omega(u(x)) + Q(x) = R(x) \neq 0, \quad x \in 0, 1.$$

To get the exact solution, we must force the residual error, $R(x)$, to be zero. However, that requires our guess to be exact, which can happen but is very unlikely. Let the approximate solution be constructed from a set of constants, D_k, multiplying a set of spatial functions, $h_k(x)$, which exactly satisfy the boundary conditions in advance, say

$$u(x) = \sum_{k=1}^{n} D_k h_k(x) = \boldsymbol{h}(x)\,\boldsymbol{D} = \boldsymbol{D}^{\boldsymbol{T}} \boldsymbol{h}^{\boldsymbol{T}}(x).$$

To obtain the best set of constants, D_k, we require that a weighted integral of the residual must vanish:

$$\int_0^1 R_\Omega(x)\, w_k(x)\, dx = 0_k, \quad 1 \le k \le n,$$

$$\int_0^1 R_\Omega(x)\, \boldsymbol{w}(x)\, dx = \boldsymbol{0}.$$

This yields a set of linear equations for the n unknown constants, and defines the MWRs. There are several common choices for the weighting function, $w_k(x)$, and each choice has its own name in the literature.

Assume that the above-mentioned example has only the two essential boundary conditions that $u^*(0) = 0 = u^*(1)$. A choice for the approximate solution with two constants, that satisfies the essential boundary conditions in advance, might be

$$\begin{aligned} u(x) &= D_1(x - x^2) + D_2(x^2 - x^3) \equiv D_1 h_1(x) + D_2 h_2(x) \\ &\equiv \boldsymbol{h}(x)\boldsymbol{D} = \boldsymbol{D}^{\boldsymbol{T}} \boldsymbol{h}(x)^{\boldsymbol{T}}, \end{aligned}$$

with $\boldsymbol{h}(x)$ being a row matrix containing all the contributions to the assumed global spatial shape of the solution,

$$\boldsymbol{h}(x) = \begin{bmatrix} h_1(x) & h_2(x) \end{bmatrix} = \begin{bmatrix} (x - x^2)\ (x^2 - x^3) \end{bmatrix},$$

and $\boldsymbol{D}$ being a vector of constants to be determined,

$$\boldsymbol{D} = \begin{Bmatrix} D_1 \\ D_2 \end{Bmatrix}.$$

This guess satisfies both boundary conditions and defines the residual error in terms of the unknown constants as

$$R_\Omega(x) \equiv \frac{d^2 \boldsymbol{h}(x)}{dx^2} \boldsymbol{D} + \boldsymbol{h}(x)\boldsymbol{D} + Q(x) \neq 0, \quad x \in \,]0, 1[$$

or

$$R_\Omega(x) = Q(x) + \left[\frac{d^2\boldsymbol{h}(x)}{dx^2} + \boldsymbol{h}(x)\right]\boldsymbol{D}.$$

In general, the residual error in a *global* approximation can be defined symbolically as

$$R_\Omega(x) = Q(x) + \boldsymbol{b}(x)\boldsymbol{D},$$

so here the differential operator acting on the assumed spatial shapes yields a $1 \times n$ row matrix:

$$\boldsymbol{b}(x) = \left[\frac{d^2\boldsymbol{h}(x)}{dx^2} + \boldsymbol{h}(x)\right].$$

Therefore, the general residual error depends on any source term approximation, $Q(x)$, and the differential operator acting on the assumed spatial form:

$$\boldsymbol{b}(x)\boldsymbol{D} = \boldsymbol{D}^{\boldsymbol{T}}\boldsymbol{b}(x)^{\boldsymbol{T}} = \sum_{j=1}^{n} b_j(x)D_j,\ b_j(x) \equiv h_j''(x) + h_j(x).$$

Thus, the general form of a weighted residuals method is

$$\int_0^1 \left[Q(x) + \sum_{j=1}^{n} b_j(x)D_j\right] w_k(x)dx = 0_k, \quad 1 \le k \le n.$$

This can written in a linear matrix system as

$$\left[\sum_{j=1}^{n}\int_0^1 w_k(x)\, b_j(x)\, dx\right] D_j = -\int_0^1 w_k(x)Q(x)\, dx, \quad 1 \le k \le n,$$

which defines a known square matrix, $\boldsymbol{S}$, and a known source vector, $\boldsymbol{C}$, as

$$\sum_{j=1}^{n} S_{kj}D_j = C_k, \quad 1 \le k \le n,$$

$$\boldsymbol{S}\boldsymbol{D} = \boldsymbol{C},$$

with known square matrix and known column matrix entries

$$S_{kj} = \int_0^1 w_k(x)\, b_j(x)\, dx, \quad 1 \le j,\ k \le n,$$

$$C_k = -\int_0^1 w_k(x)\, Q(x)\, dx.$$

Since the essential boundary conditions were satisfied in advance, the matrix square system can be inverted and solved for the best constants

$$\boldsymbol{D} = \boldsymbol{S}^{-1}\boldsymbol{C}.$$

In practice, matrix inversion is usually replaced with a factorization process or an iterative solution.

5.5 Global Methods

Method 1: Least Squares Method (Global):

The least squares weight function is defined as the derivative of the residual with respect to the unknown constants:

$$w_k(x) = \frac{\partial R_\Omega(x)}{\partial D_k} = b_k(x),$$

$$\boldsymbol{w}(x) = \frac{\partial R_\Omega(x)}{\partial \boldsymbol{D}} = \boldsymbol{b}(x).$$

The name of the method comes from the fact that it corresponds to minimizing the integral of the square of the error:

$$\frac{1}{2}\int_0^{L=1} R_\Omega(x)^2\, dx \to \text{minimum}$$

or

$$\frac{2}{2}\int_0^{L=1} R_\Omega(x)\, \frac{\partial R_\Omega(x)}{\partial D_k}\, dx = 0 = \int_0^{L=1} R_\Omega(x) w_k(x)\, dx, \quad 1 \le k \le n.$$

The square and column matrix entries for the least square choice are

$$S_{jk} = S_{kj} = \int_0^1 b_k(x)\, b_j(x)\, dx, \quad \text{and} \quad C_k = -\int_0^1 b_k(x)\, Q(x) dx,$$
$$1 \le jk \le n,$$
$$\boldsymbol{S} = \int_0^1 \boldsymbol{b}(x)^T \boldsymbol{b}(x) dx, \quad \text{and} \quad \boldsymbol{C} = \int_0^1 \boldsymbol{b}(x)^T\, Q(x) dx.$$

Note that this approach always makes the square matrix symmetric.

Method 2: Galerkin Method (Global):

The concept here is to make the residual error orthogonal (perpendicular) to the approximate spatial functions in the solution:

$$w_k(x) = \frac{\partial u}{\partial D_k} = h_k(x).$$

This method has an important mathematical advantage wherein, in a specific mathematical norm, it yields the 'best' solution. It can also be shown that the exact solution of this integral form is the same as the exact solution of the differential equation. The square and column matrix entries for the choice are

$$S_{kj} = \int_0^1 h_k(x)\, b_j(x)\, dx, \quad \text{and} \quad C_k = -\int_0^1 h_k(x)\, Q(x) dx,$$
$$1 \le jk \le n,$$
$$\boldsymbol{S} = \int_0^1 \boldsymbol{h}^T(x) \boldsymbol{b}(x)\, dx, \quad \boldsymbol{C} = \int_0^1 \boldsymbol{h}^T(x) Q(x)\, dx.$$

This appears to give $S_{kj} \neq S_{jk}$, which means that the Galerkin method can yield a non-symmetric square matrix, but the square matrix can often be made symmetric by integration by parts. It also becomes symmetric for self-adjunct differential operators (such as stress analysis and heat transfer). It will be symmetric in most of the applications to be considered in this book.

Recall that the differential operator acting on the assumed spatial form, R_Ω, in this example, gives

$$b_j(x) \equiv h_j''(x) + h_j(x),$$

so for the two specific spatial contributions assumed earlier,

$$b_1(x) = -2 + x - x^2, \quad b_2(x) = 2 - 6x + x^2 - x^3,$$
$$b(x) = \left[(-2 + x - x^2)(2 - 6x + x^2 - x^3)\right],$$

and the Galerkin form integrals yield the symmetric matrix system

$$\begin{bmatrix} \frac{3}{10} & \frac{3}{20} \\ \frac{3}{20} & \frac{13}{105} \end{bmatrix} \begin{Bmatrix} D_1 \\ D_2 \end{Bmatrix} = \begin{Bmatrix} \frac{1}{12} \\ \frac{1}{20} \end{Bmatrix}.$$

Since the essential boundary conditions were satisfied in advance, the system is non-singular and inversion yields the coefficients

$$\begin{Bmatrix} D_1 \\ D_2 \end{Bmatrix} = \begin{Bmatrix} 71/369 \\ 7/41 \end{Bmatrix} = \begin{Bmatrix} 0.1924 \\ 0.1707 \end{Bmatrix}.$$

When this Galerkin approximate global solution is graphed along with the exact solution,

$$u^*(x) = \sin(x)/\sin(1) - x$$

The two curves appear visually the same. The accuracy is clearly good, but that may be because the guessed spatial form included the first three terms of the series representation of the sine function.

As another global example, with a Neumann condition, consider the ODE

$$\frac{d^2u^*}{dx^2} + cx = 0, \quad x =]0, L[,$$

with the non-essential boundary condition

$$\frac{du^*}{dx} - b = 0, \quad x = L,$$

and the essential boundary condition $u^*(x = 0) = 0$. (Note that if u^* has units of length, then the constant, c, in the ODE must have the units of length^{-2} to remain a dimensionally correct equation.)

To guess an approximate solution, start with a term that satisfies both the NBC and the EBC and add other terms, multiplied by the unknown constants, that are zero at the location of the EBC and have a zero slope at the location of the NBC. Try

$$\begin{aligned} u(x) &= bx + D_1\left[2\left(\frac{x}{L}\right) - \left(\frac{x}{L}\right)^2\right] + D_2\left[3\left(\frac{x}{L}\right) - \left(\frac{x}{L}\right)^3\right] \\ &= bx + D_1 h_1(x) + D_2 h_2(x). \end{aligned}$$

Then, the domain residual is

$$R_\Omega(x) = \frac{d^2u}{dx^2} + cx = -\frac{2}{L^2}D_1 - \frac{6x}{L^3}D_2 + cx.$$

The weighted integrals of the domain residual are

$$\int_0^L R_\Omega(x)\, h_1(x)\, dx = \frac{5cL^3}{12} - \frac{4}{3}D_1 - \frac{5}{2}D_2 = 0,$$

$$\int_0^L R_\Omega(x)\, h_2(x)\, dx = \frac{4cL^3}{5} - \frac{5}{2}D_1 - \frac{24}{5}D_2 = 0,$$

which are rewritten in the governing matrix system (when multiplied by 60) as

$$\begin{bmatrix} 80 & 150 \\ 150 & 288 \end{bmatrix} \begin{Bmatrix} D_1 \\ D_2 \end{Bmatrix} = cL^3 \begin{Bmatrix} 25 \\ 48 \end{Bmatrix}, \quad \boldsymbol{SD} = \boldsymbol{C}.$$

Since the EBC was satisfied in advance, this system is non-singular and the square matrix can be inverted to give the desired constants,

$$\boldsymbol{D} = \boldsymbol{S}^{-1}\boldsymbol{C}, \begin{Bmatrix} D_1 \\ D_2 \end{Bmatrix} = \frac{cL^3}{540}\begin{bmatrix} 288 & -150 \\ -150 & 80 \end{bmatrix}\begin{Bmatrix} 25 \\ 48 \end{Bmatrix} = \frac{cL^3}{6}\begin{Bmatrix} 0 \\ 1 \end{Bmatrix}.$$

This means that $h_1(x)$ is excluded from the approximation, which is

$$u(x) = bx + \frac{cL^3}{6}\left[3\left(\frac{x}{L}\right) - \left(\frac{x}{L}\right)^3\right].$$

This happens to be the exact solution. This illustrates that the MWR will select the exact solution if it happens to be a combination of the $h_k(x)$ terms. Later, it will be shown that the above-mentioned ODE can represent displacement of an axial bar or the temperature distribution along a bar.

When the IGA extension to the MWR is introduced, there will be a few changes, but they yield a drastic increase in the power of weighted residual methods. When the iElement mesh is introduced, the global spatial approximation, $h_k(x)$, will be replaced by a set of piecewise spatial basis functions, $\boldsymbol{N}_k(x)$. The differential operator will act on them and will be represented as $\boldsymbol{B}_k(x)$ in place of the global $b_k(x)$. Another important change is that the spatial approximations are not required to satisfy the essential boundary conditions in advance. This means that the matrix system will just satisfy the integral form and it must be modified to include the essential boundary conditions before the system can be solved for the unknown constants (and reactions).

5.6 Summary

Galerkin weighted residuals:

$$\int_\Omega \frac{\partial u(x)}{\partial D_k} R_\Omega(x)\, dx = 0_k, \quad 1 \le k \le n.$$

Least squares weighted residuals:

$$\int_\Omega \frac{\partial R_\Omega(x)}{\partial D_k} R_\Omega(x)\, dx = 0_k, \quad 1 \le k \le n.$$

Isogeometric analysis piecewise extension:

$$\Omega = \cup_e \Omega^e, \quad \text{the union of non-overlapping patches}$$

$$u(x) = \sum_k N_k(x)\, D_k^e \quad \text{in } \Omega^e \subset \Omega.$$

$$\int_\Omega R_\Omega(x)\, w_k(x)\, dx = \sum_e \int_{\Omega^e} R_{\Omega^e}(x) w_k^{\Omega^e}(x)\, dx.$$

Chapter 6

Numerical Integration

6.1 Introduction

The previous chapter showed that a user has a few choices for selecting non-dimensional coordinates for carrying out an IGA or an FEA study. Both types of studies require that physical derivatives have to be determined in whichever system a user selects. Also, integrations over the domain of the coordinate system are always required. In practical applications of IGA and FEA, the integrals are evaluated by numerical integration using tabulated 'quadrature rule' data for the type of functions being integrated. There are rules for logarithmic, exponential, polynomial, and so on, but in IGA, the basis functions are the polynomials of choice.

In FEA, Gaussian quadrature exactly integrates iElement polynomial basis functions, which are generally C^0 at element boundaries. In IGA, the NURBS and T-splines are ratios of polynomials, so classic Gaussian quadrature over a knot span is not exact and the iElement integrations are approximate. Due to the IGA basis functions having $C^{>0}$ continuity between iElements, practice shows that using the classic Gaussian rule with the theoretical number of quadrature points, n_q, is too high and iElements are over-integrated and their numerical integration can be highly inefficient. It has been found that the practical number of Gauss points needed to integrate B-spline basis functions depends more on the number of basis functions (i.e., the number of interior knots) over the patch than on the basis degree, p.

This is probably due to the higher (than FEA) smoothness across iElement boundaries, which causes fewer degrees of freedom over the entire patch. This fact is exploited in the 'half point rule' given by Hughes *et al.* (2010). This rule spans the entire patch, instead of knot spans within a patch, and typically requires about half the number of integration points compared to the classis Gaussian number, n_q.

Alternative numerical integration rules have been proposed, including using the Lobatto rules, which always include the knot span end points, unlike the Gauss rules.

Herein, efficiency is less important than education and Gaussian quadratures are used for most numerical integrations at the knot span (iElement) level.

6.2 Natural coordinate quadrature

The first dataset for exactly integrating a polynomial over a line in the 'natural' coordinates of $-1 \leq a \leq +1$ was developed by Gauss and is usually called Gaussian quadrature. Gauss proved that the sum of the polynomial evaluated at n weighted locations, a_k, would exactly integrate a polynomial of degree $p \leq (2n-1)$:

$$\int_{-1}^{+1} a^p \, da \equiv \sum_{k=1}^{n} (a_k)^p,$$

where, as a check, for any polynomial degree, the sum of the weights must equal the 'measure' of the non-dimensional space being used. Set $p = 0$, then the constant $a^0 = 1$ and its integral requires

$$\int_{-1}^{+1} 1 \, da = a \Big|_{-1}^{+1} = (+1 - (-1)) = +2 \equiv \sum_{k=1}^{n} 1 w_k.$$

In other words, to help find the proper table of tabulated numerical integration data for natural coordinates on a line, the sum of the weights must be two, because that is the 'measure or non-dimensional length' of the line. Gauss's derivation yielded a set of points and weights for each polynomial degree, p. The locations are symmetric

Table 6.1. First three Gauss rules for evaluating an integral on $-1 \leq a \leq +1$

$$\int_{-1}^{+1} a^p da \equiv \sum_{k=1}^{n} (a_k)^p \quad \text{where} \sum_{k=1}^{n} w_k = 2$$

Polynomial degree	n	Location, a_k	Weights, w_k
$p = 0$	1	0	2
$p = 0, 1, 2, 3$	2	$-\sqrt{1}/3$, $+\sqrt{1}/3$	1, 1
$p = 0, 1, 2, 3, 4, 5$	3, etc.	$-\sqrt{3}/5$, 0, $+\sqrt{3}/5$	5/9, 8/9, 5/9

about the center and include the center point ($a = 0$) for odd polynomial degrees. The first three analytic results are given in Table 6.1. Since evaluating square roots and similar functions is slow, the coefficients are evaluated once to high precision and stored in a function for fast retrieval of the data. Here, those data, for up to $p = 13\,(n = 7)$ in the Matlab function *qp_rule_Gauss.m* where the top three cases are shown in Script 6.1.

Physical quadrilaterals are mapped to a non-dimensional square in natural coordinates $-1 \leq a, b \leq +1$, so each side ranges from -1 to $+1$. The 'measure' (local area) of the square is thus four:

$$\int_{-1}^{+1} \int_{-1}^{+1} 1\, da\, db = (+2)(+2) = 4 = \left(\sum_{j=1}^{n} w_j\right) \left(\sum_{k=1}^{n} w_k\right)$$

$$\equiv \sum_{q=1}^{n^2} w_q.$$

As mentioned, the 2D quadrilateral data points are the tensor product of the above-stated 1D natural coordinate data. They can be placed in a similar script or evaluated from the 1D rule using the tensor product form in *qp_rule_nat_quad.m*, which is shown in Script 6.2. Likewise, hexahedra in 3d natural coordinates $-1 \leq a, b, c \leq +1$ have their data evaluated by a triple tensor product of the 1D data, and the natural coordinates' weights must sum to 8.

Script 6.1. Tabulated Gauss quadrature data to degree $p = 7$

```
function [r_q, w_q] = qp_rule_Gauss (n_q) % === ====
%  tables of quadrature point locations and weights
%  for lines interpolated in natural coordinates
%             -1 <= r <= 1,    sum (w_q) == 2
% n_q = number of quadrature points required
% r_q = all of first parametric quadrature coordinates
% w_q = weight at all quadrature points in n_p
 switch n_q; % identity of data set to gather
  case 1 % precision: exact for polynomial of degree 1
    r_q = [0] ; w_q = [2];

  case 2 % precision: exact for polynomial of degree 3
    r_q = [-0.57735026918962576450914 ...
            0.57735026918962576450914] ;
    w_q = [1 1]                       ;

  case 3 % precision: exact for polynomial of degree 5
    r_q = [-0.77459666924148337703585 ...
            0.00000000000000000000000 ...
            0.77459666924148337703585];
    w_q =  [0.55555555555555555555556 ...
            0.88888888888888888888889 ...
            0.55555555555555555555556] ;

  case 4 % precision: exact for polynomial of degree 7
    r_q = [-0.86113631159405257522394 ; ...
           -0.33998104358485626480266 ; ...
            0.33998104358485626480266 ; ...
            0.86113631159405257522394] ;
    w_q =  [0.34785484513745385737306 ; ...
            0.65214515486254614262694 ; ...
            0.65214515486254614262694 ; ...
            0.34785484513745385737306] ;
. . .
  otherwise                           % update the tables
   fprintf ('Requested data where n_q = %i \n', n_q)
   error ('\n ERROR rule not tabulated for n_q > 7')
 end                  % switch number of quadrature points
% end qp_rule_Gauss % ================================
```

Script 6.2. Tensor product rule for quadrilateral integration on $-1 \leq a,\ b \leq +1$

```
function [r_q, s_q, w_q] = qp_rule_nat_quad (n_q) % ==========
%  tables of quadrature point locations and weights for
%     quadrilaterals interpolated in natural coords,
%                       -1 <= r,s <= 1

% n_1 = number of 1-D points in each direction
% n_q = 2-D number of quadrature points required, = n_1^2
% r_q = all of first parametric quadrature coordinates
% s_q = all of second parametric quadrature coordinates
% w_q = weight at all quadrature points in n_p

r_q = zeros(n_q, 1); s_q = zeros(n_q, 1); w_q = zeros(n_q, 1);

if ( n_q == 1 )    ; % exact for constant or linear polynomial
  r_q = [0]; s_q = [0]; w_q = [4];       % centroid point data

%                       tensor products of 1-D rule
elseif (n_q==4 | n_q==9 | n_q==16 | n_q==25);  % are tabulated
  n_1 = fix ( sqrt ( n_q ) )                   ; % size of 1-D rule
  [r_1, w_1] = qp_rule_Gauss (n_1)            ; % get 1-D rule data
  k = 0                                  ; % initialize point number
  for i = 1 : n_1             ; % loop over points in s-direction
    for j = 1 : n_1           ; % loop over points in r-direction
      k = k + 1              ; % point number in the quadrilateral
      w_q (k) = w_1 (i) * w_1 (j)    ; % product of 1-D weights
      r_q (k) = r_1 (j)        ; % r-coordinate in quadrilateral
      s_q (k) = r_1 (i)        ; % s-coordinate in quadrilateral
    end                         ; % for j point in quadrilateral
  end                           ; % for i point in quadrilateral
else                         ; % update the tables or elseif above
  error ('\nERROR quad rule not tabulated for these points')
end          ; % if number of quadrature points in quadrilateral
% end qp_rule_nat_quad % ====================================
```

6.3 Unit coordinate quadrature

In IGA and FEA, the analytic basis functions are often defined in unit coordinate space of $0 \leq u \leq +1$. This choice causes some example studies to be presented in exact analytic form, but in practice almost all integrals are evaluated by numerical integration. Since the linear relation between natural and unit coordinates is simply $u = (a+1)/2$, the numerical values in Script 6.1 can easily be transformed to the equivalent values in unit coordinates. Of course, the weights must change because the measure of the unit space is $(+1-0) = 1$, so the weights in that set of units must always sum to 1 for a line, square, and hexahedra. Script 6.3 gives the 1D unit coordinate data.

> **Looking Ahead**: In IGA studies, the basis functions are done in unit coordinates. When they need to be integrated numerically, using unit coordinates instead of natural coordinates makes the Jacobian a unity matrix, and thus avoids one minor product that is often needed millions of times in the integrations.

Script 6.3. Gauss 1D data converted to unit coordinates

$$\int_0^1 a^p da \equiv \sum_{k=1}^{n} (a_k)^p \quad \text{where} \sum_{k=1}^{n} w_k = 1$$

```
function [r_q, w_q] = qp_rule_unit_Gauss (n_q) % ======
% tables of quadrature point locations and weights
% for lines interpolated in unit coordinates.
%                0 <= r <= 1,    sum (w_q) == 1
% n_q = number of quadrature points required
% r_q = all of first parametric quadrature coordinates
% w_q = weight at all quadrature points in n_p
switch n_q             ; % identity of data set to gather
 case 1 ; % precision: exact for polynomial of degree 1
   r_q = [0.5] ; w_q = [1] ;

 case 2 ; % precision: exact for polynomial of degree 3
   r_q = [2.11324865405187117745 43e-01 ...
          7.88675134594812882254 57e-01] ;
   w_q = [5.00000000000000000000 00e-01 ...
          5.00000000000000000000 00e-01] ;
 case 3 ; % precision: exact for polynomial of degree 5
   r_q = [1.12701665379258311482 08e-01 ...
          5.00000000000000000000 00e-01 ...
          8.87298334620741688517 92e-01] ;
   w_q = [2.77777777777777777777 78e-01 ...
          4.44444444444444444444 45e-01 ...
          2.77777777777777777777 78e-01] ;
```

```
  case 4 % precision: exact for polynomial of degree 7
    r_q = [-0.86113631159405257523946 ; ...
           -0.33998104358485626480266 ; ...
            0.33998104358485626480266 ; ...
            0.86113631159405257523946] ;
    w_q =  [0.3478548451374538573730 6 ; ...
            0.6521451548625461426269 4 ; ...
            0.6521451548625461426269 4 ; ...
            0.3478548451374538573730 6] ;
. . .
otherwise                        ; % update the tables
  fprintf ('Requested data where n_q = %i \n', n_q)
  error ('\n ERROR  rule not tabulated for n_q > 7')
end                   % switch number of quadrature points
% end qp_rule_unit_Gauss % ============================
```

6.4 Lobatto

By evaluating the integrand at n points, two of which are the end points, Lobatto's method exactly integrates polynomials of degree $2n-3$.

The Lobatto data are listed in Script 6.4 for polynomials to degree 5.

Script 6.4. Lobatto quadrature data in natural coordinates

$$\int_{-1}^{+1} a^p da \equiv \sum_{k=1}^{n} (a_k)^p \quad \text{where} \sum_{k=1}^{n} w_k = 2$$

```
function [pt, wt] = qp_Lobatto_nat_rule (n_qp)
% ==============================================
% Lobatto quadrature abscissae and weight coeffs
%    in natural coordinates -1 <= pt <= +1
% using end points, unlike Gaussian quadratures
% ==============================================
% n_qp = number of quadrature points
% pt   = abscissae of Lobatto points
% wt   = weights   of Lobatto points
%
% max_pt = max. no. of points tabulated
max_pt=10; pt=zeros(n_qp,1); wt=zeros(n_qp,1);
if ( n_qp > max_pt ); n_qp = max_pt; fprintf ...
```

```
 ('warning, Lobatto used n_qp = %i \n',max_pt)
end                  ; % if data are in this table
switch (n_qp)      ; % jump to the tabulated case
  case (1)                ; % 1 point for degree 0
    pt = [0.00000000000000d+00];
    wt = [2.00000000000000d+00];
  case (2)               ; % 2 points for degree 1
    pt = [1.00000000000000d+00 ...
         -1.00000000000000d+00] ;
    wt = [1.00000000000000d+00 ...
          1.00000000000000d+00] ;
  case (3)               ; % 3 points for degree 3
    pt = [1.00000000000000d+00 ...
          0.00000000000000d+00 ...
         -1.00000000000000d+00] ;
    wt = [3.33333333333333d-01 ...
          1.33333333333333d+00 ...
          3.33333333333333d-01] ;
 case (4)               ; % 4 points for degree 5
    pt = [1.00000000000000d+00 ...
          4.47213595499958d-01 ...
         -1.00000000000000d+00 ...
         -4.47213595499958d-01] ;
    wt = [1.66666666666667d-01 ...
          8.33333333333333d-01 ...
          1.66666666666667d-01 ...
          8.33333333333333d-01] ;
 . . .
otherwise                           ; % report error
  error ('Lobatto rule requested < 1 point')
end  ; % select the number of quadrature points
% end qp_Lobatto_nat_rule ======================
```

Chapter 7

Second-Order Elliptic ODEs via B-Splines

7.1 Preview

There are relatively few applications where IGA algorithms and results are compared to the huge number of similar FEA studies. The word similar is used here because many IGA applications in the literature often do not adequately discuss general considerations such as various essential (Dirichlet) boundary conditions (EBCs) and their associated reactions, which are often important in engineering studies. IGA applications often point out the advantage of higher-order continuity of the solution results compared to FEA results. That is certainly an advantage of IGA over FEA in 2D applications and in several 3D applications, and advances in both fields are continuing.

Many IGA solutions that are $C^{n>0}$ continuous are compared to C^0 FEA results, implying that FEA always involves C^0 continuity. That is usually true in higher dimensions, but in 1D the FEA can easily produce $C^{n>0}$ solutions that can match or exceed the benefits of IGA. So, some 1D comparisons hide important benefits for IGA in higher dimensions. Yet, some challenges related to Dirichlet BCs, non-homogeneous material region interfaces, etc., can be demonstrated in 1D.

When viewing results from IGA and FEA, it is important to have some insight into the nature of the exact solution of the PDE or ODE. Both FEA and IGA use piecewise polynomials to approximate their solutions. If the unknown exact solution is known to be a piecewise polynomial, the user should be aware of that. If the exact solution is expected to include exponential, hyperbolic, or trigonometric functions, then the differences, if any, between IGA and FEA formulations are more significant.

7.2 Self-Weight column IGA and FEA studies

Consider a linearly elastic vertical bar, of length L, supporting its own uniform weight per unit length, $w(x)$, and an axial force, F, at a non-restrained point. In engineering jargon, this structure is often called a column and the distributed axial load is $w(x) = \rho g A(x)$. To be a non-singular model, the bar must have an axial displacement, δ, specified at least at one location. The equilibrium of the bar is defined by the second-order ordinary differential equation:

$$-\frac{d}{dx}\left[E(x)A(x)\frac{d\delta(x)}{dx}\right] - w(x) = 0. \tag{7.1}$$

For a unique solution, two boundary conditions (BCs) are needed, giving the essential displacement (Dirichlet) value, $\delta(x_D) \equiv \delta_D$, and/or the non-essential displacement gradient (Neumann) value, $d\delta(x_N)/dx$, at one or more locations along the domain. At a point, x_D, with a Dirichlet boundary condition, there is a reaction of

$$R_D = E(x_D)A(x_D)\,\frac{d\delta(x_D)}{dx}, \tag{7.2}$$

which is usually recovered as part of the overall solution. The interpretations of this ODE for some different fields of application are listed in the Summary.

If there is a discontinuity in any of the (piecewise) constant data, E, A, or ρ due to a material change interface (for E or ρ) or a step change in size (A), then the exact solution must have a discontinuity in its slope ($d\delta/dx$) there, but would still be a quadratic polynomial between such points. A point force, F, can be introduced at some point x_F as a special case of the distributed line load by using the Dirac Delta Distribution, $\Delta(x - x_F)$, to define the distributed load as $w(x) \equiv \Delta(x - x_F)F$. Such point discontinuities in the data are very common and numerical solutions should be able to easily include them.

If the data E, A, ρ and g are constant, then two integrations would show the exact solution to be a quadratic polynomial. Likewise, if $g = 0$ (a horizontal bar), this becomes a homogeneous ODE having just a linear polynomial for its exact solution.

There are several ways that the required Dirichlet displacement can be imposed in a physical model, as sketched in Fig. 7.1, along with the support reactions:

(a) The top end of a hanging bar can be restrained (second most common), which puts the entire bar in tension (if F is downward).
(b) The bottom end of a structural column can be restrained (most common), which puts the entire bar in compression (if F is downward).
(c) Both ends of the bar, which puts all of the bar in either compression or tension (if $F = 0$).
(d) The displacement restraint can be anywhere along the bar, $0 > x < L$, which puts part of the bar in compression and part in tension (if $F = 0$).

For Case (1), the leftmost image in Fig. 7.1, with $F = 0$ and constant properties, integrating twice and applying the Dirichlet displacement condition $\delta = 0$ at $x = 0$ and the Neumann condition that $d\delta(L)/dx = 0$ at $x = L$ yields the exact solution, *along the x-axis*, of

$$\delta(x) = \frac{W}{EA}\left(x - \frac{x^2}{2L}\right),$$

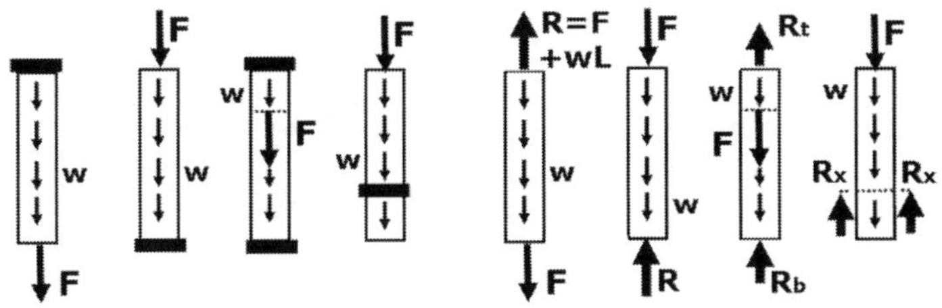

Figure 7.1. Bars supported at one end, both ends, or between ends, and their reaction forces, R

where the total weight of the bar is $W = w\,L$. The mechanical strain, ε, *along the x-axis* is the displacement gradient

$$\varepsilon(x) = \frac{\partial \delta(x)}{\partial x} = \frac{W}{\text{EA}}\left(1 - \frac{x}{L}\right),$$

and the mechanical stress *along the x-axis*, σ, from Hooke's Law is

$$\sigma(x) = E\,\varepsilon(x) = \frac{W}{A}\left(1 - \frac{x}{L}\right).$$

At the lower end of the bar at $x = L$, the force per unit area goes to zero, while at the top, $x = 0$, the force per unit area is $\sigma(0) = W/A$.

Since the exact solution is just a quadratic polynomial, a single quadratic Lagrangian FEA element with three nodes gives the exact results everywhere. Likewise, a single IGA quadratic B-spline, with three control points (CPs), should give the exact results everywhere when projected *along the x-axis*. Thus, the following discussion is just about illustrating the slightly different procedures required to implement a 1D IGA model.

FEA+

Tong's Theorem states that if the 1D FEA interpolations include the homogeneous solution, then the computed 1D nodal values will be exact.
Thus, quadratic 1D element interpolations will yield the exact 1D result everywhere. Here, IGA & FEA procedures are compared, not their accuracy.

Looking Ahead: The FEA procedure for enforcing Dirichlet BCs is the same for zero and non-zero values, but IGA often requires a different procedure for non-zero Dirichlet BCs especially on surfaces and solids.

After developing an equivalent governing integral form over the domain $[0,\ L]$, the bar will be divided into physical segments for evaluating the integrals. An IGA model will be divided into non-overlapping sub-regions spanned by B-splines. In an FEA, a mesh of non-overlapping elements is created. In both approaches, the solution will be approximated with the corresponding piecewise polynomials and IGA CPs (or FEA node interpolation points).

Note: Currently, the displacement is considered to have only one component, but in general it has three components. In the solid mechanics literature, the components are often written as $\{\boldsymbol{\delta}\}^T = [\delta_x(x,y,z)\ \delta_y(x,y,z)\ \delta_z(x,y,z)]$. For the 1D cases, $\delta \equiv \delta_x$ is used for simplicity. A second notation of $\{\boldsymbol{\delta}\}^T = [u(x,y,z)\ v(x,y,z)\ w(x,y,z)]$ is avoided here to avoid confusion with the parametric coordinate, u, used here to define the IGA basis functions.

7.3 From ODE to integral form

By applying the Galerkin weighted residual method, the equivalent integral form is found by multiplying the residual by the approximating spatial forms and setting their integral equal to zero (i.e., making the residual error in the differential equation orthogonal to the solution):

$$I = \int_0^L \delta(x)\left[-\frac{d}{dx}\left[E(x)A(x)\frac{d\delta(x)}{dx}\right] - w(x)\right] dx \equiv 0. \tag{7.3}$$

> **Looking Ahead**: This will end with an equivalent matrix equilibrium form, $\{\boldsymbol{\delta}\}^T([\boldsymbol{K}_{\mathrm{EA}}]\{\boldsymbol{\delta}\} - \{\mathbf{c}_w\} - \{\mathbf{c}_{\mathrm{EBC}}\}) \equiv 0$, and beginners will be confused by the change from the exact scalar $\delta(x)$ to its approximate value interpolated from the numerical degrees of freedom, $\{\boldsymbol{\delta}\}$.

First, assume that the displacement restraint will occur at one end of the bar. The first term can be transformed using Green's Theorem from calculus (integration by parts) as

$$I = -\left[\delta(x)\left(E(x)A(x)\frac{d\delta(x)}{dn}\right)\right]_0^L + \int_0^L \left(\frac{d\delta(x)}{dx}E(x)A(x)\frac{d\delta(x)}{dx}\right)dx - \int_0^L \delta(x)\, w(x)\, dx \equiv 0, \tag{7.4}$$

where the normal gradient of the solution, $d\delta/dn$ is the gradient vector dotted with the boundary unit normal vector. Here, that is $d\delta/dn = n_x d\delta/dx$ with $n_x = \pm 1$.

Examining the boundary contribution, the quantity in the curved brackets usually has a physical meaning. Here, the displacement gradient is the mechanical strain:

$$d\delta(x)/dx = \varepsilon(x). \tag{7.5}$$

For a linear material, Hooke's Law gives the axial stress (force per unit area) as

$$E(x)\varepsilon(x) = \sigma(x), \tag{7.6}$$

and multiplying the stress by the cross-sectional area yields the axial

$$F(x) = A(x)\sigma(x) = A(x)E(x)\varepsilon(x) = E(x)A(x)\frac{d\delta(x)}{dn}. \tag{7.7}$$

Therefore, the boundary contributions can be viewed as

$$-\left[\delta(x)\left(E(x)A(x)\frac{d\delta(x)}{dn}\right)\right]_0^L = -[\delta(x)F_n(x)]_0^L = -[\delta(x)n_xF(x)]_0^L.$$

At either end of the domain, specifying the displacement, $\delta(x)$, is a Dirichlet condition and specifying the normal force, F_n, would be a Neumann boundary condition. Mathematics prevents both types of BCs from being applied to the same boundary region. In both IGA and FEA, either one of the two terms can be specified. If the force is specified, then the displacement will be computed. If the displacement is specified (as essential), then the force there can be optionally calculated as the reaction needed to maintain the Dirichlet condition. If neither of those two is specified, then the boundary condition defaults to a stress-free section having a zero displacement gradient, $d\delta(x)/dx = 0$.

This transformed integral is called the 'weak form' because any approximation now has to have only a non-vanishing first derivative instead of the second derivative in the ODE. Thus, a piecewise linear, C^0, approximation is the lowest-order polynomial that can be used to obtain a solution with this equivalent form. Generally, that would not be desirable in IGA since it is usually better to have an C^1 approximation, which is always easy with B-splines in an IGA, but is only easy for 1D FEA.

7.4 From integral to matrix forms

The integration-by-parts term often causes confusion when it is converted to matrix notation. Recall that the scalar boundary condition term is a sum of displacements multiplied by forces:

$$\begin{aligned} -[\delta(x)F(x)]_0^L &= -[\delta(L)F_n(L) - \delta(0)F_n(0)] \\ &= [\delta(0)F_n(0) - \delta(L)F_n(L)], \end{aligned} \tag{7.8}$$

which shows that the Galerkin method of weighted residuals, combined with integration by parts, brings in the product of the Dirichlet BC and the Neumann BC on the boundary of a domain. In an FEA study, there are always nodes on the boundary, at $x = 0$ and $x = L$. But, in an IGA study, the control point locations are often not on a body at the locations where BCs are to be imposed. In the special 1D case, it is usually practical to place a control point location at the location of an essential boundary condition (for at least three of the four cases in Fig. 7.1).

The independent displacements to be computed will either be

(a) values at the CPs generally off a body, having units of displacements, in an IGA system vector of displacement control values, $\{\boldsymbol{\delta}\}^T = [\delta_1\ \delta_2\ \delta_3 \cdots \delta_{n_{\text{CP}}}]$, where n_{CP} is the number of CPs for the piecewise B-splines, or

(b) the displacement values at all nodes in a mesh on a body in an FEA system displacement vector, $\{\boldsymbol{\delta}\}^T = [\delta_1\ \delta_2\ \delta_3 \cdots \delta_{n_{\text{NM}}}]$, where n_M is the number of analysis control points (APs) in the domain.

> An extremely important difference between the two approaches is that for an FEA study the solution vector, $\{\boldsymbol{\delta}\}$, contains the values of the displacements on a body, while in an IGA study the values in $\{\boldsymbol{\delta}\}$ are control values off the body, with the units of displacements. The actual body displacements are recovered from the product of the analytic basis functions multiplied by the control point displacements, evaluated at locations on the body. The location of a displacement on the body is the product of the analytic basis functions multiplied by the control point spatial coordinates.

In either case, let the last term in the displacement vector be denoted as δ_L. For the first three cases in Fig. 7.1, it is assumed that a CP is at the location of an essential boundary condition. Also, in either case, it is possible that there is an external force applied at every displacement point and each force is a component in a system force vector, $\{\boldsymbol{F}\}^T = [F_1,\ F_2,\ F_3,\ \ldots,\ F_L]$, even though in practice most of those external load entries are identically zero.

> **Looking Ahead:** Recall that in an FEA the nodes are always on the body (the mesh), but in an IGA the CPs (especially in 2D and 3D) are usually not on the body and thus may not be located at the positions where physical BCs must be applied. Later, additional IGA procedures will be required when CP positions do not correspond to BC positions.

All of the terms in (7.8) will eventually appear as a matrix product with the system displacement vector, $\{\boldsymbol{\delta}\}$, and another matrix (of a consistent size). The choice of matrix notations requires the scalar boundary condition work terms in (7.8) to also be rewritten as a vector inner product (dot product) of all of the displacement degrees of freedom and the two normal force terms:

$$- [\delta(x)F(x)]_0^L = -[\delta(L)F_n(L) - \delta(1)F_n(1)],$$

$$- [\delta(x)F(x)]_0^L = -[\delta(1)\ \delta(L)] \begin{Bmatrix} -F_n(1) \\ F_n(L) \end{Bmatrix}.$$

From experience with matrix IGA and FEA forms, it is necessary to expand the displacements to include all of the system displacements. Thus, this simple scalar boundary condition work product, in terms of the system displacement vector, is expanded (with the two sign changes) to

$$- [\delta(x)F(x)]_0^L \equiv -\{\boldsymbol{\delta}\}^T\{\boldsymbol{c}_{EBC}\}, \tag{7.9}$$

where the system displacement vector, $\{\boldsymbol{\delta}\}$, is always fully populated, while currently the forces associated with the two BCs only contain two forces, so the above is not a valid matrix produce. To use the above-mentioned notation, as will be done throughout this book, it is necessary to pad the center of the defined force vector, $\{\boldsymbol{c}_{\mathrm{EBC}}\}$, with $(L-2)$ zeros. Thus, *assuming* that the BCs are associated with the first and last displacements, the EBC force vector is

$$\{\boldsymbol{c}_{\mathrm{EBC}}\}^T \equiv [-F_n(x_0)\ 0\ 0 \cdots 0\ F_n(x_L)].$$

Or, thinking in terms of Neumann conditions,

$$\{\boldsymbol{c}_{\text{EBC}}\}^T \equiv \left[-\left(E(0)A(0)\frac{d\delta(0)}{dn}\right)\ 0\ 0\cdots 0\ \left(E(L)A(L)\frac{d\delta(L)}{dn}\right)\right]. \tag{7.10}$$

The remaining terms in the integral form require spatial integrals over the physical domain, $\Omega = [0, L]$, of the spatial approximation of the unknown displacement $\delta(x)$ and its gradient, $d\delta(x)/dx$. The numerical approximation for IGA splits the analysis domain into non-overlapping finite sub-regions (iElements), while the FEA uses non-overlapping finite element regions. However, the IGA regions are controlled by overlapping solution control values (degrees of freedom). Thus, an FEA solution has less local support than an IGA solution.

To address the remaining integrals, a common notation, $[\boldsymbol{h}(u)]$, will be used to denote a set of piecewise polynomial analysis basis functions, whether from IGA or FEA, which are restricted to the non-dimensional parameter domain of $0 \leq u \leq 1$. Initially, the parameter domain symbol, u, is one that is commonly used in the computer graphics community. Let the spatial basis functions approximation of the unknown solution in a finite sub-region, $\Omega^r \subset_r \Omega$, be denoted as $[\boldsymbol{h}(u)]$. They are connected of a sub-set of n_i independent degrees of freedom, $\{\boldsymbol{\delta}^r\}$, of the total number of system degrees of freedom, n_d. That is, $\boldsymbol{\delta}^r \subset_r \boldsymbol{\delta}$, and the local unknown is

$$\delta(x(u)) = [\boldsymbol{h}(u)]\{\boldsymbol{\delta}^r\} = \{\boldsymbol{\delta}^r\}^T[\boldsymbol{h}(u)]^T = \sum_{i=1}^{i=n_i} h_i(u)\delta_i^r. \tag{7.11}$$

Combined with the isoparametric method, the spatial position is also defined by

$$x(u) = [\boldsymbol{h}(u)]\{\boldsymbol{x}^r\} = \{\boldsymbol{x}^r\}^T[\boldsymbol{h}(u)]^T = \sum_{i=1}^{i=n_i} h_i(u)\mathrm{x}_i^r, \tag{7.12}$$

where the x_i^r are the coordinates of the analysis degrees of freedom, δ_i^r, controlling sub-region r. That is, a control point can have associated with it both known coordinates and unknown displacement values.

Here, the basis functions can be the IGA B-spline basis functions, $h_i(u) = N_i(u)$, the FEA Lagrange basis functions, $h_i(u) = L_i(u)$, the FEA Hermite basis functions, $h_i(u) = H_i(u)$, or any other useful analysis basis function, like the IGA Bernstein basis functions, $h_i(u) = B_i(u)$, or NURBS.

Having assumed spatial forms for the piecewise polynomial basis functions and the geometric mapping of the sub-region, the solution and location local derivatives are

$$\frac{d\delta(u)}{du} = \left[\frac{d\boldsymbol{h}(u)}{du}\right]\{\boldsymbol{\delta}^r\} = \{\boldsymbol{\delta}^r\}^T \left[\frac{d\boldsymbol{h}(u)}{du}\right]^T = \sum_{i=1}^{i=n_i} \frac{dh_i(u)}{du}\delta_i^r, \tag{7.13}$$

$$\frac{dx(u)}{du} = \left[\frac{d\boldsymbol{h}(u)}{du}\right]\{\boldsymbol{x}^r\} = \{\boldsymbol{x}^r\}^T \left[\frac{d\boldsymbol{h}(u)}{du}\right]^T = \sum_{i=1}^{i=n_i} \frac{dh_i(u)}{du}\mathrm{x}_i^r. \tag{7.14}$$

In calculus, the term $dx(u)/du$ accounts for the relation between physical derivatives and parametric derivatives due to the change of variables that map the physical space [0, L] into the non-dimensional space [0, 1]. That derivative is called the Jacobian matrix of that mapping. For one-to-one mappings (like this straight line), it is a square matrix of the size of the spatial dimensions. Here, the physical space and the basis function space are both one-dimensional, so the Jacobian matrix is (1×1) and appears as a scalar:

$$[\mathbb{J}_{x:u}(u)] = \left[\frac{dx(u)}{du}\right]. \tag{7.15}$$

Also from calculus, it is known that the scalar determinant of the Jacobian, $|\mathbb{J}_{x:u}(u)|$, relates the differential length in the first space, dx, to the differential length in the second space, du:

$$dx(u) = |\mathbb{J}_{x:u}(u)|du = \det([\mathbb{J}_{x:u}(u)])\, du. \tag{7.16}$$

Note that in general the Jacobian matrix varies with u, but in special cases it is a constant when both sets of geometric data are uniformly spaced. Thus, a general IGA or FEA algorithm using the Jacobian matrix must assume the Jacobian to be spatially dependent. To have the correct units, remember that dx has the units of length, and in this case, so does the determinant of the Jacobian.

When the Jacobian matrix is square, like here, the inverse Jacobian matrix relates parametric derivatives, du, to the physical derivatives, dx:

$$[\mathbb{J}_{x:u}(u)]^{-1} = \text{inv}([\mathbb{J}_{x:u}(u)]) = \left[\frac{du}{dx}\right]. \tag{7.17}$$

Recall from the calculus definition of a derivative with respect to x,

$$\frac{df(x)}{dx} = \underset{\Delta x \to 0}{\text{limit}} \frac{f(x + \Delta x) - f(x)}{\Delta x}.$$

Since Δx in the denominator has the unit of length, taking a derivative also introduces an entity with a new set of units. In this example, they are the units of $f(x)$ divided by the unit of length.

The governing integral for the desired approximate solution involves the physical derivatives integrated over physical space:

$$\begin{aligned} I = &- \left[\delta(x)\left(E(x)A(x)\frac{d\delta(x)}{dx}\right)\right]_0^L + \int_0^L \left(\frac{d\delta(u)}{dx}E(u)A(u)\frac{d\delta(u)}{dx}\right) dx \\ &- \int_0^L \delta(u)\, w(x)\, dx \equiv 0. \end{aligned} \tag{7.18}$$

These integrals must now be converted to the sum of integrals over the physical space consisting of the non-overlapping sub-region lengths:

$$\begin{aligned} I = &- \left[\delta(x)\left(E(x)A(x)\frac{d\delta(x)}{dx}\right)\right]_0^L \\ &+ \sum_r \int_{L^r} \left(\frac{d\delta(u)}{dx}E(u)A(u)\frac{d\delta(u)}{dx}\right) dx \\ &- \sum_r \int_{L^r} \delta(u)\, w(x)\, dx \equiv 0. \end{aligned} \tag{7.19}$$

Next, map the physical length integrals, over each sub-region physical length, from x in L^r to integrals in the dimensionless space [0,1], to get the Jacobian matrix:

$$[\mathbb{J}_{x:u}(u)] = \left[\frac{dx(u)}{du}\right], \quad x \in L^r,$$

its determinant, $dx(u) = |\mathbb{J}_{x:u}(u)|du$, relates the differential lengths, and its inverse to converts the physical derivatives

$$\frac{d\delta(u)}{dx} = \frac{d\delta(u)}{du}\left[\frac{du}{dx}\right] = \frac{d\delta(u)}{du}[\mathbb{J}_{x:u}(u)]^{-1}.$$

Finally, generalize the isoparametric concept to use basis functions to approximate the spatial nature of data if they are not constant. For example, the elastic modulus could be represented as $E(u) = [\boldsymbol{h}(u)]\{\boldsymbol{E^r}\} = \{\boldsymbol{E^r}\}^T[\boldsymbol{h}(u)]^T = \sum_{i=1}^{i=n_i} h_i(u)E_i^r$. Then, all of the data are considered to have known spatial forms inside the integrals.

$$\begin{aligned}
I = &-\left[\delta(x)\left(E(x)A(x)\frac{d\delta(x)}{dx}\right)\right]_0^L \\
&+\sum_r \int_0^1 \left(\frac{d\delta(u)}{dx}E(u)A(u)\frac{d\delta(u)}{dx}\right)|\mathbb{J}_{x:u}(u)|du \\
&-\sum_r \int_0^1 \delta(u)\,w(x)\,|\mathbb{J}_{x:u}(u)|du \equiv 0.
\end{aligned}$$

Remember, the Jacobian and its determinant, $|\mathbb{J}_{x:u}(u)|$, must be considered as a variable in each sub-region.

First, consider the integral of the force (source) per unit length:

$$I_w^r = -\int_{L^r} \delta(u)w(x)dx = -\int_0^1 \delta(u)w(x)|\mathbb{J}_{x:u}^r(u)|du.$$

In other words, the size (length) of any sub-region vector $\{\boldsymbol{\delta^r}\}$ is smaller than the size (length) of the total system's CP displacements, $\{\boldsymbol{\delta}\}$ (except for the trivial case where there is only one region). Thus, after evaluating any sub-region vector, it is scattered (assembled) into integer rows in the system vector where those locations are defined by the CP connection data array for the entire system.

Inside each such knot span

$$\delta^r(u) = [\boldsymbol{h}^r(u)]\{\boldsymbol{\delta}^r\} = \{\boldsymbol{\delta}^r\}^T[\boldsymbol{h}^r(u)]^T,$$

the rightmost choice gives

$$I_w^r = -\int_{u_r}^{u_{r+1}} \{\boldsymbol{\delta}^r\}^T[\boldsymbol{h}^r(u)]^T\, w(x)|\mathbb{J}_{x:u}^r(u)|du,\ 0 \le [u_r, u_{r+1}) < 1,$$

$$I_w^r = -\{\boldsymbol{\delta}^r\}^T \int_{u_r}^{u_{r+1}} [\boldsymbol{h}^r(u)]^T\, w(x)\,|\mathbb{J}_{x:u}^r(u)|du \equiv -\{\boldsymbol{\delta}^r\}^T\{\boldsymbol{c}_w^r\}. \tag{7.20}$$

Thus, the distributed axial load, $w(u)$, over a sub-region is converted to a resultant force vector, $\{\boldsymbol{c}_w^r\}$, acting at the IGA APs (or FEA nodes) for the current sub-region. The force vector is defined as

$$\{\boldsymbol{c}_w^r\} = \int_{u_r}^{u_{r+1}} [\boldsymbol{h}^r(u)]^T\, w(x)\,|\mathbb{J}_{x:u}^r(u)|du. \tag{7.21}$$

Proceeding in a similar manner, the stiffness integral always incudes the determinant of the Jacobian, but it also brings in the physical gradient of the solution:

$$\begin{aligned} I_k &= \int_{L^r} \left(\frac{d\delta(u)}{dx} E(x)A(x)\frac{d\delta(u)}{dx}\right) dx \\ &= \int_0^1 \left(\frac{d\delta}{dx} E(x)A(x)\frac{d\delta}{dx}\right) |\mathbb{J}_{x:u}^r(u)|du. \end{aligned}$$

But, the chain rule of calculus also brings in the inverse of the Jacobian matrix to address the physical derivative relations. Introducing a common notation from FEA, and usually continued in IGA, let the rectangular matrix, $[\boldsymbol{B}^r(u)]$, be defined as the physical derivative(s) of the basis functions within a sub-region (or an iElement):

$$[\boldsymbol{B}^r(u)] \equiv \left[\frac{d\boldsymbol{h}^r(u)}{dx}\right] = \left[\frac{d\boldsymbol{h}^r(u)}{du}\right] [\mathbb{J}_{\boldsymbol{x:u}}^{\boldsymbol{r}}(u)]^{-1} \tag{7.22}$$

and

$$\frac{d\delta(u)}{dx} = [\boldsymbol{B}^r(u)]\{\boldsymbol{\delta}^r\} = \{\boldsymbol{\delta}^r\}^T[\boldsymbol{B}^r(u)]^T. \tag{7.23}$$

Substituting both notations into the quadratic stiffness integral gives the definition of a sub-region stiffness matrix:

$$I_k^r = \int_{L^r} (\{\boldsymbol{\delta^r}\}^T[\boldsymbol{B}^r(x)]^T E(x)A(x)[\boldsymbol{B}^r(x)]\{\boldsymbol{\delta^r}\})dx, \tag{7.24}$$

$$\begin{aligned} I_k^r &= \{\boldsymbol{\delta^r}\}^T \int_{u_r}^{u_{r+1}} ([\boldsymbol{B}^r(u)]^T E(u)A(u)[\boldsymbol{B}^r(u)]|\mathbb{J}_{x:u}^r(u)|du)\{\boldsymbol{\delta^r}\} \\ &\equiv \{\boldsymbol{\delta^r}\}^T[\boldsymbol{K}_{EA}^r]\{\boldsymbol{\delta^r}\}. \end{aligned}$$

A typical sub-region stiffness matrix is

$$[\boldsymbol{K}_{EA}^r] = \int_{u_r}^{u_{r+1}} [\boldsymbol{B}^r(u)]^T E(u)A(u)[\boldsymbol{B}^r(u)]|\mathbb{J}_{x:u}^r(u)|du. \tag{7.25}$$

Substituting the matrix expressions for the EBCs, $\{\boldsymbol{c}_{\mathrm{EBC}}\}$, the assembly of the line load resultants, $\{\boldsymbol{c}_w^r\}$, and the sub-region stiffness matrix, $[\boldsymbol{K}_{\mathrm{EA}}^r]$, into the governing integral form and scattering (assembling) the sub-region matrices into the corresponding larger system matrices, using the connection data that define each sub-region sub-set $\{\boldsymbol{\delta^r}\} \subset_r \{\boldsymbol{\delta}\}$, lead to the symbolic governing matrix form:

$$\{\boldsymbol{\delta}\}^T([\boldsymbol{K}_{EA}]\{\boldsymbol{\delta}\} - \{\boldsymbol{c}_w\} - \{\boldsymbol{c}_{\mathrm{EBC}}\}) \equiv 0.$$

To obtain a non-trivial solution, $\{\boldsymbol{\delta}\} \neq \{\boldsymbol{0}\}$, for arbitrary geometry, data, and BCs, the quantity within the curved brackets must vanish:

$$[\boldsymbol{K}_{\mathrm{EA}}]\{\boldsymbol{\delta}\} - \{\boldsymbol{c}_w\} - \{\boldsymbol{c}_{\mathrm{EBC}}\} = \{\boldsymbol{0}\}.$$

Or,

$$[\boldsymbol{K}_{\mathrm{EA}}]\{\boldsymbol{\delta}\} = \{\boldsymbol{c}_w\} + \{\boldsymbol{c}_{\mathrm{EBC}}\} \tag{7.26}$$

is the system to be solved, after satisfying the Dirichlet BCs, for the displacement controls, $\{\boldsymbol{\delta}\}$. In other words, the system stiffness matrix is singular, before being modified to include the Dirichlet BCs, because the system model is not unique before that.

However, unlike in FEA, in IGA, the domain [0,1] is further sub-divided into analysis knot spans defined by the analysis knot vector, $\mathbb{A}u$. In each knot span, the basis functions, $[\boldsymbol{h}(u)]$, are controlled by a different set of control point data. Generally, an IGA solution uses NURBS as the basis functions. For now, the examples will use the simpler B-splines so that simpler closed-form examples can be used. The matrix symbols used are valid for either choice.

To illustrate these matrices with a quadratic B-spline analysis knot vector with two knot spans with unequal divisions on [0,1], consider the analysis knot vector $\mathbb{A}u = \{0, 0, 0, 0.6, 1, 1, 1\}$. There, the first knot span is $[u_1 = 0, u_2 = 0.6)$ and the second knot span is $[u_2 = 0.6, u_3 = 1)$. These split of the parametric space $0 \leq u \leq 1$, into a union of two domains, and changes the line load resultant force vector from into two sub-region force resultants. The first sub-region, controlled by CP 1, 2, and 3,

$$\{\boldsymbol{c}_w^{r=1}\} = \int_{u_1=0}^{u_2=0.6} [\boldsymbol{h}(u)]^T\, w(x)\, |\mathbb{J}_{x:u}^r(u)| du$$

will add (scatter) into those same row numbers of the system matrix equilibrium equations. Likewise, the second knot contribution

$$\{\boldsymbol{c}_w^{r=2}\} = \int_{u_2=0.6}^{u_3=1} [\boldsymbol{h}(u)]^T\, w(x)\, |\mathbb{J}_{x:u}^r(u)| du$$

controlled by CP 2, 3, and 4 will scatter into those system rows. The three basis functions in each sub-span, $[\boldsymbol{h}(u)]$, join with slope continuity at $u_2 = 0.6$. This contrasts with an FEA where it remains a single integral over the parametric space. Likewise, the iElement (sub-region) stiffness matrix changes to the pair of integrals

$$[\boldsymbol{K}_{EA}^{r=1}] = \int_{u_1=0}^{u_2=0.6} [\boldsymbol{B}^r(u)]^T E(u) A(u) [\boldsymbol{B}^r(u)] |\mathbb{J}_{x:u}^r(u)| du,$$

$$[\boldsymbol{K}_{EA}^{r=2}] = \int_{u_2=0.6}^{u_3=1} [\boldsymbol{B}^r(u)]^T E(u) A(u) [\boldsymbol{B}^r(u)] |\mathbb{J}_{x:u}^r(u)| du.$$

In practice, both the knot span integrals are numerically integrated, usually with Gaussian quadratures, which convert the integral into a summation. Such quadrature data are tabulated over another non-dimensional space $-1 \leq a \leq 1$, or $0 \leq r \leq 1$, which are called natural and unit coordinates, respectively. The numerical integration approach will be illustrated later.

Table 7.1. Quadratic B-Spline bases from Knot vector $\mathbb{A}u = \{0, 0, 0, 0, 5, 1, 1, 1\}$

Length = 7, Degree = 2, Number of CP = 4,
Number of spans = 2 on [0, 0.5), [0.5, 1) with CP 1-3, 2-4

Count	Basis Functions $N(u)$	Derivatives: $dN(u)/du$
1, 0	$[(1-2u)^2, 0]$	$[(8u-4), 0]$
2, 4	$[2u(2-3u), 2(1-u)^2]$	$[(4-12u), (4u-4)]$
3, 5	$[2u^2, 8u-2-6u^2]$	$[4u, (8-12u)]$
0, 6	$[0, (1-2u)^2]$	$[0, (8u-4)]$

IGA Case (1) B-Spline Example: A similar, but more common approach is to select a symmetric quadratic analysis knot vector $\mathbb{A}u = \{0, 0, 0, 1/2, 1, 1, 1\}$ to make the two knot spans [0, 0.5) and [0.5, 1) symmetric about the parametric center.

The analytic forms of the three quadratic B-spline basis functions of each of the two knot spans from Appendix A1 are listed ($[\boldsymbol{h}(u)] \to [\boldsymbol{N}(u)]$) in Table 7.1.

First, evaluate and check the resultant source vector due to the weight per unit length. The resultant force vector associated with the first three CPs and the first set of blending functions is

$$\{\boldsymbol{c}_w^{r=1}\} = \int_{u_1=0}^{u_2=0.5} [\boldsymbol{N}^{\boldsymbol{r}}(u)]^T \, w(ux) \, |\mathbb{J}_{x:u}^r(u)| du.$$

The resultant force vector associated with the last three CPs and the second set of blending functions is

$$\{\boldsymbol{c}_w^{r=2}\} = \int_{u_2=0.5}^{u_3=1} [\boldsymbol{N}^{\boldsymbol{r}}(u)]^T \, w(x) \, |\mathbb{J}_{x:u}^r(u)| du.$$

Each will make three contributions to the resultant 4×1 system weight vector, $\{\boldsymbol{c_w}\}$. Because each span basis function array omitted one identically zero term in the knot span contributions, $\{\boldsymbol{c}_w^r\}$, each is only a 3×1 vector. Matrix algebra does not allow the addition of matrices of different sizes. The scattering into the resultant system matrices uses the associated control point numbers to identify the row where each term is to be added. That assembly can be viewed as

$$\{\boldsymbol{c_w}\} = \left\{ \begin{matrix} \{\boldsymbol{c}_w^{r=1}\} \\ 0 \end{matrix} \right\} + \left\{ \begin{matrix} 0 \\ \{\boldsymbol{c}_w^{r=2}\} \end{matrix} \right\},$$

but, in practice, the addition of zeros is skipped.

To evaluate the example case with analytic integration, assume a constant weight per unit length and the same analysis knot vector as mentioned earlier, with the four CPs located at $\mathrm{CP}_x = [0, 1/4, 3/4, 1]L$ to cause a constant Jacobian of $[\mathbb{J}_{\boldsymbol{x:u}}^{\boldsymbol{r=1}}(u)] = |\mathbb{J}_{x:u}^{r=2}(u)| = L$, so the *matrix* contribution from CP 1, 2, 3 is

$$\{\boldsymbol{c}_w^{r=1}\} = \int_{u_1=0}^{u_2=0.5} \left\{ \begin{matrix} (1-2u)^2 \\ 2u(2-3u) \\ 2u^2 \end{matrix} \right\} w(L\, du) = \frac{wL}{12} \left\{ \begin{matrix} 2 \\ 3 \\ 1 \end{matrix} \right\}$$

and from CP 2, 3, 4 is

$$\{\boldsymbol{c}_w^{r=2}\} = \int_{u_2=0.5}^{u_2=1} \left\{ \begin{matrix} 2(1-u)^2 \\ -6u^2+8u-2 \\ (1-2u)^2 \end{matrix} \right\} w\,(L\, du) = \frac{wL}{12} \left\{ \begin{matrix} 1 \\ 3 \\ 2 \end{matrix} \right\}.$$

For the full region, the resultant force vector, $\{\boldsymbol{c}_w\}$, from the line load has four rows corresponding to the four CPs. Scattering these two arrays into their system positions gives

$$\{\boldsymbol{c}_w\} = \frac{wL}{12} \left\{ \begin{matrix} 2 \\ 3+1 \\ 1+3 \\ 2 \end{matrix} \right\} = \frac{wL}{6} \left\{ \begin{matrix} 1 \\ 2 \\ 2 \\ 1 \end{matrix} \right\}.$$

The sum of the force resultants at all four of the CPs is $6wL/6 = wL = W$ the full weight of the bar, as required.

Again making the above-mentioned simplifying assumptions (for an easier example only), the two stiffness matrix sub-region contributions can be written in closed form. Recalling that

$$\begin{aligned}[\boldsymbol{B}^r(u)] &\equiv \left[\frac{d\boldsymbol{N^r}(u)}{dx} \right] = \left[\frac{d\boldsymbol{N^r}(u)}{du} \right] \frac{du}{dx} = \left[\frac{d\boldsymbol{N^r}(u)}{du} \right] [\mathbb{J}_{\boldsymbol{x:u}}^{\boldsymbol{r}}(u)]^{-1} \\ &= \left[\frac{d\boldsymbol{N^r}(u)}{du} \right] \frac{1}{L},\end{aligned}$$

for CP 1, 2, 3, controlling the first knot span $[\boldsymbol{B}^1(u)] = [(8u - 4)\ (4 - 12u)\ 4u]/L$, then

$$[\boldsymbol{K}^1_{EA}] = \int_{u_1=0}^{u_2=0.5} [\boldsymbol{B}^1(u)]^T EA [\boldsymbol{B}^1(u)](L\,du),$$

so

$$[\boldsymbol{K}^1_{EA}] = EA \int_{u_1=0}^{u_2=0.5} \frac{1}{L} \begin{Bmatrix} 8u - 4 \\ (4 - 12u) \\ 4u \end{Bmatrix} \frac{1}{L} \times [(8u - 4)\ (4 - 12u)\ 4u]\,(L\,du)$$

$$[\boldsymbol{K}^1_{EA}] = \frac{EA}{3L} \begin{bmatrix} 8 & -6 & -2 \\ -6 & 6 & 0 \\ -2 & 0 & 2 \end{bmatrix}.$$

Note that this IGA result follows the nature of FEA element stiffness matrices in that the sum of each row is zero, the sum of each column is zero, and the matrix is singular. For CP 2, 3, 4, the sub-region integral is

$$[\boldsymbol{K}^2_{\mathrm{EA}}] = \frac{\mathrm{EAL}}{L^2} \int_{u_2=0.5}^{u_3=1} \begin{Bmatrix} (4u - 4) \\ (8 - 12u) \\ (8u - 4) \end{Bmatrix} [(4u - 4)\ \ (8 - 12u)\ \ (8u - 4)]du,$$

$$[\boldsymbol{K}^2_{\mathrm{EA}}] = \frac{\mathrm{EA}}{3L} \begin{bmatrix} 2 & 0 & -2 \\ 0 & 6 & -6 \\ -2 & -6 & 8 \end{bmatrix}$$

Assembling each sub-region 3×3 stiffness matrix into the 4×4 system stiffness matrix,

$$[\boldsymbol{K}_{\mathrm{EA}}] = \frac{\mathrm{EA}}{3L} \begin{bmatrix} 8 & -6 & -2 & 0 \\ -6 & (6+2) & (0+0) & -2 \\ -2 & (0+0) & (2+6) & -6 \\ 0 & -2 & -6 & 8 \end{bmatrix}.$$

Again, this IGA result follows the nature of FEA system stiffness matrix in that the sum of each row is zero, the sum of each column is zero, and the system matrix is singular before any Dirichlet boundary condition(s) is enforced.

Currently, without BCs, the governing matrix form of equilibrium in terms of the CP displacement controls is

$$[\boldsymbol{K}_{\mathrm{EA}}]\{\boldsymbol{\delta}\} = \{\boldsymbol{c}_w\} + \{\boldsymbol{c}_{\mathrm{EBC}}\} + \{\boldsymbol{c}_F\},$$

$$\frac{\mathrm{EA}}{3L}\begin{bmatrix} 8 & -6 & -2 & 0 \\ -6 & 8 & 0 & -2 \\ -2 & 0 & 8 & -6 \\ 0 & -2 & -6 & 8 \end{bmatrix}\begin{Bmatrix} \delta_1 \\ \delta_2 \\ \delta_3 \\ \delta_4 \end{Bmatrix}$$

$$= \frac{wL}{6}\begin{Bmatrix} 1 \\ 2 \\ 2 \\ 1 \end{Bmatrix} + \begin{Bmatrix} -\mathrm{EA}d\delta(0)/dn \\ 0 \\ 0 \\ \mathrm{EA}d\delta(L)/dn \end{Bmatrix} + \begin{Bmatrix} F_1 \\ F_2 \\ F_3 \\ F_4 \end{Bmatrix}.$$

Here, the externally applied force vector at each CP, $\{\boldsymbol{c}_F\}$, is identically zero for the current assumptions.

The boundary condition vector, $\{\boldsymbol{c}_{\mathrm{EBC}}\}$, in this IGA model is misleading because it is a 1D example and it was trivial to require that CP1 was exactly at $x = 0$ and CP4 was exactly at $x = L$. Therefore, a Dirichlet boundary condition can be applied at CP1, or CP2, or both. Referring back to Fig. 7.1, this IGA matrix form can be used to solve the first three cases of restraint conditions, but would be a challenge to place a CP exactly at a random support point along the length. That challenge is higher in 2D and 3D IGA applications. In contrast, the FEA can easily handle all four of the illustrated support options.

Case 1: For the first case in Fig. 7.1, CP1 *happens to exactly correspond* to $x = 0$; it can be used to impose a known Dirichlet displacement, δ_1, and the Neumann condition is $d\delta(L)/dx = 0$. With those two conditions, the four matrix equations will be unique and will yield the four remaining unknowns of $\delta_2, \delta_3, \delta_4$ and the reaction force $\mathrm{EA}d\delta(0)/dx$. Rearranging the four equations would destroy the symmetry and other useful numerical properties of a system stiffness matrix.

A matrix partition can be used to retain the important symmetry of the stiffness matrix, allow for the solution constraints, and yield a smaller and non-singular system of equations to be solved. In any field of application, the initially singular governing FEA or IGA matrix system can be written in a general form that more clearly defines what must be done to reduce the system to a solvable (non-singular) form by enforcing known essential boundary condition values.

Let the system equilibrium equations have a square matrix, $[\boldsymbol{S}]$, and a combined column (source) matrix, $\{\boldsymbol{c}\}$, such that the initially singular matrix system is

$$[\boldsymbol{S}]\{\boldsymbol{\delta}_{\mathrm{CP}}\} = \{\boldsymbol{c}\}.$$

In other words, this is a singular matrix form to be modified for the required EBCs to be imposed to render a modified and smaller system non-singular. One way to do that is by partitioning the rows and columns to be associated either with the known CP control values, $\{\boldsymbol{\delta}_k\}$, or the unknown displacement control values, $\{\boldsymbol{\delta}_u\}$, so that $\{\boldsymbol{\delta}_{\mathrm{CP}}\}^T \equiv [\boldsymbol{\delta}_u\ \boldsymbol{\delta}_k]$ creates a matrix partitioned form:

$$[\boldsymbol{S}]\{\boldsymbol{\delta}_{\mathrm{CP}}\} = \{\boldsymbol{c}\} \Leftrightarrow \begin{bmatrix} [\boldsymbol{S}_{uu}] & [\boldsymbol{S}_{uk}] \\ [\boldsymbol{S}_{ku}] & [\boldsymbol{S}_{kk}] \end{bmatrix} \begin{Bmatrix} \{\boldsymbol{\delta}_u\} \\ \{\boldsymbol{\delta}_k\} \end{Bmatrix} = \begin{Bmatrix} \{\boldsymbol{c}_u\} \\ \{\boldsymbol{c}_k\} + \{\boldsymbol{r}_k\} \end{Bmatrix}.$$

The sub-matrices $[\boldsymbol{S}_{uu}]$ and $[\boldsymbol{S}_{kk}]$ are square, whereas $[\boldsymbol{S}_{uk}]$ and $[\boldsymbol{S}_{ku}]$ are rectangular, in general. In an IGA or FEA formulation, all of the coefficients in the $[\boldsymbol{S}]$ and $\{\boldsymbol{c}\}$ matrices are known. This means that in general after the EBCs, $\{\boldsymbol{\delta}_k\}$, are prescribed, the remaining unknowns are $\{\boldsymbol{\delta}_u\}$ and the associated reactions $\{\boldsymbol{r}_k\}$ at the Dirichlet BCs. The unknown degrees of freedom, $\{\boldsymbol{\delta}_u\}$, are always calculated, but the calculation of the reaction unknowns is optional.

For future reference, it should be noted that there are some fields of application, like heat transfer with convection and structures supported on elastic foundations, where the system contains the sum of two square matrices, $[\boldsymbol{S}] = [[\boldsymbol{K}] + [\boldsymbol{C}]]$, where the second square matrix, $[\boldsymbol{C}]$, is non-singular and no explicit Dirichlet boundary condition is given. In such cases, the above-mentioned matrix equations are solved for the unknowns, $\{\boldsymbol{\delta}_{\mathrm{CP}}\}$, without any further modifications. For such applications, there are Dirichlet conditions which are satisfied indirectly through the mathematics. An elastic foundation could be supplied by soil which has zero displacement at infinity. With heat convection, the surrounding fluid temperature at infinity has a specified value.

Continuing with the current structure, the essential boundary condition, $\delta_1 = 0$, can be enforced using the above-mentioned matrix partitioning operations (and other methods). Specifically, the currently singular matrix equilibrium equations for the hanging bar are

$$[\boldsymbol{S}] \Leftrightarrow \frac{\mathrm{EA}}{3L}\begin{bmatrix} 8 & \vdots & -6 & -2 & 0 \\ \cdots & \vdots & \cdots & \cdots & \cdots \\ -6 & \vdots & 8 & 0 & -2 \\ -2 & \vdots & 0 & 8 & -6 \\ 0 & \vdots & -2 & -6 & 8 \end{bmatrix}, \quad \{\boldsymbol{\delta}_{\mathrm{CP}}\} = \begin{Bmatrix} \delta_1 \\ \cdots \\ \delta_2 \\ \delta_3 \\ \delta_4 \end{Bmatrix},$$

$$\{\boldsymbol{c}\} = \frac{wL}{6}\begin{Bmatrix} 1 \\ \cdots \\ 2 \\ 2 \\ 1 \end{Bmatrix} + \begin{Bmatrix} r_1 \\ \cdots \\ 0 \\ 0 \\ 0 \end{Bmatrix}.$$

Since this application yields a symmetric square matrix, the rectangular partitions are the transpose of each other, $[\boldsymbol{S}_{ku}] \equiv [\boldsymbol{S}_{uk}]^T$.

The top partitioned row of unknowns requiring

$$[\boldsymbol{S}_{uu}]\{\boldsymbol{\delta}_u\} + [\boldsymbol{S}_{uk}]\{\boldsymbol{\delta}_k\} = \{\boldsymbol{c}_u\}$$

is solved for the CP unknowns

$$\{\boldsymbol{\delta}_u\} = [\boldsymbol{S}_{uu}]^{-1}\{\{\boldsymbol{c}_u\} - [\boldsymbol{S}_{uk}]\{\boldsymbol{\delta}_k\}\}.$$

Then, if desired, the unknown reactions are

$$\{\boldsymbol{r}_k\} = [\boldsymbol{S}_{ku}]\{\boldsymbol{\delta}_u\} + [\boldsymbol{S}_{kk}]\{\boldsymbol{\delta}_k\} - \{\boldsymbol{c}_k\}.$$

For simplicity here, it has been assumed that the equations have been numbered in a manner that places the prescribed CP analysis values (EBCs) at the beginning of the system equations, but in practice this is not required. In the algorithms used here, the row numbers that have known Dirichlet values assigned are stored in a *vector subscript* named [Fixed]. Of course, all other degree-of-freedom numbers must be identified and they are stored in a vector subscript named [Free]. After imposing the essential boundary condition, the reduced non-singular matrix equilibrium equations are

$$\frac{\text{EA}}{3L}\begin{bmatrix} 8 & 0 & -6 \\ 0 & 8 & -2 \\ -6 & -2 & 8 \end{bmatrix}\begin{Bmatrix} \delta_2 \\ \delta_3 \\ \delta_4 \end{Bmatrix}$$

$$= \frac{wL}{6}\begin{Bmatrix} 2 \\ 2 \\ 1 \end{Bmatrix} + \begin{Bmatrix} 0 \\ 0 \\ 0 \end{Bmatrix} + \begin{Bmatrix} 0 \\ 0 \\ 0 \end{Bmatrix} - \frac{EA}{3L}\begin{Bmatrix} 6 \\ -2 \\ 0 \end{Bmatrix}\delta_1.$$

Here, the essential boundary condition value can be any real number. The cited figure implies a rigid foundation with $\delta_1 \equiv 0$ thus reducing the load vector to just the partition of the line load resultant. This 3×3 partition of the original CP stiffness is now non-singular and symmetric. Its inverse matrix is

$$[\boldsymbol{S}_{uu}]^{-1} = \frac{L}{16\text{EA}}\begin{bmatrix} 7 & 3 & 4 \\ 3 & 15 & 12 \\ 4 & 12 & 16 \end{bmatrix}.$$

Multiplying both sides by this inverse matrix gives the CP displacement control values defining the displacements on the bar: $\{\boldsymbol{\delta}_u\} = [\boldsymbol{S}_{uu}]^{-1}\{\{\boldsymbol{c}_u\} - [\boldsymbol{S}_{uk}]\{\boldsymbol{\delta}_k\}\}$,

$$\begin{Bmatrix} \delta_2 \\ \delta_3 \\ \delta_4 \end{Bmatrix} = \frac{L}{16\text{EA}}\begin{bmatrix} 7 & 3 & 4 \\ 3 & 15 & 12 \\ 4 & 12 & 16 \end{bmatrix}\left(\frac{wL}{6}\begin{Bmatrix} 2 \\ 2 \\ 1 \end{Bmatrix} - \begin{Bmatrix} 0 \\ 0 \\ 0 \end{Bmatrix}\right)$$

$$= \frac{wL^2}{96\text{EA}}\begin{Bmatrix} 48 \\ 24 \\ 48 \end{Bmatrix} = \frac{wL}{4\text{EA}}\begin{Bmatrix} 1 \\ 2 \\ 2 \end{Bmatrix}.$$

Inserting the displacement boundary condition, the CP displacement control values of

$$\{\boldsymbol{\delta}_{CP}\}^T = \begin{bmatrix} 0 & 1 & 2 & 2 \end{bmatrix} wL/4\text{EA}$$

are NOT the bar displacements.

Since the CP subject to the EBC is actually on the bar, the reaction computed there is the actual reaction. In this application, the reaction force at the top Dirichlet boundary condition is useful information and can be obtained in two ways. The most accurate way is to use the remaining equation in the matrix equilibrium system. The second, less accurate, way is to evaluate the gradient of the adjacent knot span displacements at the point of the Dirichlet boundary condition.

First, use the remaining rows of the original matrix equations of equilibrium. Substituting into the first row to recover the reaction,

$$\{\boldsymbol{r}_k\} = [\boldsymbol{S}_{ku}]\{\boldsymbol{\delta}_u\} + [\boldsymbol{S}_{kk}]\{\boldsymbol{\delta}_k\} - \{\boldsymbol{c}_k\},$$

$$\left\{-\mathrm{EA}\tfrac{d\delta(0)}{dn}\right\} = \frac{\mathrm{EA}}{3L}\begin{bmatrix}8 & -6 & -2 & 0\end{bmatrix}\frac{w\,L^2}{4\mathrm{EA}}\begin{Bmatrix}0\\1\\2\\2\end{Bmatrix} + \{0\} - \frac{wL}{6}\{1\}.$$

In this example, there is just one equation (one row) and one unknown:

$$-\mathrm{EA}\frac{d\delta(0)}{dn} = \frac{\mathrm{EA}}{3L}\frac{wL^2}{4\mathrm{EA}}(8*0-6*1-2*2+0*2) - \frac{wL}{6},$$

$$-(-1)\mathrm{EA}\frac{d\delta(0)}{dx} = \frac{-5w\,L}{6} - \frac{wL}{6} = -wL = -W.$$

Thus, the reaction force at the essential boundary condition of $u(0) = 0$ is the weight of the bar acting upward, as expected.

For future reference, the above-mentioned partitioning of the matrices, enforcing the Dirichlet boundary equations, solving the reduced equations, and recovering the reactions at each of the Dirichlet conditions can be carried out with the function ***Solve_Free_DOF.m***, given in the next section, with arguments $[\boldsymbol{\delta}_{\mathrm{CP}}, \{\boldsymbol{r}\}]$ = Solve_Free_DOF ($[\boldsymbol{S}], \{\boldsymbol{c}\}$, Free, Fixed, Given), where for the above-mentioned example [Free] = [2, 3, 4], [Fixed] = [1], and Given = [0.0].

Applying the usually less accurate way (when the $\{\boldsymbol{\delta}_{\mathrm{CP}}\}$ are usually approximate) to estimate the normal reaction force by extrapolating the adjacent iElement strain to the boundary can be done using the first knot span. Evaluating the gradient of the basis functions using the first three CP control values gives the normal mechanical strain at the support point of

$$\varepsilon_n = \frac{\partial\delta(u=0)}{\partial n} = n_x\frac{\partial\delta(u=0)}{\partial x} = (-1)\frac{\delta\partial(u=0)}{\partial x},$$

$$\varepsilon_n = -\frac{\partial\delta(u=0)}{\partial x} = -\frac{\partial\delta(u=0)}{\partial u}\frac{\partial u}{\partial x} = -\frac{\partial \boldsymbol{N}(u=0)}{\partial u}\frac{1}{L}\{\boldsymbol{\delta}_{\mathrm{CP}}\},$$

$$\varepsilon_n = -[-4 \quad 4 \quad 0]\frac{1}{L}\frac{\mathrm{WL}}{4\mathrm{EA}}\begin{Bmatrix}0\\1\\2\end{Bmatrix} = \frac{-W}{\mathrm{EA}}.$$

Multiplying by the elastic modulus gives the normal stress, $\sigma_n = E\varepsilon_n = -W/A$, and finally multiplying by the end area gives the reaction force, $F = A\sigma = -W$ to be equal and opposite to the hanging weight, as shown earlier. In general, the matrix equilibrium approach is always much more accurate when the displacement control values are approximate.

7.5 Matrix operations

The process of partitioning a set of matrix equilibrium equations, with Dirichlet BCs, using vector subscripts is easily automated. The process is shown in Algorithm 7.1. The key restriction is that the sequential lists of Fixed and Free must reference all the degrees of freedom without duplicating any. Usually, the Fixed list is defined by the input data defining the Dirichlet requirements. For a small number of unknowns, like the one above, the 'Fixed' list can be set by inspection. In very rare cases, like structures on an elastic foundation or heat transfer defined only by convection, there may be no direct Dirichlet BCs. Of course, after identifying the dof with Dirichlet conditions, all other dofs must go into the 'Free' list. Algorithm 7.2 illustrates one way to build that list of equation numbers.

Practical applications often involve a huge number of degrees of freedom with only a very few dofs subject to Dirichlet conditions. The approach shown here assembles all of the equations and only imposes the Dirichlet conditions at the end of the process to make the solution unique. Others, like Hughes (2000), solve FEA studies with reduced memory requirements by just assembling the smaller Free matrices and inserting the Fixed requirements into the right-hand side of the smaller system matrices.

Algorithm 7.1. Partitioned matrix solution for equilibrium systems

```
function [D, R] = Solve_Free_DOF (K, P, Free, Fixed, Given)
% Solve for unknowns in equilibrium equations: K*U = (P+R)
% ==========================================================
%                           Knowns
% K     = system stiffness matrix
% P     = external applied loads
% Free  = list of unknown displacement numbers
% Fixed = list of known displacement numbers
% Given = values of known Fixed displacements
%                         Unknowns
% D     = system displacements
% R     = reaction forces at the Fixed nodes
%
% n_DOF = number of degrees of freedom
% n_BC  = number of Dirichlet boundary conditions
%              Matrix System Partitioned as
% |K(Fixed,Fixed) |K(Fixed,Free)|*|D(Fixed)|=|P&R(Fixed)|
% |K(Free, Fixed) |K(Free, Free)| |D(Free )| |P(Free )  |
%         Requires n_DOF - n_BC == size(Free, 1)

n_DOF = size  (K,      1)  ;   % number of rows in K, P, U
n_BC  = size  (Fixed, 1)  ; % number of Given and R terms
D     = zeros (n_DOF, 1)  ;        % allocate displacements
R     = zeros (n_BC,  1)  ;% allocate Dirichlet reactions
if ( n_BC > 0 )                    ; % almost always true
  % Insert Given values into control displacements, U
  D(Fixed) = Given (:)       ; % copy into unknown results
  % Subtract Given values times Free columns to RHS
  P(Free) = P(Free)-K(Free, Fixed)*D(Fixed); % enforce BC
end                 ; % if Dirichlet conditions are present

%    Solve the Free partition for Free displacements
% D(Free) = inv(K(Free, Free))*P(Free);% matrix inversion
D(Free) = K(Free, Free) \ P(Free); % matrix solution
if ( n_BC > 0 )                    ; % almost always true
  % Recover the reactions at Fixed nodes
  R = K(Fixed, :) * D(:) - P(Fixed) ; % optional recovery
end                 ; % if Dirichlet conditions are present
% end Solve_Free_DOF ====================================
```

7.6 Post-processing the control values

It is very rare to terminate an FEA or IGA study after computing just the displacements and Dirichlet reactions. Generally, it is necessary to compute the strains, the stresses, and appropriate material yield criteria throughout the analysis domain. Stresses should be continuous through a domain except at interfaces where the material properties change. FEA models using Lagrange interpolation always have stresses that are discontinuous across element boundaries. After all FEA stresses are known, they are usually processed again to make them continuous at the nodes in a homogeneous region. Similar continuous stress data can be further processed to establish mathematical error estimates (Ainsworth and Oden, 1997) to guide in adapting the IGA solution to improve its accuracy.

Algorithm 7.2. Identifying degrees of freedom without dirichlet BC

```
function [Free] = get_free_from_fixed (n_d, Fixed)
% =====================================================
%   Define the free dof from the fixed dof list
% n_d   = number of st=ystem degrees of freedom
% Fixed = sequential list of dof with Dirichlet BC
n_c = size (Fixed, 2) ; % the number of Dirichlet BC
d_n = n_d - n_c         ; % number of independent dof
Free = zeros (d_n, 1)    ; % allocate memory for Free
kount = 0               ; % initialize Free dof count
for k = 1:n_d   ; % loop over all dof ==>> ==>> ==>>
  if ( any ( k == Fixed ) )       ; % is k constrained
    continue            ; % to the next dof --> --> -->
  else                  ; % k is free, put in Free list
    kount = kount + 1               ; % location in Free
    Free (kount) = k             ; % insert dof number k
  end           ; % if k is in Fixed list <-- <-- <--
end  ; % for dof k possibly free <<== <<== <<== <<==
% end get_free_from_fixed =========================
```

The steps are different in IGA studies once the CP displacement control values are known. The physical displacement components, $\boldsymbol{\delta}(\boldsymbol{x}(u))$, for any parametric coordinate, u, in a region of the body are defined by the basis functions as

$$\boldsymbol{\delta}(\boldsymbol{x}) = \boldsymbol{\delta}(\boldsymbol{x}(u)) = [\boldsymbol{N}(u)]\{\boldsymbol{\delta}_{CP}\},$$

where the corresponding physical point locations (for plotting or listing) are likewise

$$\boldsymbol{x}(u) = [\boldsymbol{N}(u)]\{\boldsymbol{x}_{CP}\}.$$

If the displacement components, $\boldsymbol{\delta}(\boldsymbol{x}_p)$, are needed at a specific physical point, $\boldsymbol{x}_p$, then the required parametric coordinate(s) can be found using a Newton–Raphson iterative solution.

7.6.1 *Post-processing Case 1*

The physical bar displacements are found from the combination of the basis functions using the CP x-coordinates to define physical locations $x(u) = [\boldsymbol{N}(u)]\{\boldsymbol{x}_{\text{CP}}\}$ and the CP displacement control values to define the actual bar displacements

$$\delta(x) = \delta(x(u)) = [\boldsymbol{N}(u)]\{\boldsymbol{\delta}_{\text{CP}}\}.$$

That combination of functions can be used to graph the solution by looping over u from 0 to 1 in small increments and connecting those point values with line segments. Determine the maximum bar displacement at the free end of the second knot span, at $u = 1$, by first evaluating the three basis functions and multiplying them by the control coordinates at CPs 2, 3, and 4 to verify the physical location:

$$x(u=1) = [\boldsymbol{N}(u=1)]\{\boldsymbol{x}_{\text{CP}}\} = \begin{bmatrix} 0 & 0 & 1 \end{bmatrix} \begin{Bmatrix} L/4 \\ 3L/4 \\ L \end{Bmatrix} = L.$$

Similarly, the end displacement is

$$\delta(u=1) = [\boldsymbol{N}(u=1)]\{\boldsymbol{\delta}_{CP}\} = [0 \quad 0 \quad 1]\frac{wL}{4\text{EA}} \begin{Bmatrix} 1 \\ 2 \\ 2 \end{Bmatrix} = \frac{wL}{2\text{EA}},$$

which is the exact analytic end displacement, as expected from any IGA or FEA study using quadratic or higher piecewise polynomials. Expressing in a non-dimensional form, the maximum displacement is $4\delta\text{EA}/wL = 2$.

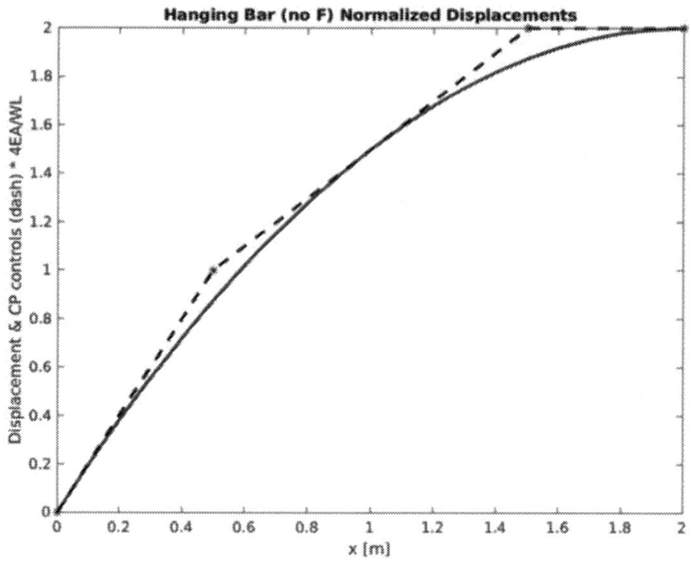

Figure 7.2. Hanging bar (solid) and AP (dashed) displacements, Case 1

The non-dimensional graph of the actual bar displacement for this case is given in Fig. 7.2. In that figure, the last two CP control values cause the displacement gradient to exactly satisfy the natural Neumann condition $(d\delta(L)/dx = 0)$ of zero slope (and force) at $x = L$. In Lagrangian FEA studies, the natural Neumann condition is satisfied weakly. Its value is mesh dependent and can be improved by having smaller elements adjacent to the Neumann boundary.

Knowing the basis functions for each knot span, $[\boldsymbol{N}(u)]$, the local parametric derivatives of the displacement and position are known as given earlier.

$$\frac{d\delta(u)}{du} = \left[\frac{d\boldsymbol{N}(u)}{du}\right]\{\boldsymbol{\delta}^r\} = \{\boldsymbol{\delta}^r\}^T\left[\frac{d\boldsymbol{N}(u)}{du}\right]^T,$$

$$\frac{dx(u)}{du} = \left[\frac{d\boldsymbol{N}(u)}{du}\right]\{\boldsymbol{x}^r\} = \{\boldsymbol{x}^r\}^T\left[\frac{d\boldsymbol{N}(u)}{du}\right]^T.$$

To recover the axial strain in the bar, the axial displacement gradient component, $\partial\delta(u)/\partial x \equiv \delta_{,x}$, is required. This requires the chain rule of calculus and the Jacobian of the mapping of the span, $[\mathbb{J}_{x:u}(u)]$:

$$\delta_{,x}(u) = \frac{\partial \delta(u)}{\partial u}\frac{\partial u}{\partial x} = \frac{\partial \delta(u)}{\partial u}\left(\frac{\partial x}{\partial u}\right)^{-1} = \frac{\partial \delta(u)}{\partial u}[\mathbb{J}_{\boldsymbol{x:u}}(u)]^{-1}$$

or

$$\delta_{,x}(u) = \left[\frac{\partial \boldsymbol{N}(u)}{\partial x}\right]\{\boldsymbol{\delta}^r\} = \left[\frac{\partial \boldsymbol{N}(u)}{\partial u}\right][\mathbb{J}_{\boldsymbol{x:u}}(u)]^{-1}\{\boldsymbol{\delta}^r\}.$$

Here, the Jacobian matrix is 1×1 in size (a scalar), and for the currently chosen control coordinates $(0, L/4, 3L/4, L)$, it is a constant, $[\mathbb{J}_{\boldsymbol{x:u}}(u)] = L$. With the special CPs, the displacement gradient component is

$$\delta_{,x}(u) = \left[\frac{\partial \boldsymbol{N}(u)}{\partial u}\right]\frac{1}{L}\{\boldsymbol{\delta}^r\} = \frac{\partial \delta_x}{\partial x}$$

within each knot span.

The displacement gradient is substituted into the 'strain–displacement relations' and that yields the axial strain as the only non-zero term:

$$\varepsilon_x = \frac{\partial \delta_x}{\partial x} = \left[\frac{\partial \boldsymbol{N}(u)}{\partial u}\right][\mathbb{J}_{\boldsymbol{x:u}}(u)]^{-1}\{\boldsymbol{\delta}^r\} = \left[\frac{\partial \boldsymbol{N}(u)}{\partial u}\right]\frac{1}{L}\{\boldsymbol{\delta}^r\}$$

for the current control point x-coordinates.

Next, the 'stress–strain law' ('constitutive equation') gives the axial stress $\sigma_x = E\varepsilon_x$. Now, all of the quantities are available for graphing across the two knot spans.

7.7 Case 1 implementation

Generally, this IGA application would be carried out using numerical integration, and assuming a variable Jacobian caused by a user's arbitrary placement of the CPs. The algorithm, hard coded for this example in Algorithm 7.3, assumes constant properties and is very similar to the one given earlier where just the length and centroid of this same line were obtained. Here, not only will the derivatives basis functions be applied to the control positions to obtain the Jacobian but they are also applied a second time to the displacement controls to form the strain–displacement matrix in each knot span, $[\boldsymbol{B}^r(u)]$, which appears twice in the integrand of the local stiffness matrix. Also, the basis functions, $[\boldsymbol{N}^r(u)]$, are used for the first time to form the line load resultant vector matrix. The same algorithm can be used to numerically solve Cases 2 and 3 simply by changing the Dirichlet boundary condition locations and values in line 14.

An important aspect of Algorithm 7.3 is establishing the required number of integration points, n_q (line 24). Also, the section where the basis functions and their local derivatives (N and dN) are calculated is only good for this analysis knot vector and it should be replaced with a function call to return the non-zero basis functions for any knot vector.

Algorithm 7.3. Numerical integration solution for the hanging bar

```
%         [L, K, c] = HangingBar_QP ()
% ==========================================================
% Integrate over a B-spline with Gaussian quadratures
%  for the length of the spans, line length, L, the
%  stiffness matrix, S, load matrix, c, and the CP
% displacement control values for displacements, v,
%         from ODE d(EA dv/dx)/dx = Fx
%   EA = Modulus*Area, Fx = Force per unit length
% ==========================================================
%                  Application Data
Au    = [0,0,0, 1/2, 1,1,1]  ; % Analysis knot vector
CP_x  = [0, 0.5, 1.5, 2]     ; % CP x-coordinates [cm]
CP_n  = [1, 2, 3 ; 2, 3, 4]           ; % CPs per span
EA    = 6; Fx = 0.5; % Modulus*Area, Force per length
Fixed = [1]; EBC = [0]  ; % Dirichlet location & value
Free  = [2, 3, 4]  ; % manually set unknown equations
n_D   = max(size(Fixed))   ; % Number of Dirichlet BC
%                Control Integers
A_L  = length (Au)               ; % Knot vector length
p    = sum( Au == 1 ) - 1         ; % Polynomial degree
n_CP = A_L - p - 1         ; % Number of control points
n_ks = n_CP - p                 ; % Number of knot spans
n_bf = n_ks + 1           ; % Number of basis functions
n_g  = 1                        ; % number of dof per CP
```

```
n_q  = 3                  ; % Number of quadrature pts
n_i  = n_g*n_bf      ; % # of independent dof per span
n_d  = n_g*n_CP               ; % number of system dof
%            Allocate Array Memory
N  = zeros(1,  n_bf)    ; % basis functions of a span
dN = zeros(1,  n_bf)       ; % basis derivative dN/du
S  = zeros(n_d, n_d)             ; % stiffness matrix
c  = zeros(n_d,   1)                ; % source vector
R  = zeros(n_D,   1)   ; % recovered reactions at EBC
v  = zeros(n_d,   1);% CP displacement control values
v(Fixed) = EBC(1:n_D) ; % set the known displacememts
%        Echo given data for two spans
disp ('Analysis Knot Vector');           disp(Au   )
disp ('Control Point x-Coords');         disp(CP_x )
disp ('Control Points per Span');        disp(CP_n )
disp ('Number of Dirichlet BC');         disp(n_D  )
disp ('Given displacement CP numbers'); disp(Fixed)
disp ('Given CP displacement values');  disp(EBC  )

% Gather the tabulated quadrature data in unit coords
[r_q, w_q] = qp_rule_unit_Gauss (n_q)     ; % 1D table

L = 0  ; % initialize length. Ans. CP_x(n_CP)-CP_x(1)
for K = 1:n_ks % loop over all knot spans ===>> ===>>
  S_r  = zeros(n_bf, n_bf)  ; % span stiffness matrix
  c_r  = zeros(n_bf,    1)      ; % span source matrix
  sL = 0                       ; % initialize span length

  %  Extract two coordinates on span in 0 <= u <= 1
  U_k  = Au(p+K)       ; % where span begins in 0<=u<=1
  U_kp = Au(p+K+1)       ; % where span ends in 0<=u<=1

  for Q = 1:n_q;% loop over integration pts ---> --->
    r = r_q(Q)  ; % where in quadrature space 0<=r<=1
    w = w_q(Q)                      ; % weight at r, 0<w<1
    u    = (1 - r)*U_k + r*U_kp  ; % in span for pt Q
    J_ur = U_kp - U_k           ; % Jacobian  du/dr at Q
```

```
% Evaluate N and dN for this span at this point
    if ( K == 1 )              ; % then first span of two
     N  = [(2*u - 1)^2, -2*u*(3*u - 2), 2*u^2]  ; % N
     dN = [(8*u - 4), (4 - 12*u), 4*u]       ; % dN/du
    else                     ; % second possible case here
     N  = [(2*u - 2)*(u - 1),(-6*u^2 + 8*u - 2), ...
           (2*u - 1)^2]                            ; % N
     dN = [(4*u - 4), (8 - 12*u), (8*u - 4)]; % dN/du
    end                     ; % if which span K of two ***

    %  Gather CP_x and evaluate dx/du & du/dx at r_Q
    s_CPn = CP_n (K, :)       ; % CP numbers, this span
    x_ks  = CP_x (s_CPn)        ; % x-coords at span CP
    x_u   = N*x_ks'                     ; % x(u) at r_Q
    dx_du = dN*x_ks'               ; % J = dx/du at r_Q
    du_dx = 1 / dx_du      ; % inverse Jacobian at r_Q
    dN_dx = dN * du_dx                  ; % dN(r_Q)/dx

    %   Update the span stiffness and source vector
    S_r = S_r + EA*dN_dx'*dN_dx*dx_du*w ; % stiffness
    c_r = c_r + Fx*N'           *dx_du*w       ; % load

    % Add qp contribution to span length (debug aid)
    sL  = sL + dx_du*J_ur*w ; % increment span length
  end ; % for all quadrature points <-- <-- <--- <---

  fprintf ('Span %i length = %g cm \n', K, sL) % list
  L    = L+sL  ; % increment total length (debug aid)
  rows = s_CPn            ; % gather system row numbers
  S(rows, rows) = S(rows, rows) + S_r   ; % stiffness
  c(rows)       = c(rows)       + c_r          ; % load
end ; % for all 2 knot spans  <<=== <<=== <<=== <<===
fprintf ('Total line length = %g cm \n', L) % list

%  Modify system matrices for Dirichlet BC and solve
%    Subtract EBC values times Free columns to RHS
c(Free) = c(Free)-S(Free,Fixed)*v(Fixed);% enforce BC
%  Solve the Free partition for Free displacements
v(Free) = S(Free,Free)\c(Free);% matrix factorization
```

```
%        Recover the reactions at Fixed nodes
R = S(Fixed, :)*v(:) - c(Fixed);  % optional recovery
disp('Final CP displacement values');          disp(v)
disp('For Dirichlet BC at CPs');    disp(CP_n(Fixed))
disp('The reaction forces are');               disp(R)
% end of HangingBar_QP ===============================
```

7.8 External force

Next, consider the additional requirement where the external end force, F, is not zero. Here, that can be easily done only because the fourth control point is at $x = L$. Assume that the bottom point also supports an object weighing three times the weight of the bar, $F = 3W$. It converts the prior natural Neumann condition into a non-zero Neumann boundary condition, $\mathrm{EA}d\delta(L)/dn = F = 3W$, and adds one new term into the system equilibrium equations before the Dirichlet condition is enforced:

$$\frac{\mathrm{EA}}{3L}\begin{bmatrix} 8 & -6 & -2 & 0 \\ -6 & 8 & 0 & -2 \\ -2 & 0 & 8 & 6 \\ 0 & -2 & -6 & 8 \end{bmatrix}\begin{Bmatrix} \delta_1 \\ \delta_2 \\ \delta_3 \\ \delta_4 \end{Bmatrix}$$

$$= \frac{wL}{6}\begin{Bmatrix} 1 \\ 2 \\ 2 \\ 1 \end{Bmatrix} + \begin{Bmatrix} -\mathrm{EA}d\delta(0)/dn \\ 0 \\ 0 \\ \mathrm{EA}d\delta(L)/dn = 3W \end{Bmatrix} + \begin{Bmatrix} F_1 \\ F_2 \\ F_3 \\ F_4 \end{Bmatrix}.$$

After imposing the essential boundary condition, the reduced non-singular matrix equilibrium equations are now

$$\frac{\mathrm{EA}}{3L}\begin{bmatrix} 8 & 0 & -6 \\ 0 & 8 & -2 \\ -6 & -2 & 8 \end{bmatrix}\begin{Bmatrix} \delta_2 \\ \delta_3 \\ \delta_4 \end{Bmatrix}$$

$$= \frac{\mathrm{wL}}{6}\begin{Bmatrix} 2 \\ 2 \\ 1 \end{Bmatrix} + \begin{Bmatrix} 0 \\ 0 \\ 3W \end{Bmatrix} + \begin{Bmatrix} 0 \\ 0 \\ 0 \end{Bmatrix} - \frac{\mathrm{EA}}{3L}\begin{Bmatrix} 6 \\ -2 \\ 0 \end{Bmatrix}\delta_1 = \frac{W}{6}\begin{Bmatrix} 2 \\ 2 \\ 19 \end{Bmatrix}.$$

Multiplying both sides by the previously computed inverse of the stiffness matrix gives the new control displacements:

$$\begin{Bmatrix} \delta_2 \\ \delta_3 \\ \delta_4 \end{Bmatrix} = \frac{L}{16\text{EA}} \begin{bmatrix} 7 & 3 & 4 \\ 3 & 15 & 12 \\ 4 & 12 & 16 \end{bmatrix} \frac{W}{6} \begin{Bmatrix} 2 \\ 2 \\ 19 \end{Bmatrix} = \frac{\text{WL}}{96\text{EA}} \begin{Bmatrix} 96 \\ 264 \\ 336 \end{Bmatrix}$$

$$= \frac{\text{WL}}{\text{EA}} \begin{Bmatrix} 1 \\ 2.75 \\ 3.5 \end{Bmatrix}.$$

Again, evaluating the second knot span basis functions at $u = 1$ and multiplying them by the CPs 2, 3, and 4 displacement control values gives the end displacement of

$$\delta(u = 1) = [\boldsymbol{N}(u = 1)]\{\boldsymbol{\delta}_{\text{CP}}\} = [0 \quad 0 \quad 1] \frac{\text{WL}}{\text{EA}} \begin{Bmatrix} 1 \\ 2.75 \\ 3.5 \end{Bmatrix} = 3.5 \frac{\text{WL}}{\text{EA}},$$

which is the analytic exact end displacement, as expected from any IGA or FEA study using quadratic or higher piecewise polynomials. But, these results were obtainable only because it was easy to put a CP at the end of a line (or at a corner in higher dimensions).

The top reaction force should now be increased to $W + F = 4\text{W}$. Determine it by substituting the new displacements into the unused first row of the matrix equilibrium equations,

$$\{\boldsymbol{r}_{\text{k}}\} = [\boldsymbol{S}_{ku}]\{\boldsymbol{\delta}_u\} + [\boldsymbol{S}_{kk}\{\boldsymbol{\delta}_k\} - \{\boldsymbol{c}_k\} :$$

$$\left\{ -\text{EA}\frac{d\delta(0)}{dn} \right\} = \frac{\text{EA}}{3L} [8 \quad -6 \quad -2 \quad 0] \frac{\text{WL}}{\text{EA}} \begin{Bmatrix} 0 \\ 1 \\ 2.75 \\ 3.5 \end{Bmatrix} + \{0\} - \frac{W}{6}\{1\}.$$

In this example, there is just one equation (one row) and one unknown:

$$-\text{EA}\frac{d\delta(0)}{dn} = \frac{\text{EA}}{3L}\frac{\text{WL}}{\text{EA}}(8 * 0 - 6 * 1 - 2 * 2.75 + 0 * 3.5) - \frac{W}{6},$$

$$-\text{EA}\frac{d\delta(0)}{dn} = -(-1)\text{EA}\frac{d\delta(0)}{dx} = \frac{-11.5\,W}{3} - \frac{W}{6} = -\frac{24W}{6} = -4W$$

as expected.

7.9 Case 2

For the second case in Fig. 7.1, CP 4, the last control point coordinate, *happens to exactly correspond* to $x = L$. It can be used to impose a known Dirichlet displacement, δ_4, and the default Neumann condition would be $d\delta(0)/dx = 0$. This is identical to Case 1, except for the control displacements changing their signs. This changes the signs of all of the displacements, strains, stresses, and forces.

7.10 Case 3

For the third case in Fig. 7.1, the analysis will be even easier because CP 1 and CP 4 *happen to exactly correspond* to $x = 0$, and $x = L$, respectively. Thus, those displacement controls can be used to impose two zero Dirichlet BCs. All of the calculations in Case 1 are unchanged up to the point where the singular matrix equilibrium equations are modified to reflect the two EBCs.

By inspection, this case should cause the bottom half of the bar to be in varying compression, while the top half has varying compression; and the two reaction forces will be upward with half of the weight, W, each. Preparing to enforce the EBCs, the singular matrix equilibrium equations, with no externally applied point loads, are

$$[\boldsymbol{K}_{\mathrm{EA}}]\{\boldsymbol{\delta}\} = \{\boldsymbol{c}_w\} + \{\boldsymbol{c}_{\mathrm{EBC}}\} + \{\boldsymbol{c}_F\},$$

$$\frac{\mathrm{EA}}{3L}\begin{bmatrix} 8 & -6 & -2 & 0 \\ -6 & 8 & 0 & -2 \\ -2 & 0 & 8 & 6 \\ 0 & -2 & -6 & 8 \end{bmatrix}\begin{Bmatrix} \delta_1 \\ \delta_2 \\ \delta_3 \\ \delta_4 \end{Bmatrix}$$

$$= \frac{wL}{6}\begin{Bmatrix} 1 \\ 2 \\ 2 \\ 1 \end{Bmatrix} + \begin{Bmatrix} -\mathrm{EA}\dfrac{d\delta(0)}{dn} \\ 0 \\ 0 \\ \mathrm{EA}\dfrac{d\delta(L)}{dn} \end{Bmatrix} + \begin{Bmatrix} F_1 = 0 \\ F_2 = 0 \\ F_3 = 0 \\ F_4 = 0 \end{Bmatrix}.$$

Partitioning the equations for $\delta_1 = 0$ and $\delta_4 = 0$ and retaining the middle two rows of the system and the middle two columns of the square matrix give the non-singular unknown displacement control value relations

$$\frac{\text{EA}}{3L}\begin{bmatrix}8 & 0\\0 & 8\end{bmatrix}\begin{Bmatrix}\delta_2\\\delta_3\end{Bmatrix} = \frac{wL}{6}\begin{Bmatrix}2\\2\end{Bmatrix} + \begin{Bmatrix}0\\0\end{Bmatrix} + \begin{Bmatrix}F_2 = 0\\F_3 = 0\end{Bmatrix}$$
$$- \frac{\text{EA}}{3L}\begin{Bmatrix}-6\\-2\end{Bmatrix}\delta_1 - \frac{\text{EA}}{3L}\begin{Bmatrix}-2\\-6\end{Bmatrix}\delta_4$$

for any values of δ_1 and δ_4. Assigning zero to both Dirichlet conditions gives CP displacement control values of

$$\begin{Bmatrix}\delta_2\\\delta_3\end{Bmatrix} = \frac{3L}{\text{EA}}\frac{1}{8}\begin{bmatrix}1 & 0\\0 & 1\end{bmatrix}\frac{\text{wL}}{6}\begin{Bmatrix}2\\2\end{Bmatrix} = \frac{\text{wL}^2}{8\text{EA}}\begin{Bmatrix}1\\1\end{Bmatrix} = \frac{\text{WL}}{8\text{EA}}\begin{Bmatrix}1\\1\end{Bmatrix}.$$

7.11 Post-processing Case 3

Those displacements give the exact physical bar displacements shown in the non-dimensional graph for this case is in Fig. 7.3.

From anti-symmetry, the maximum displacement is at the center of the bar, at $u = 0.5$, and can be evaluated at that common point

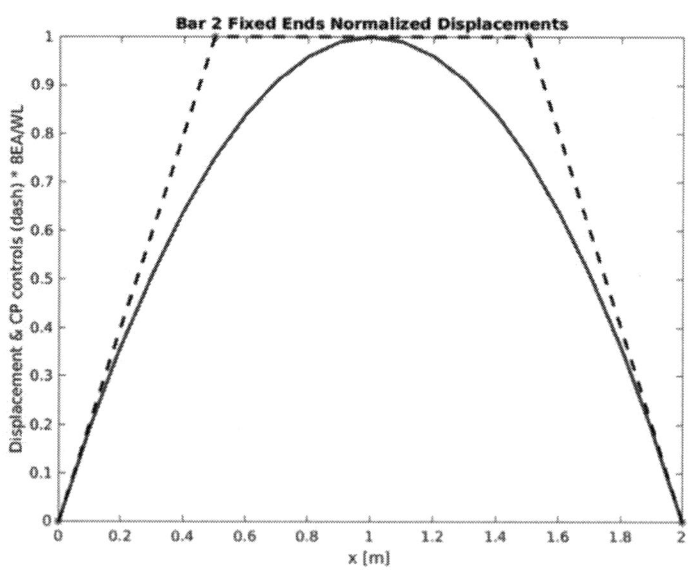

Figure 7.3. **Bar (solid) and control point (dashed) displacements, Case 3**

on either knot span. From the first knot span, using CP 1, 2, 3 gives

$$\delta(u = 0.5) = [\boldsymbol{N}(u = 0.5)]\{\boldsymbol{x}_{CP}\} = [0 \quad 0.5 \quad 0.5]\frac{\mathrm{WL}}{8\mathrm{EA}}\begin{Bmatrix} 0 \\ 1 \\ 1 \end{Bmatrix} = \frac{\mathrm{WL}}{8\mathrm{EA}}.$$

Or, from the second knot span and CP 2, 3, and 4, the same result is obtained:

$$\delta(u = 0.5) = [\boldsymbol{N}(u = 0.5)]\{\boldsymbol{x}_{CP}\} = \begin{bmatrix}0.5 & 0.5 & 0\end{bmatrix}\frac{\mathrm{WL}}{8\mathrm{EA}}\begin{Bmatrix} 1 \\ 1 \\ 0 \end{Bmatrix} = \frac{\mathrm{WL}}{8\mathrm{EA}}.$$

Both are the analytically exact midpoint deflections.

The first and fourth rows of the matrix equilibrium equation establish the two reaction forces:

$$\begin{aligned}
&\frac{\mathrm{EA}}{3L}\begin{bmatrix} 8 & -6 & -2 & 0 \\ 0 & -2 & -6 & 8 \end{bmatrix}\frac{\mathrm{wL}^2}{8\mathrm{EA}}\begin{Bmatrix} 0 \\ 1 \\ 1 \\ 0 \end{Bmatrix} = \frac{\mathrm{wL}}{6} \\
&\quad \begin{Bmatrix} 1 \\ 1 \end{Bmatrix} + \begin{Bmatrix} -\mathrm{EA}d\delta(0)/dn \\ \mathrm{EA}d\delta(L)/dn \end{Bmatrix} + \begin{Bmatrix} F_1 = 0 \\ F_4 = 0 \end{Bmatrix}, \\
&\frac{\mathrm{EA}}{3L}\frac{\mathrm{wL}^2}{8\mathrm{EA}}\begin{Bmatrix} -8 \\ -8 \end{Bmatrix} \\
&\quad = \frac{wL}{6}\begin{Bmatrix} 1 \\ 1 \end{Bmatrix} + \begin{Bmatrix} -\mathrm{EA}d\delta(0)/dn \\ \mathrm{EA}d\delta(L)/dn \end{Bmatrix} + \begin{Bmatrix} 0 \\ 0 \end{Bmatrix}, \\
&-\frac{wL}{3}\begin{Bmatrix} 1 \\ 1 \end{Bmatrix} - \frac{wL}{6}\begin{Bmatrix} 1 \\ 1 \end{Bmatrix} \\
&\quad = -\frac{wL}{2}\begin{Bmatrix} 1 \\ 1 \end{Bmatrix} = \begin{Bmatrix} -\mathrm{EA}d\delta(0)/dn \\ \mathrm{EA}d\delta(L)/dn \end{Bmatrix} = \begin{Bmatrix} \mathrm{EA}d\delta(0)/dx \\ \mathrm{EA}d\delta(L)/dx \end{Bmatrix}.
\end{aligned}$$

That is, the reaction forces are each upward at half of the total bar weight, as expected.

7.12 Case 3 non-homogeneous EBC

Since the CPs were easily placed at $x = 0$ and $x = L$, it would be easy to impose non-zero support displacements at either of those locations. If the bottom control displacement settles downward (+), it will reduce the bottom reaction and eventually turn that reaction force into a tension (−) force. Recall that the maximum free hanging bar had a maximum displacement of $\delta_{\text{max}} = \text{WL}/2\text{EA}$. A numerical example illustrates the effects of changing the lower Dirichlet boundary condition from zero to a downward settlement displacement proportional to the maximum displacement in the free-hanging bar of $\delta_4 = \text{WL}/\text{n EA}$, so that $n > 2$ gives a smaller value (the fixed-fixed case occurs as $n \to \infty$). The two reaction forces caused by such a settlement are easily found from the mechanics of materials, and will be used to check the following IGA solution.

The settlement will only change the right-hand side of the previous second and third rows of the equilibrium equations to

$$\frac{\text{EA}}{3L}\begin{bmatrix}8 & 0\\0 & 8\end{bmatrix}\begin{Bmatrix}\delta_2\\\delta_3\end{Bmatrix} = \frac{\text{wL}}{6}\begin{Bmatrix}2\\2\end{Bmatrix} + \begin{Bmatrix}0\\0\end{Bmatrix} + \begin{Bmatrix}0\\0\end{Bmatrix} + \begin{Bmatrix}0\\0\end{Bmatrix}$$

$$+\frac{\text{EA}}{3L}\begin{Bmatrix}2\\6\end{Bmatrix}\left(\frac{\text{WL}}{\text{n EA}}\right) = \frac{W}{6}\begin{Bmatrix}2\\2\end{Bmatrix} + \frac{W}{3n}\begin{Bmatrix}2\\6\end{Bmatrix}$$

$$\frac{\text{EA}}{3L}\begin{bmatrix}8 & 0\\0 & 8\end{bmatrix}\begin{Bmatrix}\delta_2\\\delta_3\end{Bmatrix} = \frac{W}{6}\left(\begin{Bmatrix}2\\2\end{Bmatrix} + \frac{1}{n}\begin{Bmatrix}4\\12\end{Bmatrix}\right).$$

This is starting to get messy, and the reaction calculations will be worse. The results for both were obtained by Matlab Script 7.1. (The above-mentioned fixed-fixed results can be verified by changing line three to d4 = 0.)

Script 7.1. Symbolic displacements and reaction recovery

```
%        Fixed (d1)& Settled (d4) Hanging Bar
syms W n L EA d1 d2 d3 d4 r1 r2 real
d1 = 0; d4 = W*L/(n*EA)          ; % Dirichlet BC
d = [d1; d2; d3; d4]              % Displacements
%             Reduced stiffness matrix
k = 8*EA*[1, 0; 0, 1]/(3*L)        ; % after EBC
%       Net forces from weight & BC 4
F = W*[2; 2]/6 + d4*EA*[2; 6]/(3*L)  % after EBC
```

```
%         Solve for free displacements
[free] = inv(k)*F                    ; % Get free dof
d(2) = free(1); d(3) = free(2)     ; % update dof
disp('System displacements');disp(d)      % list
%           Get BC Reaction Forces
%       Two reaction rows from K 1 and 4
R = EA*[8 -6 -2 0;0 -2 -6 8]/(3*L)  ; % 2 K rows
LHS = R*d                     ; % multiply by all dof
r  = LHS - W*[1; 1]/6       ; % subs. For reactions
disp('Reactions at BC');disp(r)      % list disp
%                 Running gives
% System displacements: d1 set, d2 d3 d4 set
%                              0            % EBC
% (3*L*(W/3 + (2*W)/(3*n)))/(8*EA)    % solution
%     (3*L*(W/3 + (2*W)/n))/(8*EA)    % solution
%                       (L*W)/(EA*n)        % EBC
% Reactions at BC 1 and 4
% - W/2 - W/n                 % required force at d1
%   W/n - W/2                 % required force at d4
```

Reaction forces of $R_{\text{top}} = -W(1/n + 1/2)$ and $R_{\text{bot}} = W(1/n - 1/2)$ add up to a net reaction of $-W$, which is equal and opposite to the weight of the bar. These two reaction forces are also obtainable from mechanics of materials and illustrated in Fig. 7.4. Referring to that figure, from Case 2 and the mechanics of materials, it is known that the two reactions (on the left) are both half the weight acting upward, $-W/2$. Also from mechanics of materials, it is known that to elongate a bar of length L by an amount Δ requires a force $P = \Delta\,\text{EA}/L$ (in the middle). When both conditions are active, the reactions on the right are obtained. In the current example, the elongation has been set to $\Delta \equiv \text{WL}/n\,\text{EA}$, which makes the reactions become $R_{\text{top}} = -W(1/n+1/2) = R_1$, and $R_{\text{bot}} = W(1/n - 1/2) = R_4$, which match the IGA result obtained from the symbolic Matlab solution.

Varying the amount of the relative elongation (that is, n) and/or its direction (sign) covers an infinite number of situations. Recall that the end displacement of the above free bar (with no end force) was $\delta(u = 1) = \text{WL}/2\text{EA}$, which should correspond to $n = 2$. Indeed, inserting that value gives $R_{\text{top}} = -W$ and $R_{\text{bot}} = 0$ as expected.

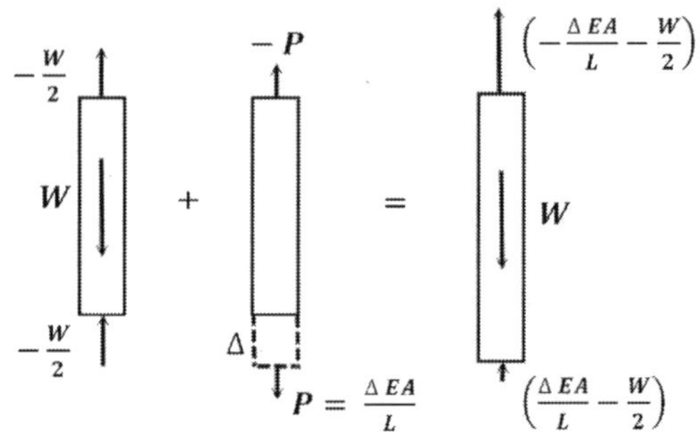

Figure 7.4. Combinations of a Fixed-Fixed bar and an elongated bar

7.13 Case 4

An FEA study with at least quadratic C^0 Lagrange element interpolation (basis) functions would also give the same exact answers everywhere for the above-mentioned three cases, and for this case. Now, the procedures of the two analysis methods become more notably different. The support point is now at an arbitrary point along the bar. In FEA, it is easy to place a node at such a support point and obtain the exact deflections and stresses everywhere.

In an IGA study, the process of imposing a Dirichlet boundary condition anywhere in the domain is not as simple, especially in higher dimensions. There are at least three ways to accomplish this task: Change the analysis knot vector, introduce a second sub-domain (iElement), or introduce a constraint condition on the assembled system matrices.

First, consider the physical insight from Figure 7.5. The thin vertical line in Fig. 7.5 represents zero values of axial force (dashed) and axial displacement (solid) along the length of the bar. The prior results illustrate that the displacements vary quadratically with axial position and the axial forces vary linearly. The displacement values in Case 4 are positive (downward) except at the displacement constraint. There, it goes to zero, *and has a discontinuity* (jump) in its slope. In other words, at the Dirichlet condition reaction force, the sign of the slope changes.

Addressing the axial force, above a restraint point, the top of the bar is stress-free (connected to air). Moving downward, the weight of

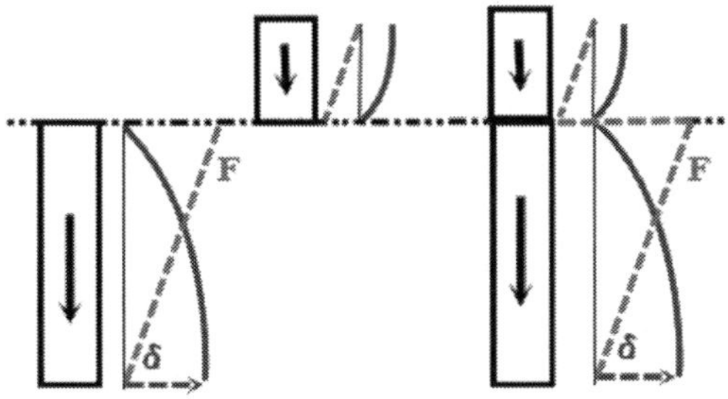

Figure 7.5. Case 4 internal support causes a kink in the displacement's slope

the above material causes an increasing compression force (and axial stress), which is maximum at the support location. Conversely, the bottom end of the bar is stress-free (connected to air), and moving up the bar causes a tension force to support the lower weight. That tension force becomes maximum at the point below the restrained displacement. In other words, at the Dirichlet condition reaction force, the sign of the slope changes.

This case will be addressed here by changing the analysis knot vector to include a slope discontinuity by repeating a knot, here at the center. To understand why a slope discontinuity is needed in this case, the displacement and axial force shapes from Cases 1 and 2 are sketched in Fig. 7.5 in comparison to this case. In each case, the displacement is zero at the support and increases from there, reaching their maximum value, with a zero slope (and thus no force) at their distal end. The part above the support is in compression, while the lower portion is in tension. The discontinuity in the displacement slope corresponds to a discontinuity in the axial strain, the axial stress, and the internal axial force.

There are other common changes in the data that likewise require inserting a repeating knot (change in slope) at such points in an IGA study. In an FEA study, allowing for a possible slope discontinuity is automatic when common C^0 Lagrange interpolations are used and a node is placed at such locations. Think of the governing second-order ODE with piecewise constant properties as being

$$-\frac{d}{dx}\left[E(x)A(x)\frac{d\delta(x)}{dx}\right] - w(x) = 0 \rightarrow \frac{d}{dx}[\text{strain}] = \frac{w}{EA}.$$

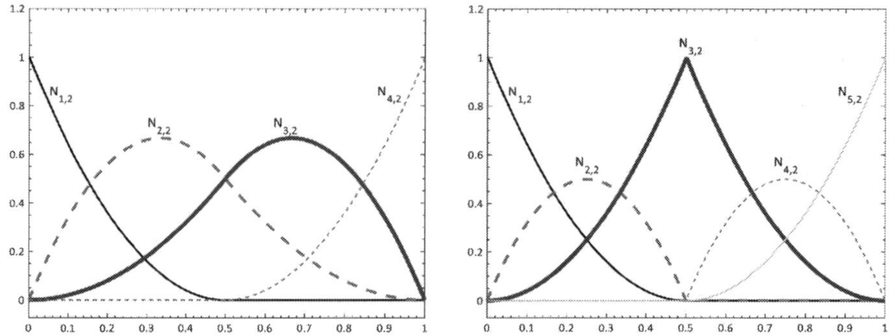

Figure 7.6. Quadratic basis functions for knot vector $\mathbb{A}u$ = {0, 0, 0, 0.5,1,1,1} and with a repeated center knot, $\mathbb{A}u$ = {0, 0, 0, 0.5, 0.5, 1, 1, 1}

If there is a jump in the value of E, A, or w, there will be a kink in the slope of the displacement, $d\delta/dx$. In an IGA study, this means a repeated knot needs to occur at the control point where any such data discontinuity is present. In the current case, the single repeated knot is at the control point where the internal support point is located. This is because the reaction force at that point, R, is a singular line load (mathematically, $w(x)$ is the reaction force, R, acted upon by a Dirac Delta distribution).

Up to this point, the analysis has used the four quadratic basis functions (three per knot span), with continuous slopes, shown on the left of Fig. 7.6. Now, with the repeated knot, the recursion relations create the five basis functions (four per knot span) shown on the right of that figure. All five of those basis functions have a kink in their slope at the repeated knot at $u = 0.5$ due to the two repeated center knots.

Before looking at the new basis functions, it is easier to review the geometrical effects of repeated knots on a planar curve shown in Fig. 7.7. There, the top image uses the computer graphics counting system (beginning with 0) and the lower two images use the engineering counting system. Think of the location on the planar curve as the magnitude of the solution in a scalar IGA study. If the fifth knot had not been repeated, the curve location would have smoothly passed near control point six. Instead, the location goes exactly through that point but with a jump in its direction (slope). If the geometric knot vector is thought of as an analysis knot vector, it would mean that the solution on either side of knot five start with the same value, but not

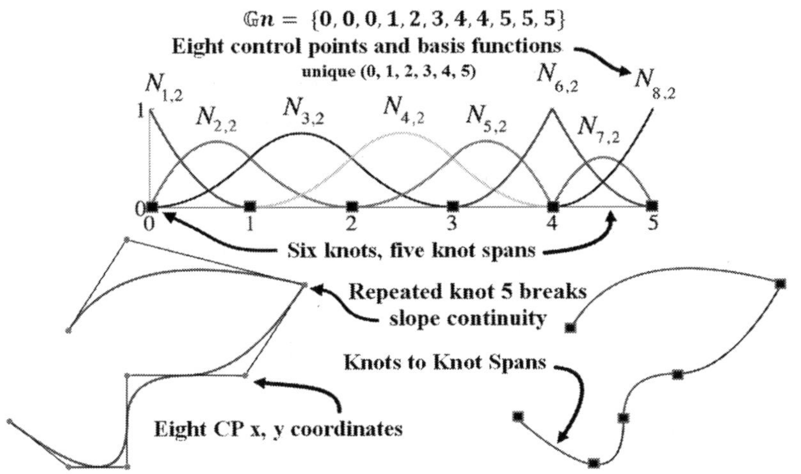

Figure 7.7. Effect of two repeated knots on locations on a planar curve

Table 7.2. A symmetric Knot vector with one repeated Knot

Analysis knot vector $\mathbb{A}u = \{0, 0, 0, 1/2, 1/2, 1, 1, 1\}$, Length = 8, Degree = 2, Number of CP = 3, Number of spans = 2 on [0, 1/2), [1/2,1); CP 1-2, 2-3.

Count	Basis Functions N(u)	Derivatives: dN(u)/du
1, 0	$[(1-2u)^2,\ 0]$	$[(-4+8u),\ 0]$
2, 4	$[4u(1-2u),\ 4(1-u)^2]$	$[(4-16u),\ (-8+8u)]$
3, 5	$[4u^2,\ 4(2u-1)(1-u)]$	$[8u,\ (12-16u)]$
0, 6	$[0,\ (2u-1)^2]$	$[0,\ (-4+8u)]$

with the same slope. In math jargon, the solution is C^1 continuous at knots 1, 2, 4, 5, but at knot 3, it is only C^0 continuous (has a jump in its slope).

For Case 4, the needed analytic forms of those five basis functions (Fig. 7.6 right) are given in Table 7.2.

Let the five CPs be placed at the Greville points: $\mathrm{CP}_x = [0 \quad 1/4 \quad 2/4 \quad 3/4 \quad 1]L$. Now, the Jacobian of the two mappings has to be determined in each knot span. The Jacobian matrix in a quadratic knot span is

$$[\mathbb{J}^r_{x:u}(u)] = \left[\frac{dN_j(u)}{du} \ \frac{dN_{j+1}(u)}{du} \ \frac{dN_{j+2}(u)}{du}\right] \{\boldsymbol{x}^r\}.$$

For knot span 1,

$$[\mathbb{J}_{x:u}^{r=1}(u)] = [(-4+8u)\ (4-16u)\ (8u)] \begin{Bmatrix} 0 \\ 1/4 \\ 2/4 \end{Bmatrix} L = L.$$

For knot span 2,

$$[\mathbb{J}_{x:u}^{r=2}(u)] = [(-8+8u)\ (12-16u)\ (-4+8u)] \begin{Bmatrix} 2/4 \\ 3/4 \\ 1 \end{Bmatrix} L = L.$$

As before, check the resultant source vectors due to the weight per unit length.

$$\{\boldsymbol{c}_w^r\} = \int_{u_j}^{u_{j+1}} [\boldsymbol{N}^r(u)]^T\, w(u)\, |\mathbb{J}_{x:u}^r(u)|\, du.$$

But, here, the geometric Jacobian matrix, $[\mathbb{J}_{\boldsymbol{x:u}}(u)]$, and its determinant are the same in each span where $\mathrm{dx} = L\,du$.

$$\{\boldsymbol{c}_w^{r=1}\} = \int_{u_1=0}^{u_2=0.5} \begin{Bmatrix} (1-2u)^2 \\ 4u(1-2u) \\ 4u^2 \end{Bmatrix} w\,(L\,du) = \frac{wL}{3} \begin{Bmatrix} 1 \\ 1 \\ 1 \end{Bmatrix}$$

and

$$\{\boldsymbol{c}_w^{r=1}\} = \int_{u_1=0.5}^{u_2=1} \begin{Bmatrix} 4(1-u)^2 \\ 4(2u-1)(1-u) \\ (2u-1)^2 \end{Bmatrix} w\,(L\,du) = \frac{\mathrm{wL}}{3} \begin{Bmatrix} 1 \\ 1 \\ 1 \end{Bmatrix}.$$

As required, the full weight of the bar, $W = \mathrm{wL}$, is contained in the assembled resultant load (column) vector, $\{\boldsymbol{c}_w\}$, with a third being above the support point. Likewise, the strain–displacement matrix, $[\boldsymbol{B}(u)] \equiv [\frac{\partial \boldsymbol{N}(u)}{\partial u}][\mathbb{J}_{\boldsymbol{x:u}}(u)]^{-1}$, has the same inverse Jacobian matrix (here a scalar) in each span. Evaluating each knot span stiffness from

$$[\boldsymbol{K}_{\mathrm{EA}}^j] = \int_{u_j}^{u_{j+1}} [\boldsymbol{B}^j(u)]^T \mathrm{EA} [\boldsymbol{B}^j(u)](|\mathbb{J}_{x:u}^r(u)|\, du)$$

gives

$$[\boldsymbol{K}_{\mathrm{EA}}^{j=1}] = \mathrm{EA} \int_{u_1=0}^{u_2=0.5} \left[\frac{1}{L}\right] \begin{Bmatrix} (-4+8u) \\ (4-16u) \\ (8u) \end{Bmatrix} \left[\frac{1}{L}\right]$$

$$\times [(-4+8u)\ (4-16u)\ (8u)]\,(Ldu),$$

$$[\boldsymbol{K}_{\mathrm{EA}}^{1}] = \frac{4}{3}\frac{\mathrm{EA}}{L} \begin{bmatrix} 2 & -1 & -1 \\ -1 & 2 & -1 \\ -1 & -1 & 2 \end{bmatrix}.$$

Similarly, $[\boldsymbol{K}_{\mathrm{EA}}^{2}] = [\boldsymbol{K}_{\mathrm{EA}}^{1}]$ and the assembled matrices, for the third point fixed, no external point loads, and with pending essential boundary condition modifications are

$$\frac{4}{3}\frac{\mathrm{EA}}{L} \begin{bmatrix} 2 & -1 & -1 & 0 & 0 \\ -1 & 2 & -1 & 0 & 0 \\ -1 & -1 & 4 & -1 & -1 \\ 0 & 0 & -1 & 2 & -1 \\ 0 & 0 & -1 & -1 & 2 \end{bmatrix} \begin{Bmatrix} \delta_1 \\ \delta_2 \\ \delta_3 = 0 \\ \delta_4 \\ \delta_5 \end{Bmatrix}$$

$$= \frac{\mathrm{wL}}{6} \begin{Bmatrix} 1 \\ 1 \\ 2 \\ 1 \\ 1 \end{Bmatrix} + \begin{Bmatrix} -d\delta(0)/dn = 0 \\ 0 \\ 0 \\ 0 \\ d\delta(0)/dn = 0 \end{Bmatrix} + \begin{Bmatrix} 0 \\ 0 \\ R_3 \\ 0 \\ 0 \end{Bmatrix}.$$

Modifying these matrices to enforce the essential boundary condition on the support on the third point of the bar, CP value at δ_3, gives the four independent CP control value matrices:

$$\frac{4}{3}\frac{\mathrm{EA}}{L} \begin{bmatrix} 2 & -1 & 0 & 0 \\ -1 & 2 & 0 & 0 \\ 0 & 0 & 2 & -1 \\ 0 & 0 & -1 & 2 \end{bmatrix} \begin{Bmatrix} \delta_1 \\ \delta_2 \\ \delta_4 \\ \delta_5 \end{Bmatrix}$$

$$= \frac{\mathrm{wL}}{6} \begin{Bmatrix} 1 \\ 1 \\ 1 \\ 1 \end{Bmatrix} + \begin{Bmatrix} 0 \\ 0 \\ 0 \\ 0 \end{Bmatrix} - \delta_3 \frac{4\mathrm{EA}}{L} \begin{Bmatrix} -1 \\ -1 \\ -1 \\ -1 \end{Bmatrix}.$$

Multiplying both sides by the stiffness inverse gives the control displacement values of

$$\begin{Bmatrix} \delta_1 \\ \delta_2 \\ \delta_4 \\ \delta_5 \end{Bmatrix} = \frac{3}{4}\frac{L}{\mathrm{EA}}\frac{1}{9}\begin{bmatrix} 2 & 1 & 0 & 0 \\ 1 & 2 & 0 & 0 \\ 0 & 0 & 2 & 1 \\ 0 & 0 & 1 & 2 \end{bmatrix} \frac{\mathrm{wL}}{6}\begin{Bmatrix} 1 \\ 1 \\ 1 \\ 1 \end{Bmatrix}$$

$$= \frac{\mathrm{wL}^2}{8\,\mathrm{EA}}\begin{Bmatrix} 3 \\ 3 \\ 1 \\ 1 \end{Bmatrix} = \frac{\mathrm{wL}^2}{27\,\mathrm{EA}}\begin{Bmatrix} 1 \\ 1 \\ 2 \\ 2 \end{Bmatrix},$$

which result in a total CP displacement values matrix of

$$\{\boldsymbol{\delta}_{CP}\} = \mathrm{wL}^2[1\ 1\ 0\ 1\ 1]/(8\,\mathrm{EA}).$$

Since the fourth CP is at $x = L$, the physical displacement there in the second span is

$$\delta_{\mathrm{bot}} = [4(1-u)^2,\ 4(2u-1)(1-u),\ (2u-1)^2]\begin{Bmatrix} \delta_3 \\ \delta_4 \\ \delta_5 \end{Bmatrix}.$$

At $u = 1$, $\delta_{\mathrm{bot}} = [0\ 0\ 1]\frac{\mathrm{wL}^2}{8\,\mathrm{EA}}\begin{Bmatrix} 0 \\ 1 \\ 1 \end{Bmatrix} = \frac{2\,\mathrm{wL}^2}{8\mathrm{EA}}$.

To check this, compare to the bottom end displacement using the exact answer for Case 1 of

$$\delta_{\mathrm{bot}} = \frac{wL}{2\mathrm{EA}} = \frac{(wL/2)(L/2)}{2\mathrm{EA}} = \frac{\mathrm{wL}^2}{8\mathrm{EA}},\ \text{as expected.}$$

7.14 Multiple Patches

So far, the IGA studies have only used a single spatial patch with basis functions in various knot spans. In higher dimensions, different material domains are modeled with different patches. When one such patch is joined to an adjacent patch, the basis functions only maintain C^o continuity of the primary unknown (unless numerous multipoint constraints, MPCs, are added to the system matrices). In other words, at patch interfaces, there is a kink in the slope of the solution in the direction normal to the patch interface.

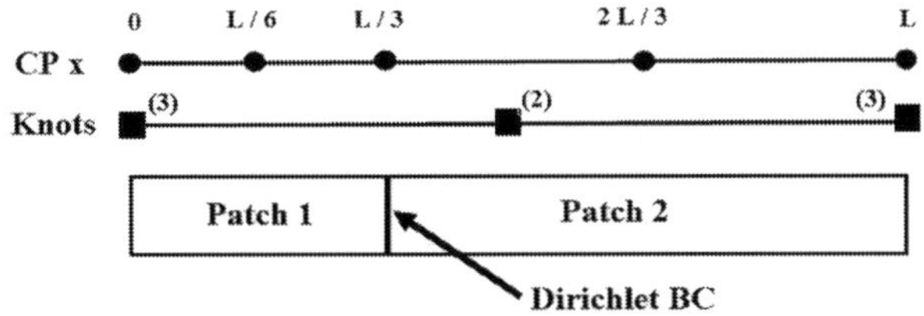

Figure 7.8. Two patches can meet at an interface with essential BC

This provides an alternate approach to Case 4 of Fig. 7.5 of having a bar supported anywhere along its length, as sketched in Fig. 7.8. One patch can contain the upper bar segment, with multiple knot spans, while a second patch contains the lower bar segment, with multiple knot spans. At the control point, dividing the two patches at the support interface, the bar displacement there, and on either side, will be continuous, but the slope (strain) will be discontinuous, which is needed for Case 4.

7.14.1 *Resolving Case 4 with two quadratic patches*

The assembled stiffness and load vectors for each panel are known from the solution of Case 1 since they depend on the length of the bar when all properties are constant. Let the support section be at $x = L/3$. Then, the lengths of the upper (patch 1) and lower segment (patch 2) are $L/3$ and $2L/3$, respectively. Substituting those lengths into the matrices in Case 1 makes the assembled equilibrium matrices for the first panel

$$\frac{\mathrm{EA}}{3(\mathrm{L}/3)}\begin{bmatrix} 8 & -6 & -2 & 0 \\ -6 & 8 & 0 & -2 \\ -2 & 0 & 8 & -6 \\ 0 & -2 & -6 & 8 \end{bmatrix}\begin{Bmatrix} \delta_1 \\ \delta_2 \\ \delta_3 \\ \delta_4 \end{Bmatrix} = \frac{\mathrm{w}(\mathrm{L}/3)}{6}\begin{Bmatrix} 1 \\ 2 \\ 2 \\ 1 \end{Bmatrix}$$

and for the lower panel (which shares CP 4)

$$\frac{\mathrm{EA}}{3(2\mathrm{L}/3)}\begin{bmatrix} 8 & -6 & -2 & 0 \\ -6 & 8 & 0 & -2 \\ -2 & 0 & 8 & -6 \\ 0 & -2 & -6 & 8 \end{bmatrix}\begin{Bmatrix} \delta_4 \\ \delta_5 \\ \delta_6 \\ \delta_7 \end{Bmatrix} = \frac{\mathrm{w}(2\mathrm{L}/3)}{6}\begin{Bmatrix} 1 \\ 2 \\ 2 \\ 1 \end{Bmatrix}.$$

Assembling (scattering) the two panels into the system matrix equilibrium equations, before enforcing the Dirichlet condition at CP 4 that $\delta_4 = 0$, will cause an external reaction force, R_4, to give

$$\frac{\mathrm{EA}}{L}\begin{bmatrix} 8 & -6 & -2 & 0 & 0 & 0 & 0 \\ -6 & 8 & 0 & -2 & 0 & 0 & 0 \\ -2 & 0 & 8 & -6 & 0 & 0 & 0 \\ 0 & -2 & -6 & 12 & -3 & -1 & 0 \\ 0 & 0 & 0 & -3 & 4 & 0 & -1 \\ 0 & 0 & 0 & -1 & 0 & 4 & -3 \\ 0 & 0 & 0 & 0 & -1 & -3 & 4 \end{bmatrix}\begin{Bmatrix} \delta_1 \\ \delta_2 \\ \delta_3 \\ \delta_4 = 0 \\ \delta_5 \\ \delta_6 \\ \delta_7 \end{Bmatrix}$$

$$= \frac{\mathrm{wL}}{18}\begin{Bmatrix} 1 \\ 2 \\ 2 \\ 3 \\ 4 \\ 4 \\ 2 \end{Bmatrix} + \begin{Bmatrix} 0 \\ 0 \\ 0 \\ R_4 \\ 0 \\ 0 \\ 0 \end{Bmatrix}.$$

Setting the vector subscripts to partition the solving steps with [Fixed] = [4], for the interface support condition at δ_4, along with the essential boundary condition value(s) of [EBC] = [0], and making all other dofs unknown [Free] = [1, 2, 3, 5, 6, 7], yields the free control point displacement values of the two panels of

$$\{\boldsymbol{\delta}\,(\text{Free})\} = [\boldsymbol{K}\,(\text{Free, Free})]^{-1}\{\boldsymbol{c}\,(\text{Free})\} \text{ or } \begin{Bmatrix} \delta_1 \\ \delta_2 \\ \delta_3 \\ \delta_5 \\ \delta_6 \\ \delta_7 \end{Bmatrix} = \frac{w\,L^2}{36\,EA}\begin{Bmatrix} 2 \\ 2 \\ 1 \\ 4 \\ 8 \\ 8 \end{Bmatrix}.$$

The unused fourth row of the stiffness is multiplied by all seven of the now known displacements to yield the support reaction (which must equal the total weight of the bar):

$$\frac{\mathrm{EA}}{L}[0 \quad -2 \quad -6 \quad 12 \quad -3 \quad -1 \quad 0]\{\boldsymbol{\delta}\} - \frac{\mathrm{wL}}{18}\{3\} = \{R_4\}.$$

As expected, $-\mathrm{wL}\{1\} = -W = \{R_4\}$. Of course, in most practical applications, the above-mentioned integrals would have to be evaluated numerically.

Looking Ahead: Often, the dynamics and/or vibration studies of a structure require its 'mass matrix' in addition to its stiffness matrix. Without proof, the knot span mass matrix is

$$[\boldsymbol{M^r}] = \int_{u_r}^{u_{r+1}} [\boldsymbol{N^r}(u)]^T \rho \, [\boldsymbol{N^r}(u)] \, |\mathbb{J}^r_{x:u}(u)| \, du,$$

where ρ is the mass density per unit length. For the first knot span of the bar,

$$[\boldsymbol{M}^{r=1}] = \int_{u_1=0}^{u_2=0.5} \begin{Bmatrix} (1-2u)^2 \\ 2u(2-3u) \\ 2u^2 \end{Bmatrix} \rho \, [(1-2u)^2 \; 2u(2-3u) \; 2u^2](L \, du),$$

$$[\boldsymbol{M}^{r=1}] = \frac{\rho L}{120} \begin{bmatrix} 12 & 7 & 1 \\ 7 & 17 & 6 \\ 1 & 6 & 3 \end{bmatrix}.$$

Note that only half of the total mass of the bar, $m = \rho L$, is present as the sum of the coefficients in the mass matrix because this knot span only covers half of the bar length. The second knot span accounts for the remaining mass:

$$[\boldsymbol{M}^{r=2}] = \frac{\rho L}{120} \begin{bmatrix} 3 & 6 & 1 \\ 6 & 17 & 7 \\ 1 & 7 & 12 \end{bmatrix}.$$

7.15 Variable force data

Bars are often components of rotating equipment. Consider Case 3 where the bar is pinned at both ends, but is now rotating at about $x = 0$ with an angular velocity of ω (rad/sec). This causes a radial acceleration of $a_x = x\,\omega^2$, which when multiplied by the mass per unit length, ρ changes the weight (force) per unit length to $w(x) = x\,\rho\omega^2$. That, in turn, changes the load vector in each span:

$$\{\boldsymbol{c}^r_w\} = \omega^2 \int_{u_r}^{u_{r+1}} [\boldsymbol{h^r}(u)]^T \rho(x)\, x(u) \; |\mathbb{J}^r_{x:u}(u)| du.$$

But, the isoparametric concept uses the same (exact) geometric description:

$$x(u) = [\boldsymbol{h^r}(u)]\{\boldsymbol{x}\}^r,$$

which creates the spinning 'mass matrix'

$$\{\boldsymbol{c}_w^r\} = \omega^2 \left[\int_{u_r}^{u_{r+1}} [\boldsymbol{h^r}(u)]^T \rho(x) [\boldsymbol{h^r}(u)] \, |\mathbb{J}_{x:u}^r(u)| du \right] \{\boldsymbol{x}\}^r$$
$$\equiv \omega^2 [\boldsymbol{M^r}] \{\boldsymbol{x}\}^r,$$

where

$$[\boldsymbol{M^r}] = \int_{u_r}^{u_{r+1}} [\boldsymbol{h^r}(u)]^T \rho(x) [\boldsymbol{h^r}(u)] \, |\mathbb{J}_{x:u}^r(u)| du$$

is the symmetric square iElement mass matrix. Usually, the mass density per unit length, ρ, is constant and can be pulled outside the integral. This integral appears many times in different fields of application of IGA. When the integral relates to a constant coefficient, a new phrase is used known as the generalized mass matrix. Its expression is

$$[\boldsymbol{M^r}] = \rho [\boldsymbol{GM^r}] = \rho \left[\int_{u_r}^{u_{r+1}} [\boldsymbol{h^r}(u)]^T \, [\boldsymbol{h^r}(u)] \, |\mathbb{J}_{x:u}^r(u)| du \right].$$

If the constant coefficient, ρ, is replaced in a thermal analysis with a convection coefficient, h_c, then the symmetric square matrix is the 'convection matrix' to be added to the 'conduction matrix' (the above-mentioned stiffness matrix numbers multiplied by thermal properties). A convection matrix is one of two integrals arising from boundary heat flow due to convection, $q_h = h_c(T - T_\infty)$.

Consider the above-mentioned second-order ODE with constant properties with the axial body force written in terms of its radial acceleration:

$$-\frac{d}{dx}\left[\frac{d\delta(x)}{dx}\right] - \frac{x\,\rho\omega^2}{\text{EA}} = 0.$$

Note that integrating this equation twice means that the exact solution must be cubic, regardless of the BCs. It also means that the all-important stresses must be quadratic in x, in contrast to the linear estimate from the above-mentioned solution. This suggests that selecting a cubic basis function would be better than a quadratic one in an IGA study if there was hope of accidentally obtaining an exact solution.

7.16 Conduction

As a minimal thermal analysis example, consider the steady state heat transfer through the wall of a chimney made of three materials. The wall is large enough that as a first approximation (except in the corners) a one-dimensional model gives a very good model of the temperature through a unit area of the wall. The chimney has an internal temperature of $1,500°\text{F}$ with 9 inches of firebrick with a thermal conductivity of $k_f = 0.72$ BTU/(hr ft F) against 5 inches of insulation with $k_i = 0.08$ BTU/(hr ft °F) against 7.5 inches of redbrick with $k_r = 0.50$ BTU/(hr ft °F), which has an exterior surface temperature of 150° F.

Without any internal heat generation or convection, the differential equation is

$$k\,A\,\frac{d^2T}{dx^2} = 0,$$

where k is the thermal conductivity, A is the area normal to x, and T is the temperature. Integrating this equation twice shows that the solution should be piecewise linear between the two Dirichlet BCs. Having three different materials means this model will require three patches to capture the interface temperatures and the kinks in the temperature gradient at each interface (or, a single knot vector with a pair of repeated knots to deal with the temperature gradient kinks at each material interface).

In addition to the temperature, the heat flux per unit area, $q(x)$, through each layer of material is also of interest. It is defined by Fourier's law as $q(x) = -k\; dT/dx$.

Use a three-panel model (the minimum for three materials) with the simplest single-span quadratic B-Spline. From Appendix A1, the quadratic analysis knot vector, $\mathbb{A}u = 0, 0, 0, 1, 1, 1\}$, creates the following basis functions having three CPs per span:

Count	Basis Functions $N(u)$	Derivatives $dN(u)/du$
1	$[(u-1)\,\hat{}\,2]$	$[(2u-2)]$
2	$[2u(1-u)]$	$[(2-4u)]$
3	$[u\,\hat{}\,2]$	$[2u]$
Span	$[0, 1)$	

When the APs are equally spaced, these have a constant Jacobian of $|\mathbb{J}_{u:L}| = L_x$ in each panel. This leads to the conduction matrix of

$$[K^p] = \frac{k^p A^p}{3L^p}\begin{bmatrix} 4 & -2 & -2 \\ -2 & 4 & -2 \\ -2 & -2 & 4 \end{bmatrix}$$

in each panel. The connection lists for the panel iElements are

IElement	AP List		
1	1	2	3
2	3	4	5
3	5	6	7

The consistent input data are as follows:

Panel	A (ft^2)	L (ft)	k $(BTU/hr\,ft\,°\mathrm{F})$	Interface x (ft)
1	1	$9/12 = 0.75$	0.72	0.75
2	1	$5/12 = 0.41667$	0.08	1.1667
3	1	$7.5/12 = 0.625$	0.50	1.7917

These data were implemented in Script 7.2 to generate the three iElement conduction matrices, assemble (scatter) those into the system matrix equations for thermal equilibrium, enforce the Dirichlet BCs, and solve for the unknown temperatures and the heat flow reactions to maintain the specified temperatures. The system was post-processed to gather each iElement's temperatures, their properties, and then obtain the physical temperature gradient which was used, via Fourier's Law, to verify that every iElement conducted the same amount of heat flux as entered at the hot end.

Script 7.2. Solution for a three-layer conducting wall

```
%                three_layer_wall.m
%             |   P1   |   P2   |   P3   |
%    1,500 F *---*---*---*---*---*---* 150 F
%         AP 1   2   3   4   5   6   7
%              k1,L1   k2,L2   k3,L3   A=1, Q=0
k = [0.72, 0.08, 0.5]; L = [0.75, 0.4167, 0.625];
K = zeros (7,7), cQ = zeros (7,1)  ; % allocate sys
%             AP connections list
AP_list = [1 2 3; 3 4 5; 5 6 7];% system eq numbers
Fixed   = [1, 7]; Given = [1500., 150.]     ; % EBC
Free    = [2, 3, 4, 5, 6]                   ; % tbd
%             Conduction matrix kernel
[KP] = [4 -2 -2;-2 4 -2;-2 -2 4]/3     ; % constant
%   Loop over panels, form Ki, assemble, solve
for i = 1:3        % Loop over 3 single span panels
  [Ki] = [KP]*k(i)/L(i)                ; % A(i) = 1
  % Scatter into the system conduction eqs
  dof = AP_list (i, :)         ; % gather iEl AP list
  K(dof, dof) = K(dof, dof) + Ki       ; % assemble
end                             ; % for all iElements

%   Solve for temperatures & heat flow reactions
[T, R] = Solve_Free_DOF(K, cQ, Free, Fixed, Given);
disp(T'), disp(R') % Running gives the exact result
% T=[1500 1406.3 1312.5 843.7 375.0 262.5 150] F
% q1 = 180. q7 = -180. BTU/hr ft^2    % - means out
%       Post-process for mid-point heat flux
u = 0.5; dN_du =[(2*u-2),(2-4*u),(2*u)] % at mid-pt

for i = 1:3        % Loop over 3 single span panels
  % Gather iEl temps from system temperatures
  dof = AP_list (i, :)         ; % gather iEl AP list
  T_e = T(dof)              ; % iElement temperatures
  dT_dx = (dN_du*T_e)/L(i) ; % Temperature gradient
  q_e = -k(i) * dT_dx                ; % Fourier's Law
end                             ; % for all iElements% Gives
exact q_e = [180, 180, 180] BTU/hr ft^2
```

As expected, the heat flow entering the chimney wall is conserved as it flows through each area and exits the inside wall. In other words, in a one-dimensional model, different materials have different temperature gradients, but the same heat flow (if no internal heat generation is present). Also, note that quadratic basis functions were used in each wall layer, but each degenerated to capture the exact linear polynomial temperature distribution for a homogeneous material.

7.17 Conduction with Convection

A fin conducts heat away from a hot wall and also convects heat away through its axial perimeter surface area and end surface. The governing one-dimensional ODE for the temperature, $T(x)$, is

$$k\,A\,\frac{d^2T}{dx^2} - h\,P\,T(x) + h\,P\,T_\infty = 0,$$

where $k = 3\,W/\text{cm}\ ^\circ\text{C}$ is the thermal conductivity of the material, $A = 4\,\text{cm}^2$ is the fin cross-sectional area (and free end area), which has a perimeter of $P = 10\,\text{cm}$, and the temperature of the surrounding fluid is $T_\infty = 20^\circ\text{C}$. The convection coefficient along the length and on the end area is $h = 0.1\ W/\text{cm}^2\,^\circ\text{C}$. The Dirichlet (essential) condition is that the wall temperature is $T(x = 0) = 80^\circ\text{C}$. The convection loss at the free end area at $x = L = 8\,\text{cm}$ is given by a Robin condition of the mixed kind:

$$q_c(L) = -k\,AdT/dx = h\,P(T(L) - T_\infty).$$

The presence of the middle term, $h\,P\,T(x)$, changes the classification of the ODE because now the exact solution can no longer be polynomials. Depending on the EBCs, and constant coefficients, the solution is usually a combination of hyperbolic trigonometric functions, cosh (x) and sinh (x). Thus, an IGA study utilizing B-splines or NURBS can no longer accidentally produce an exact solution.

In addition to determining the axial temperature distribution, $T(x)$, the effectiveness of the cooling fin is to be displayed with a graph of the heat convected away. That quantity is

$$\begin{aligned}
\text{Loss}\,(x = d \leq L) &= \int_{x=0}^{d} h\,(T(x) - T_\infty)(P\,dx) \\
&= h\,P \int_{x=0}^{d} (T(x) - T_\infty)\,dx, \\
\text{Loss}\,(d) &= h\,P \int_{x=0}^{d} ([\boldsymbol{H}(u)]\,\{\boldsymbol{T}\}^r - T_\infty)dx \\
&= hP \int_{x=0}^{d} [\boldsymbol{H}(u)]\,dx\,(\{\boldsymbol{T}\}^r - \{T_\infty\}),
\end{aligned}$$

where again $dx = |\mathbb{J}^r_{x:u}(u)|du$ and the integral is done in parametric space. There is also a heat loss at the end of the bar of

$$\begin{aligned}
\text{Loss}_L &= \int_A h(T(L) - T_\infty)\,dA = h\,(T(L) - T_\infty) \int_A dA \\
&= h\,A\,(T(L) - T_\infty).
\end{aligned}$$

Here, the simplest cubic B-spline will be used for an approximate solution. From Appendix 1, the analysis knot vector of $\mathbb{A}u = \{0,\ 0,\ 0,\ 0,\ 1,\ 1,\ 1,\ 1\}$ creates the following basis functions and derivatives:

Spans = 1 on [0, 1) with CP 1–4

Count	Basis Functions $N(u)$	Derivatives: $dN(u)/du$
1	$[-(u-1)^3]$	$[-3(u-1)^2]$
2	$[3u(u-1)^2]$	$[9u^2 - 12u + 3]$
3	$[-3u^2(u-1)]$	$[-3u(3u-2)]$
4	$[u^3]$	$[3u^2]$

So far, the physical coordinates of the geometric control points (CPs) have not been assigned. The APs are at the same physical locations but have the yet unknown control temperatures that will define the temperatures on the physical body. An important reason for establishing the CP coordinates is that they will define the geometric Jacobian that must be used in evaluating the governing integral form. In practice, the Jacobian is evaluated numerically. Here, the goal is to choose a layout of the two interior CPs so as to make the Jacobian matrix a constant so the example integrals can be done in closed form.

Try a symmetric but non-uniform spacing of geometric CPs with $\{\boldsymbol{x}\}^r = [0\ \alpha\ (1-\alpha)\ 1]'L$ and obtain the geometric Jacobian by multiplying by the above-mentioned parametric gradient:

$$|\mathbb{J}^r_{x:u}(u)| = \left[\frac{d\boldsymbol{N}(u)}{du}\right]\{\boldsymbol{x}\}^r = 3L(\alpha + u(2-6\alpha) + u^2(6\alpha - 2)).$$

In practice, the Jacobian would be of second degree and would increase the polynomial degree of any integral by 2, thereby requiring more integration points and a longer run time. But, today (2024), the number of iElements in super-complicated applications runs into the high millions, and at least one is in the low billions. So, having a constant Jacobian (degree 0) is the ideal case.

Ideally, the Jacobian can be forced to be a constant by eliminating the u and u^2 terms. Here, that is done by inspection by choosing $= 1/3$, making $|\mathbb{J}^r_{x:u}(u)| = L$, which is used here to make the example even simpler. This choice means, as usual, the calculated temperatures will be, for APs, at the same locations and not on the physical body, except for the end points, which are always interpolatory for open knot vectors.

Of course, in a very fine mesh, an iElement size, L, is so small that $dx = |\mathbb{J}^r_{x:u}(u)|\, du$ is well approximated by assuming that the Jacobian is constant during the integrations.

By adding the center convection term to the Galerkin governing integral form used for the ODE for the axial bar and changing the interpretation of the coefficients from displacements to temperatures, the governing integral is

$$\begin{aligned}\boldsymbol{I} = -\{\boldsymbol{T}\}^{\boldsymbol{T}}\{\boldsymbol{c}_{\text{NBC}}\} + \int_\Omega \{\boldsymbol{T}\}^{\boldsymbol{T}}\left\{\frac{d\boldsymbol{N}(x)^{\boldsymbol{T}}}{dx}\right\}\left[\text{KA}\frac{d\boldsymbol{N}(x)}{dx}\right]\{\boldsymbol{T}\}dx+,\\ + \int_\Omega \{\boldsymbol{T}\}^{\boldsymbol{T}}\{\boldsymbol{N}(x)^{\boldsymbol{T}}\}\, h\, P\, [\boldsymbol{N}(x)]\{\boldsymbol{T}\}dx\\ + \int_\Omega \{\boldsymbol{T}\}^{\boldsymbol{T}}\{\boldsymbol{N}(x)^{\boldsymbol{T}}\}\, h\, P\, T_\infty\, dx = 0,\end{aligned}$$

where the natural boundary condition/thermal reaction terms are

$$\{\boldsymbol{c}_{\mathrm{NBC}}\} = \begin{Bmatrix} -\mathrm{KA}\, dT(0)/dx \\ 0 \\ 0 \\ \vdots \\ 0 \\ \mathrm{KA}\, dT(L)/dx \end{Bmatrix}.$$

The first term is the 'conduction matrix'

$$[\boldsymbol{K}^r_{\mathrm{KA}}] = \int_{L^r} \left\{ \frac{d\boldsymbol{N}(x)^{\boldsymbol{T}}}{dx} \right\} \left[\mathrm{KA} \frac{d\boldsymbol{N}(x)}{dx} \right] dx$$

or the thermal stiffness because bar properties, EA, were just replaced with thermal properties, KA. The next square matrix integral is a 'convection matrix' to be combined with the conduction matrix.

While this problem is viewed as a 1D problem, the convection surface integrals include all surfaces adjacent to the convecting fluid. They are the cylindrical surface along the length of area PL, and the area, A, at the free end of the fin. For constant coefficients, the convection matrix for the cylindrical portion is a length integral, while the end contribution is just the integral over the end area at $x = L$. The convection matrix line integral produces a generalized mass matrix times hP, and the end contribution is only one non-zero term on the diagonal in the row associated with its analysis control point number (4):

$$[\boldsymbol{C}^r_{\mathrm{cyl}}] = P \int_{L^r} \{\boldsymbol{N}(x)^{\boldsymbol{T}}\}\ [\boldsymbol{N}(x)]\, dx + h \int_A \{\boldsymbol{N}(\mathrm{L})^{\boldsymbol{T}}\}\ [\boldsymbol{N}(\mathrm{L})]\, dA.$$

Specifically, because the physical end point is the point $u \equiv 1$ in the parametric space, the end 'point' contributes

$$[\boldsymbol{C}^r_L] = h\{\boldsymbol{N}(u = 1)^{\boldsymbol{T}}\}[\boldsymbol{N}(u = 1)] \int_A dA,$$

$$[\boldsymbol{C}^r_L] = h\, A[0\,0\,0\,1]^T [0\,0\,0\,1] = h\, A \begin{bmatrix} 0 & 0 & 0 & 0 \\ 0 & 0 & 0 & 0 \\ 0 & 0 & 0 & 0 \\ 0 & 0 & 0 & 1 \end{bmatrix} W/^\circ\mathrm{C}.$$

The final source vector from the surrounding fluid on both convecting areas is

$$\{\boldsymbol{c}_\infty\} = h\,P\,T_\infty \int_{L^r} \{\boldsymbol{N}(x)^{\boldsymbol{T}}\}dx + hT_\infty\{\boldsymbol{N}(L)^{\boldsymbol{T}}\} \int_{\mathrm{A}} dA.$$

To obtain the post-processing convection heat loss, requires the integral of the basis functions

$$[\boldsymbol{iN}] = \int_{L^r} [\boldsymbol{N}(x)]dx = \ [1,\ 1,\ 1,\ 1]\ L^r/4.$$

Substituting the basis functions and their derivatives into the other integrals and integrating symbolically with Matlab give the conduction matrix (which sums to zero $W/°\mathrm{C}$) of

$$[\boldsymbol{K}] = \frac{\mathrm{kA}}{10\,L^r}\begin{bmatrix} 18 & -9 & -6 & -3 \\ -9 & 12 & 3 & -6 \\ -6 & 3 & 12 & -9 \\ -3 & -6 & -9 & 18 \end{bmatrix}.$$

The cylindrical surface convection matrix (which sums to hPL $W/°\mathrm{C}$) is

$$[\boldsymbol{C}] = \frac{h\,\mathrm{PL}^r}{140}\begin{bmatrix} 20 & 10 & 4 & 1 \\ 10 & 12 & 9 & 4 \\ 4 & 9 & 12 & 10 \\ 1 & 4 & 10 & 20 \end{bmatrix}$$

and the external fluid temperature, T_∞, source contribution along the cylindrical and end surfaces is

$$\{\boldsymbol{c}_\infty\} = h\,\mathrm{PL}^r T_\infty \frac{[1,\ 1,\ 1,\ 1]^T}{4} + h\,\mathrm{AT}_\infty[0,\ 0,\ 0,\ 1]^T\,W.$$

These assemble into the combined system matrix equations:

$$[\boldsymbol{S}]\{\boldsymbol{T}\} = [[\boldsymbol{K}] + [\boldsymbol{C}] + [\boldsymbol{M}_L]]\{\boldsymbol{T}\} = \{\boldsymbol{c}_\infty\} + \{\boldsymbol{c}_{\mathrm{EBC}}\}.$$

This example application solution, via analytic matrices, was done by Script 7.3. The numerical values, before enforcing the EBCs, yield the singular system for the four AP temperature values:

$$K = \begin{matrix} [2.7000 & -1.3500 & -0.9000 & -0.4500; \\ -1.3500 & 1.8000 & 0.4500 & -0.9000; \\ -0.9000 & 0.4500 & 1.8000 & -1.3500; \\ -0.4500 & -0.9000 & -1.3500 & 2.7000] \end{matrix}$$

$$C = \begin{matrix} [1.1429 & 0.5714 & 0.2286 & 0.0571; \\ 0.5714 & 0.6857 & 0.5143 & 0.2286; \\ 0.2286 & 0.5143 & 0.6857 & 0.5714; \\ 0.0571 & 0.2286 & 0.5714 & 1.1429] \end{matrix}$$

$$C(4,4) = 1.5429,\ c_{\text{inf}} = [40,\ 40,\ 40,\ 40]',\ c_{\text{inf}}(4) = 48\,W.$$

Running that script gives temperatures of $T = [80.0000, 36.5723, 33.0920, 30.5805]°\text{C}$. The temperature from the exact hyperbolic cosine solution at $x = Lx$ is $T = 30.6°\text{C}$.

For the Dirichlet reaction, take the first row (the unused row) of the assembled matrix equilibrium relations and multiply by all of the temperatures to give the thermal reaction at first AP (and the hot wall physical) heat inflow:

$$\begin{aligned} &[3.8429 - 0.7786 - 0.6714 - 0.3929] \\ &\quad \{80.0000, 36.5723, 33.0920, 30.5805\}' \\ &\quad = \{1.1429\} - \{k(0)A(0)dT/dx(0)\}, \end{aligned}$$

giving $-KA\,dT(0)/dx = q_1 = 204.7\,W$ flowing into the fin. The heat flow leaving at the distal end is simply

$$\begin{aligned} q_4 = hA(T_4 - T_\infty) &= \left(0.1\frac{W}{\text{cm}^2}\,°\text{C}\right)(4\,\text{cm}^2)(30.5805°\text{C} - 20°\text{C}) \\ &= 4.23\,W. \end{aligned}$$

Finally, the heat flowing out along the length of the fin is $q_L = hP\,[\boldsymbol{iN}](\{\boldsymbol{T}\} - \{\boldsymbol{T_\infty}\})$:

$$\begin{aligned} q_L &= \left(0.1\frac{W}{\text{cm}^2}°\text{C}\right)(10\,\text{cm})\left(\frac{8\,\text{cm}}{4}\right)[1,1,1,1]\begin{Bmatrix} 60.00 \\ 16.57 \\ 13.09 \\ 10.58 \end{Bmatrix}°\text{C} \\ &= 200.48\,W. \end{aligned}$$

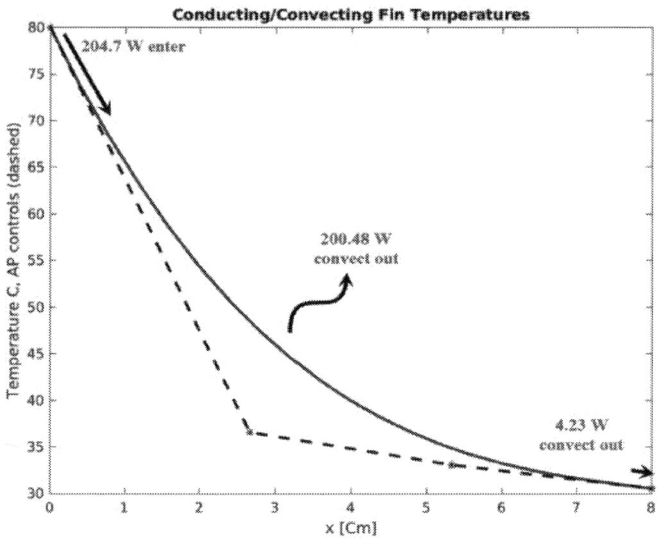

Figure 7.9. Fin temperature and convection loss summary

As expected, the entering wall heat flow of $204.7\,W$ minus the convective loss total of $(-200.48\,W - 4.23\,W) = -204.7\,W$ is balanced, as required for thermal equilibrium. The graph of the physical solution and the AP temperatures are shown in Fig.7.9.

Script 7.3. Solution of the thermal fin via analytic matrices

```
%                     thermal_fin.m
%   Single span conducting convecting fin, p=3 spline
%     with x_AP = [0,1/3,2/3,1]Lx, J = L, constant
%    AP connection list = [1, 2, 3, 4] for assembly
%                  Specific data, and units
% cm^2,    cm,  W/cm,         C, W/cm^2 C,     cm
A = 4; P = 10; k = 3; T0 = 80;  h = 0.1; Lx = 8;
T_inf = 20                     ; % Fluid temperature, C
Fixed = [1]      ; % AP with Dirichlet (EBC) condition
Given = [T0]                     ; % EBC assigned value
Free = [2 3 4]     ; % All other AP values are unknown
K = [18, -9, -6, -3; -9, 12,  3, -6;
     -6,  3, 12, -9; -3, -6, -9, 18] ...
     * k*A / Lx / 10         ; % line conduction matrix
sum(sum(K))                 %  Before EBC must = 0 always
```

```
C = [20, 10,  4,  1; 10, 12,  9,  4;
      4,  9, 12, 10;  1,  4, 10, 20] ...
      * h * P * Lx / 140     ; % Cyl. convection matrix
sum(sum(C))/h/P/Lx                          % must = 1 ?
C(4,4) = C(4,4) + h * A    ; % Add end area convection
K = K + C              ; % Combined thermal square matrix
c_inf = h*P*T_inf*Lx * [1; 1; 1; 1]/4    % Line source
c_inf (4) = c_inf (4) + h*A*T_inf     % Add end source
[T,R] = Solve_Free_DOF(K, c_inf, Free, Fixed, Given);
disp('Temperatures, C'), disp(T)     % AP temperatures
disp('Thermal reaction(s) at EBC'), disp(Fixed)
disp('are, W'), disp(R)         % Heat flow into EBC AP
```

In practice, the governing integrals would be formed by numerical integration. Gaussian quadratures should be exact for these B-splines, but not for the latter use of NURBS. The integral with the highest polynomial degree is the convection square matrix along the line (for the forced constant Jacobian):

$$[\boldsymbol{C}^r_{\text{cyl}}] = hP \int_{L^r} \{\boldsymbol{H}(x)^{\boldsymbol{T}}\} [\boldsymbol{H}(x)] dx$$

$$= hP \int_{u=0}^{u=1} \{\boldsymbol{H}(u)^{\boldsymbol{T}}\} [\boldsymbol{H}(u)] \left(|\mathbb{J}^r_{x:u}(u)|\, du\right).$$

It has the combined polynomial degree of Degree $= 3 + 3 + 0 = 6$. The 1D Gaussian rules require that Degree $\leq (2n_q - 1)$. So, $n_q = 4$ quadrature points (qps) are required in the summation:

$$[\boldsymbol{C}^r_{\text{cyl}}] = hP \sum_{q=1}^{q=4} \{\boldsymbol{H}(r_q)^{\boldsymbol{T}}\} [\boldsymbol{H}(r_q)] |\mathbb{J}_{x:u}(r_q)| |\mathbb{J}_{u:r}| w_q,$$

where the four r_q are tabulated locations in $0 \leq r \leq 1$ with associated weights of w_q. Note that the appearance of the determinant of the second Jacobian matrix, $|\mathbb{J}_{u:r}|$, is due to the necessary mapping from the spline space, $0 \leq u \leq 1$, into the tabulated integration space.

Observe, in this work, that the Gaussian data have been tabulated in the r-space, which is the same as the u-space, causing $|\mathbb{J}_{u:r}| = 1$ because only a single knot span appears in this model. If there are multiple knot spans involved, it becomes a different *constant* value of $|\mathbb{J}_{u:r}| = (u_j - u_{j+1})/1$, the knot span length divided by the length of the integration space.

Also, most IGA publications select the Gaussian data in the dimensionless space of $-1 \leq a \leq 1$. For a single knot span, the slightly different Jacobian is $|\mathbb{J}_{u:a}| = 1/2$. When multiple spans occur, the Jacobian is $|\mathbb{J}_{u:a}| = (u_j - u_{j+1})/2$.

The tabulated integration data in the unit coordinate space are extracted by the function *qp_rule_unit_Gauss.m*. The first four cases of quadrature point (qp) data in the unit space are listed in Script 7.4. For qp data in natural coordinates, use function *qp_rule_Gauss.m*. In either system, it is wise to note that the sum of correctly tabulated weights is always the measure (length, area, or volume) of the non-dimensional quadrature space.

Script 7.4. Quadrature point rules in unit coordinates

```
function [r_q, w_q] = qp_rule_unit_Gauss (n_q) % ======
%  tables of quadrature point locations and weights
%  for lines interpolated in unit coordinates.
%               0 <= r <= 1,    sum (w_q) == 1
% n_q = number of quadrature points required
% r_q = all of first parametric quadrature coordinates
% w_q = weight at all quadrature points in n_p
switch n_q            ; % identity of data set to gather
 case 1 ; % precision: exact for polynomial of degree 1
   r_q = [0.5] ; w_q = [1] ;

 case 2 ; % precision: exact for polynomial of degree 3
   r_q = [2.1132486540518711774543e-01 ...
          7.8867513459481288225457e-01] ;
   w_q = [5.0000000000000000000000e-01 ...
          5.0000000000000000000000e-01] ;

case 3 ; % precision: exact for polynomial of degree 5
   r_q = [1.1270166537925831148208e-01 ...
          5.0000000000000000000000e-01 ...
          8.8729833462074168851792e-01] ;
   w_q = [2.7777777777777777777778e-01 ...
          4.4444444444444444444445e-01 ...
          2.7777777777777777777778e-01] ;
```

```
case 4 ; % precision: exact for polynomial of degree 7
   r_q = [6.943184420297371238802 7e-02 ...
          3.300094782075718675986 7e-01 ...
          6.699905217924281324013 3e-01 ...
          9.305681557970262876119 7e-01] ;

   w_q = [1.739274225687269286865 3e-01 ...
          3.260725774312730713134 7e-01 ...
          3.260725774312730713134 7e-01 ...
          1.739274225687269286865 3e-01] ;
. . .
otherwise                          ; % update the tables
 fprintf ('See tabulated data where n_q = %i \n', n_q)
 error('ERROR quadrature rule not tabulated for n_q>7')
end                 ; % switch number of quadrature points
% end qp_rule_unit_Gauss % ============================
```

The numerical integration of this thermal problem is a somewhat longer one. It gives the exact same numerical results. In practice, the line where the analytic basis functions, and their derivative, were evaluated would be replaced by calling a function that would efficiently evaluate the NURBS needed for the same purposes. The setup segment of the script ***Fin_iga_p3_GQ.m***) is in **Script 7.5** and the numerical integration, solution, and post-processing are in **Script 7.6.**

Script 7.5. Setting up for the numerical integration of the thermal fin study

```
function [] = Fin_iga_p3_GQ ()  % conducting/convecting fin
%          The IGA 1D p=3 B-spline solution of the ODE
%              -[k A T']' + hP T - h P T_inf = 0
%       with matrix system (K + C + Ca) u = c + c_NBC via
%     numerically integrated p=3 iElement matrices from 4 AP
%    AP 1---2---3---4 set at [0 1/3  2/3 1]Lx, J_xu constant
% K = length conduction matrix, C = length convection matrix
```

```
% c  = length source due to fluid
% Ca = area end point convection matrix     (treat as scalar)
% ca = area end point source due to fluid   (treat as scalar)
%
%          ***  PHASE 1: Set control integers ***
n_B = 1;    n_e = 1;  n_g = 1;     n_n = 4; n_q = 4 ; n_S = 1;
% rows in B, iElements, DOF per AP, APs per iE,    QP, space

%  ***  PHASE 2: Set properties, coord.s, AP connections ***
% cm^2,      cm,    W/cm,       C, W/cm^2 C,      cm,          C
A = 4; P = 10; k = 3; T0 = 80;  h = 0.1; Lx = 8; T_inf = 20;
x_y = [0; 1/3; 2/3; 1]*Lx                        ; % AP coordinates
Lists (1, :) = [1, 2, 3, 4]          % list of AP per iElement
Ak =A*k; hP =h*P; hPT =h*P*T_inf; hA = h*A; hAT = h*A*T_inf;
%                    *** Dirichlet condition(s) ***
Fixed  = [1]              ; % AP with Dirichlet (EBC) condition
Free   = [2 3 4]            ; % All other AP values are unknown
Given  = [T0]                             ; % EBC assigned value
%       Allocate System Arrays  % (too small to bother width)
n_m = max(max(Lists))                          ; % number of AP
n_d = n_g * n_m             ; % system degrees of freedom (DOF)
n_i = n_g * n_n                  ; % number of DOF per iElement
S   = zeros (n_d, n_d)                  ; % system square matrix
c   = zeros (n_d,   1)                  ; % system source vector

%  *** PHASE 3: Generate matrices, scatter into system ***
[r_q, w_q] = qp_rule_unit_Gauss(n_q)  ; %Get quadrature data
span_L = 0                        ; %  initialize system length
for k = 1:n_e ; % loop over iElements    ====>> ====>> ====>>
  AP_e = Lists (k, 1:n_n)                 ; % connectivity list
  xy_e (1:n_n, 1) = x_y(AP_e(1:n_n))   ; % x coord at el APs
  Jacobian = zeros (n_S, n_S)            ; % geometry Jacobian
  c_e = zeros (n_i,   1)             ; % clear array el sources
  N_i = zeros (1,   n_n)                ; % clear integral of N
  B_q = zeros (n_B, n_i)              ; % B at quadrature point
  M_e = zeros (n_i, n_i)    ; % clear generalized mass matrix
  K_e = zeros (n_i, n_i)          ; % clear array el stiffness
  C_e = zeros (n_i, n_i)          ; % clear array el stiffness
```

Script 7.6. Numerical integration formulation of conducting/convecting fin

```
%   Numerical Integration of K_e, c_e, in 1D parametric
  span = 0                                   ; % iElement span_L
  for q = 1:n_q ; % begin quadrature loop   ---> ---> --->
    r = r_q (q) ; w = w_q (q)      ; % get integration data

%   iElement basis functions, N,    local derivatives, DLN
    N_q   = [-(r-1)^3, 3*r*(r-1)^2, -3*r^2*(r-1), r^3];
    DLN_q = [-3*(r-1)^2,9*r^2-12*r+3,-3*r*(3*r-2),3*r^2];

%   Interpolate global position of quadrature point
    xy_q = N_q * xy_e (1:n_n, :) ; % 1 x n_S, global(x,y)

%   Compute the geometry mapping data
    Jacobian = DLN_q * xy_e (1:n_n, 1)  ; % geom Jacobian
    J_det = det(Jacobian);J_Inv = inv(Jacobian);% det,inv

%   Global derivatives of N are DGN
    [B_q] = J_Inv * DLN_q    ; % Global deriv. of N, dN/dx

%   Update iElement square matrix, from K(x)
    K_e = K_e + (B_q' * Ak * B_q)*J_det*w_q(q); % conduct

%   Generalized mass matrix, from C(x)
    C_e = C_e + (N_q' * hP * N_q)*J_det*w_q(q); % convect

%   Source vector, from fluid
    c_e = c_e + N_q'*hPT*J_det*w_q(q);% line source T_inf

%   Update integral of the basis functions for post-proc.
    N_i = N_i + N_q * J_det*w_q (q);% integral for losses
    span = span + J_det * w_q (q)     ; % update iE span_L
   end  ; % for quadratures <--  <--- <--- <--- <--- <---
   C_e (4, 4) = C_e (4, 4) + hA   ; % end area convection
   c_e (4)    = c_e (4)    + hAT        ; % end area source
   K_e = K_e + C_e  ; % combine all three square matrices
```

```
 %  dof = Vector to convert elem to system eq numbers
    [dof] = Lists (k, :); % gather system equation numbers
    S (dof, dof) = S (dof, dof) + K_e ; % add to stiffness
    c (dof)      = c (dof)      + c_e ;% add to sys source
    span_L = span_L + span       ; % update iElement length
end   ; % for each k iElement in mesh   <<=== <<=== <<====
fprintf('Total knot span length = %g \n', span_L )% check
%   *** PHASE 4,5: Apply EBC, Solve, Get reaction(s) ***
[T, R]=Solve_Free_DOF (S, c, Free, Fixed, Given); % T & q

% *** PHASE 6: Post-process system and/or iElements ***
Loss_L = hP * N_i * [T - T_inf]          ; % length loss
Loss_A = hA * (T(4) - T_inf)               ; % area loss
fprintf('Total convection loss = %g \n', (Loss_L+Loss_A))
% end Fin_iga_p3_GQ ====================================
```

7.18 Numerical Evaluation of the Basis Functions

The evaluation of the basis functions at a qp by using their analytical form as previously illustrated is not practical and is inefficient. In IGA, any practical analysis means that matrix integrations are carried out numerically and this means that the basis functions and their derivative(s) must also be evaluated numerically. A practical way to do that is to evaluate the 1D r-basis functions at a tabulated 1D qp. Often, there are several (say 10–20) knot spans of the same degree in a 1D IGA analysis. An easy way to describe such regions is to create a uniform open knot vector containing a sequence of integers ranging from zero to the number of desired knot spans. For example, if the r-direction was to have five iElements, $n_{kr} = 5$, of degree two, the r-knot vector would contain 1, 2, 3, 4, 5, and would be padded with end values to declare the desired polynomial degree, like the quadratic knot vector $Ar = \{0, 0, 0, 1, 2, 3, 4, 5, 5, 5\}$.

The qps locations in an iElement (r_q) are known. At first glance, it seems like a daunting task to determine in which knot span such a point falls. Actually, given such analysis knot vector, the Cox-de Boor recursive definition does automatically generates the numerical values of the basis functions, and their parametric derivatives, at the supplied qp. Script 7.7 shows a simple script that does just that by evaluating the 1D recurrence relations in the r-direction.

Script 7.7. Numerically evaluating 1D basis functions and their derivatives

```
function [Nr, dNr] = num_basis_line (r,n_r,deg,V_r)
% This function forms basis functions & derivatives
% at parametric point (r) by first constructing the
% uniform knot vector Vr=[0,0,..,1,2,...,n_r,n_r]
%   with the end values repeated (deg+1) times.
%
% r    = parametric coordinate
% n_r  = end of sequence 1,2,3, ..., n_r
% deg  = polynomial order
% V_r  = optional pre-built knot vector, speeds up
%
% Nr  = basis functions
% dNr = derivative of basis functions
%
if (nargin == 3 )          ; % build the knot vector
  q = deg + 1;  V_r (1:q) = 0           ; % front pad
  V_r (q+1:q+n_r-1) = (1:n_r-1)        ; % middle set
  V_r (q+n_r:2*q+n_r-1) = n_r             ; % end pad
end                          ; % knot vector complete
long_r = length (V_r) - 1           ; % memory needed
Nr = zeros(1,long_r); dNr = zeros(1,long_r);% clear
%             Build basis functions
p = 0; n = length (V_r)-p-1 ; % Number of functions
for i = 1:n          ; % check if r in this knot span
  if (r >= V_r (i) && r<V_r (i+1))          ; % yes ?
    Nr(i) = 1; else; Nr(i) = 0; end         ; % end if
end                                           ; % for i
%            Build derivative values
for p = 1:deg; n = length (V_r)-p-1; % Number of BF
  for i = 1:n
    if ((V_r (i+p)-V_r (i)) == 0) ; % is i repeated
      A = 0; a = 0; else                      ; % no !
      A = (r-V_r (i))*Nr(i) / (V_r (i+p)-V_r (i));
      a = p*Nr(i) / (V_r (i+p)-V_r (i))          ; % OK
    end                           ; % end if i repeated
```

```
    if ((V_r (i+p+1)-V_r (i+1)) == 0)  ; % is i+1 ?
      B = 0; b = 0; else                     ; % no !
      B = (V_r (i+p+1)-r)*Nr(i+1) / ...
          (V_r (i+p+1)-V_r (i+1))              ; % OK
      b = p*Nr(i+1) / (V_r (i+p+1)-V_r (i+1)); % OK
    end                          ; % end if i+1 repeated
    Nr(i) = A+B; dNr(i) = a-b                ; % done
  end                                       ; % for i
end                                         ; % for p
Nr  = nonzeros(Nr) ; dNr = nonzeros(dNr)     ; % no 0
% end num_basis_line ==============================
```

In that script, the first argument gives the coordinate of a quadrature (or plotting) point, while the second gives the desired number of knot spans (iElements). The degree of the basis functions is supplied as the third argument. Those data are sufficient for the algorithm to identify the single knot span in which the coordinate falls and to return the values of the basis functions and their first derivative. The function allows for a fourth argument, which is the corresponding knot vector instead of the maximum number of iElements. The 1D function actually uses the maximum knot number to compute the knot vector and pass it along to the recurrence relations. For frequent calls to the above-mentioned function, it would be more efficient to include the optional knot vector. Post-processing graphing functions call the above-mentioned script many more times than the numerical integration of the basis functions.

7.19 NURBS basis functions

The previous sections introduced IGA procedures beyond those used in FEA because they automatically have a higher level of continuity. To keep the mathematics simple enough to illustrate those procedures in closed forms, B-splines were used as the basis functions. But, NURBS basis functions, derived from B-splines, are the true workhorses of IGA.

NURBS are rational polynomials, meaning they are the ratio of B-splines of the same order, but multiplied by different weights. They are formed in a special way to preserve the desirable mathematical property of being a 'partition of unity'. Using the notation $N_{i,p}(u)$ to represent a non-zero B-spline of degree p acting on data at analysis control point i, the weighted rational polynomial, $R_{i,p}$, defines the NURBS of the same degree at the same AP as

$$R_{i,p}(u) = \frac{N_{i,p}(u)\, w_i}{\sum_{j=1}^{i_{\max}} N_{j,p}\,(u)\, w_j},$$

which sum to

$$\sum_{i=1}^{i_{\max}} R_{i,p}(u) = \sum_{i=1}^{i_{\max}} \left(\frac{N_{i,p}(u)\, w_i}{\sum_{j=1}^{i_{\max}} N_{j,p}\,(u)\, w_j} \right) = \left(\frac{\sum_{i=1}^{i_{\max}} N_{i,p}\,(u)\, w_i}{\sum_{j=1}^{i_{\max}} N_{j,p}\,(u)\, w_j} \right) \equiv 1,$$

where the w_i are AP weights selected by the user. This means that when weighted spatial NURBS curves are to be constructed, it now requires each AP to specify its spatial coordinates and its weight value.

This section will use the non-zero quadratic B-spline basis functions defined by the analysis knot vector $\mathbb{A}n = \{0, 0, 0, 0.5, 1, 1, 1\}$ to construct the corresponding NURBS basis functions.

Quadratic B-Spline Basis Knot Vector $\mathbb{A}n = \{0, 0, 0, 0.5, 1, 1, 1\}$			
Span	Three (Non-Zero) Basis Functions per Span		
[0,0.5):	$N_1 = (1 - 2u)^2$,	$N_2 = 2u(2 - 3u)$,	$N_3 = 2u^2$
CP:	1	2	3
Limits	$u_1 = 0$		$u_2 < 0.5$
[0.5,1):	$N_4 = 2(1 - u)^2$,	$N_5 = -6u^2 + 8u - 2$,	$N_6 = (1 - 2u)^2$
CP:	2	3	4
Limits	$u_2 = 0.5$		$u_3 < 1$

Parametric derivatives $x(u)/\partial u = [\boldsymbol{N}(u)/\partial u]\{\boldsymbol{x}_{CP}\}[0, 0.5){:}\, dN_1 = (8u - 4), dN_2 = (4 - 12u), dN_3 = 4u, [0.5, 1){:}\, dN_4 = (4u - 4), dN_5 = (8 - 12u)$,

The NURBS basis function denominator polynomial is always the weighted sum of the non-zero B-spline basis functions in a knot span. Here,

$$R_{D,2}(u) = \sum_{i=1}^{i_{\max}} N_{i,p}\,(u)\, w_i = N_{1,2}(u)w_1 + N_{2,2}(u)w_2 + N_{3,2}(u)w_3 + N_{4,2}(u)w_4.$$

In the first knot span,

$$R_{D1,2}(u) = (1-2u)^2 w_1 + 2u(2-3u)w_2 + (2u^2)w_3 + 0,$$

and, in the second knot span,

$$R_{D2,2}(u) = 0 + 2(1-u)^2 w_2 + (-6u^2+8u-2)w_3 + (1-2u)^2 w_4.$$

Recall that the sum of the B-spline basis functions is always unity, so if (as given above) they are all multiplied by unity weights, this confirms that NURBS basis functions degenerate to the B-spline basis functions

$$R_{i,p}(u) = \frac{N_{i,p}(u)\, w_i}{\sum_{j=1}^{i_{\max}} N_{j,p}(u)\, w_j} \text{ for } w_j \equiv 1 \,\text{is} N_{i,p}(u).$$

Thus, the NURBS for the two knot spans are

Knot Span 1	Knot Span 2
$R_{1,2}(u) = \frac{N_{1,2}(u)w_1}{R_{D1,2}(u)} = \frac{(1-2u)^2 w_1}{R_{D1,2}(u)},$	$R_{1,2}(u) = 0,$
$R_{2,2}(u) = \frac{N_{2,2}(u)w_2}{R_{D1,2}(u)} = \frac{2u(2-3u)w_2}{R_{D1,2}(u)},$	$R_{2,2}(u) = \frac{N_{2,2}(u)w_2}{R_{D2,2}(u)} = \frac{2(1-u)^2}{R_{D2,2}(u)},$
$R_{2,2}(u) = \frac{N_{3,2}(u)w_3}{R_{D1,2}(u)} = \frac{2u^2 w_3}{R_{D1,2}(u)},$	$R_{2,2}(u) = \frac{N_{3,2}(u)w_2}{R_{D2,2}(u)} = \frac{(-6u^2+8u-2)w_3}{R_{D2,2}(u)},$
$R_{2,2}(u) = 0.$	$R_{2,2}(u) = \frac{N_{4,2}(u)w_2}{R_{D2,2}(u)} = \frac{(1-2u)^2 w_4}{R_{D2,2}(u)}.$

7.20 Summary

$n\_d \leftrightarrow n_d \equiv$ Number of system unknown degrees of freedom = $n\_g \times n\_m$,
$n\_e \leftrightarrow n_e \equiv$ Number of iElements in an IGA domain,
$n\_g \leftrightarrow n_g \equiv$ Number of generalized DOFs per node,

$n\_i \leftrightarrow n_i \quad \equiv$ Number of unknowns per sub-region or iElement $= n\_g \times n\_n$,
$n\_m \leftrightarrow n_m \equiv$ Maximum number of solution points (CP or APs),
$n\_n \leftrightarrow n_n \equiv$ Number of nodes per sub-region or iElement,
$n\_p \leftrightarrow n_p \equiv$ Dimension of parametric space,
$n\_q \leftrightarrow n_q \equiv$ Number of total quadrature points,
$n\_r \leftrightarrow n_r \equiv$ Number of sub-regions (or iElements),
$n_s \leftrightarrow n_s \equiv$ Dimension of physical space,
$\subset =$ sub − set, $\cup =$ union of sets,
Sub-region and system unknowns: $\boldsymbol{\delta}^{\boldsymbol{r}} \subset_r \boldsymbol{\delta}$,
Boolean extraction arrays: $\boldsymbol{\delta}^{\boldsymbol{r}} \equiv \boldsymbol{\beta}_{\boldsymbol{r}} \boldsymbol{\delta}$,
Geometry: $\Omega = \cup_{\boldsymbol{r}} \Omega^r \equiv$ Solution domain, $\Omega^r \equiv$ Sub−region domain.

The basis function for a NURBS curve is given by

$$R_{i,p}(u) = \frac{N_{i,p}(u)\, w_i}{\sum_{j=1}^{i_{\max}} N_{j,p}(u)\, w_j} = \frac{N_{i,p}(u)\, w_i}{W(u)},$$

$$W(u) = \sum_{j=1}^{i_{\max}} N_{j,p}(u)\, w_j.$$

The first derivative of the NURBS basis function is

$$\frac{d}{du} R_{i,p}(u) = w_i \frac{W(u) N_{i,p}^{'}(u) - W(u)^{'} N_{i,p}(u)}{W(u)^2}.$$

The basis function for a NURBS surface is given by

$$R_{i,j,p,q}(r, s) = \frac{Nr_{i,p}(r)\, Ns_{j,q}(s)\, wr_i\, ws_j}{\sum_{i=1}^{n_r} \sum_{j}^{n_s} Nr_{i,p}(r)\, Ns_{j,q}(s)\, wr_i\, ws_j}.$$

The local equation (degree of freedom) number at AP i for parameter j

$$n(i, j) = n_g * (i - 1) + j,\ 1 \le i \le n_n,\ 1 \le j \le n_g.$$

The system equation (degree of freedom) number at AP I for parameter j

$$N(I, j) = n_g * (I - 1) + j,\ 1 \leq I \leq n_m,\ 1 \leq j \leq n_g.$$

Any spatial basis function, $[\boldsymbol{N}(x)]$, over a knot span or iElement

$$\begin{aligned}\text{Displacement} : \delta(x) &= [\boldsymbol{N}(x)]\{\boldsymbol{\delta^r}\} = \{\boldsymbol{\delta^r}\}^T[\boldsymbol{N}(x)]^T\\ &= \sum_{i=1}^{i=n_i} N_i(x)\delta_i^r,\\ \text{Strain} : \frac{d\delta(x)}{dx} &= \left[\frac{d\boldsymbol{N}(x)}{dx}\right]\{\boldsymbol{\delta^r}\} = \{\boldsymbol{\delta^r}\}^T\left[\frac{d\boldsymbol{N}(x)}{dx}\right]^T\\ &= \sum_{i=1}^{i=n_i} \frac{dN_i(x)}{dx}\delta_i^r.\end{aligned}$$

Modify the system matrix equilibrium equations, 9S d = c), into sub-matrices by splitting the control displacements into known, (dk), and unknown, (du), sets

$$\{\boldsymbol{\delta}_{\text{CP}}\} \equiv \begin{Bmatrix}\boldsymbol{\delta}_u\\ \boldsymbol{\delta}_k\end{Bmatrix},$$

creating a partitioned form

$$[\boldsymbol{S}]\{\boldsymbol{\delta}_{\text{CP}}\} = \{\boldsymbol{c}\} \leftrightarrow \begin{bmatrix}[\mathrm{S}_{uu}] & [\mathrm{S}_{uk}]\\ [\mathrm{S}_{ku}] & [\mathrm{S}_{kk}]\end{bmatrix}\begin{Bmatrix}\{\boldsymbol{\delta}_u\}\\ \{\boldsymbol{\delta}_k\}\end{Bmatrix} = \begin{Bmatrix}\{\boldsymbol{c}_u\}\\ \{\boldsymbol{c}_k\} + \{\boldsymbol{r}_k\}\end{Bmatrix}.$$

The partitioned rows of unknowns requiring $[\mathrm{S}_{uu}]\{\boldsymbol{\delta}_u\} + [\mathrm{S}_{uk}]\{\boldsymbol{\delta}_k\} = \{\boldsymbol{c}_u\}$ are solved for the AP unknowns $\{\boldsymbol{\delta}_u\} = [\mathrm{S}_{uu}]^{-1}\{\{\boldsymbol{c}_u\} - [\mathrm{S}_{uk}]\{\boldsymbol{\delta}_k\}\}$.

If desired, the unknown reactions are $\{\boldsymbol{r}_k\} = [\mathrm{S}_{ku}]\{\boldsymbol{\delta}_u\} + [\mathrm{S}_{kk}]\{\boldsymbol{\delta}_k\} - \{\boldsymbol{c}_k\}$.

ODEs: Most 1D linear second-order ODEs involve a general form for $\delta(x)$ of

$$-\frac{d}{dx}\left[K(x)\,\frac{d\delta(x)}{dx}\right] + A(x)\frac{d\delta(x)}{dx} + C(x)\,\delta(x) - Q(x) = 0.$$

Typically, the coefficients in the equation represent stiffness or diffusion effects, K, advection effects, A, convection effects, C, and source or forcing terms, Q. EBCs specify the value of the unknown, $\delta(x_D) = \delta_D$, and the non-essential BCs would specify the gradient, $\partial\delta(x_N)/\partial x = \upsilon_N$.

Point sources are defined as $Q(x) \equiv \Delta(x - x_p)Q$, where

$$\Delta(x - x_p) = \begin{cases} \infty & \text{if } x = x_p \\ 0 & \text{else} \end{cases} \quad \text{and} \quad \int_{-\infty}^{\infty} \Delta(x - x_p)\, f(x)\, dx = f(x_p).$$

At a point, x_D, with a Dirichlet boundary condition, $\delta(x_D) \equiv \delta_D$, there is a reaction normal to the boundary of

$$R_D = K(x_D)\,\frac{d\delta(x_D)}{dn}.$$

Example ODEs

Elastic Bar: $-\frac{d}{dx}[E(x)A(x)\frac{d\delta(x)}{dx}] - w(x) = 0$, $\delta =$ displacement
$E =$ Elastic modulus, $A =$ area, $w =$ load per unit length, spinning $w(x) = x\,\rho\omega^2$.

Conducting Bar: $-\frac{d}{dx}[K(x)A(x)\frac{dT(x)}{dx}] - Q(x) = 0$, $T =$ Temperature
$K =$ Themal Conductivity, $A =$ Area, $Q =$ Heat generation per unit volume.

Foundation Pile: $-\frac{d}{dx}[E(x)A(x)\frac{d\delta(x)}{dx}] + k(x)(\delta - \delta_\infty) - w(x) = 0$
$k =$ Foundation stiffness, $\delta_\infty =$ Medium settlement at ∞ (usually zero).

Convecting Bar: $-\frac{d}{dx}[K(x)A(x)\frac{dT(x)}{dx}] + h(x)p(x)(T - T_\infty) - Q(x) = 0$
$h =$ Convection coeff., $p =$ Perimeter of A, $T_\infty =$ Surrounding fluid temperature.

Neutron Diffusion: $-\frac{d}{dx}\left[D(x)\frac{d\phi(x)}{dx}\right] + M(x)\phi - S(x) = 0$, $\phi =$ Neutron flux
$D =$ Diffusion Coeff., $M =$ Absorption cross-section, $S =$ Neutron source density.

Lubrication Gap: $-\frac{d}{dx}\left[\frac{h(x)^3 dP}{6\nu\, dx}\right] + \frac{d(U\,h(x))}{dx} = 0$, $P =$ Fluid pressure
$h =$ film thickness, $\nu =$ fluid viscosity, $U =$ speed of top surface.

Chapter 8

Fourth-Order Elliptic ODEs via B-Splines

8.1 Introduction

Most of the engineering applications of higher-order elliptical ordinary differential equations are of an even order and most of those are of fourth order. Fourth-order ODEs are often related to bending and buckling. A generic example is

$$\frac{d^2}{dx^2}\left[K(x)\frac{d^2v}{dx^2}\right] - \frac{d}{dx}\left[A(x)\frac{dv}{dx}\right] + C(x)v - Q(x) = 0.$$

The first and third coefficients, $K(x)$ and $C(x)$, lead to symmetric square matrix contributions, but the third term, $A(x)$ complicates the solution by giving rise to a non-symmetric square matrix contribution and is often a source of ill conditioning due to rapid changes in the solution. Of course, the scalar source term, $Q(x)$, leads to a column matrix contribution. The presence of the third term also means that the exact solution will include trigonometric or hyperbolic terms in addition to any polynomial terms. If only the first term is present, direct integration will yield a cubic polynomial as the simplest possible exact solution. The source for constant coefficients is often a linear polynomial, resulting in the exact solution of the ODE being a quintic polynomial. Those insights provide guidance on what polynomial degree should be picked in planning an FEA or IGA approximation of the solution.

When contrasting IGA and FEA approaches to this class of problems, it is important to remember the mathematical definitions of the two essential and the two non-essential boundary conditions, possibly acting at points in the domain for each application. The two possible Dirichlet (essential) conditions for a fourth-order ODE are

$$v(x_D) = v_D \quad \text{and/or} \quad \frac{dv}{dx}(x_D) = s_D$$

at one or more locations. The two possible Neumann (non-essential) conditions are

$$\frac{\partial^2 v}{\partial x^2}(x_N) = m_N \quad \text{and/or} \quad \frac{\partial^3 v}{\partial x^3}(x_N) = f_N.$$

The first Neumann leads to a jump in the slope of the solution and the second one leads to a jump in the curvature (second derivative) of the solution.

The less common sixth-order ODEs often appear when modeling a problem with a rapidly changing result at the end(s) of the solution domain (Akin and Wooten, 1978). For a sixth-order PDE, the three possible Dirichlet (essential) conditions are

$$v(x_D) = v_D \quad \text{and/or} \quad \frac{dv}{dx}(x_D) = s_D \quad \text{and/or} \quad \frac{\partial^2 v}{\partial x^2}(x_D) = m_D$$

at one or more locations. The three possible Neumann (non-essential) conditions are

$$\frac{\partial^3 v}{\partial x^3}(x_N) = f_N \quad \text{and/or} \frac{\partial^4 v}{\partial x^4}(x_N) = g_N \quad \text{and/or} \quad \frac{\partial^5 v}{\partial x^5}(x_N) = h_N.$$

FEA can easily be used to formulate solutions for this order ODEs, but in general an IGA study requires new approaches for imposing either class of boundary condition. It is easy to create an exact 1D IGA solution for simplified models where only the value of the solution is specified at two points and where the source term is chosen to make the exact result a cubic polynomial.

Next, the integral forms for a general fourth-order ODE will be developed, and a simplified 'global approximation' will used to illustrate new boundary condition issues associated with IGA studies in general.

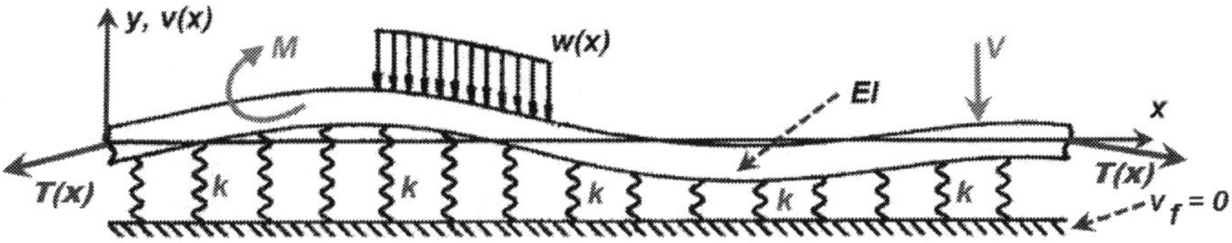

Figure 8.1. A beam column on an elastic foundation

8.2 Beam columns

The three common engineering problems of common beam deflections, beams on an elastic foundation, and the buckling of beam columns are addressed in the following ODE:

$$\frac{d^2}{dx^2}\left[\mathrm{EI}(x)\frac{d^2v}{dx^2}\right] - \frac{d}{dx}\left[N(x)\frac{dv}{dx}\right] + k(x)v - w(x) = 0,$$

which is subject to any or all of the four possible boundary conditions. This model is illustrated in Fig. 8.1 showing the transverse displacement, $v(x)$, of a beam on an elastic foundation. The beam is transversely supported by a continuous series of foundation springs, of stiffness $k(x)$ per unit length. The other end of each foundation spring is fixed (which is an implied Dirichlet boundary condition). The differential equation requires the transverse equilibrium of a beam, of flexural stiffness $\mathrm{EI}(x)$ resting on the foundation, with a transverse load per unit length of $w(x)$, and optionally subjected to an axial tensile axial load of $N(x)$. The Neumann conditions also allow for a point couple, M, and/or a transverse shear force, V. When the foundation effect is present, the structure is usually called a beam on an elastic foundation (BOEF), or a beam column on an elastic foundation if the axial load is present. This general class of fourth-order ODE has been explored by Hetenyi (1941) where he supplied numerous exact analytic solutions, generally involving a combination of hyperbolic and trigonometric functions.

Each beam has a cross section with two principle axes associated with the second moments of inertia, I_{zz}, I_{yy}, and I_{zy}. For the coordinate system shown in Figure 8.1, the principle inertia is the former: $I \equiv I_{zz} = \int_A y^2 dA$. Its defining axis is the z-axis coming out of the paper and lies in the horizontal plane. Conversely, the minor inertia y-axis lies in the vertical plane of the paper. Any changes in the two inertia axis directions becomes important if the beam is part of a space frame member located in general three-dimensional space, as will be discussed later.

The distributed load per unit length can include point transverse shear loads, V, by using the Dirac Delta distribution. Likewise, employing a doublet distribution of rotation $\ll 1$), $\theta(x) = v'(x)$, the bending moment $M(x) = \mathrm{EI}v''(x)$, and the transverse shear force $V(x) = \mathrm{EI}v'''(x)$ are defined. Engineers designing beams are usually interested in the last two quantities since they define the stress levels and the material failure criteria. This means that a model used in designing a beam should have accurate third derivatives, or a fine mesh. The sign conventions are that the position, x, is positive to the right; the deflection, v, point forces, P, and the line load, w, are positive in the y-direction (upward); and the slopes and moments are positive in the counter-clockwise direction.

The axial force, $T(x)$, is positive when it is in tension, which stabilizes the system, and negative when in compression, which destabilizes the system. Those two cases are typically called a 'tensioned beam' and a 'beam column', respectively. When the axial force is in tension, the solution to the homogeneous equation tends to smoothly decay away from the supports. Conversely, when the axial force is in compression, the solution to the homogeneous solution oscillates and decays very slowly. Frequently, the axial force of interest is an unknown global constant to be computed. Then, the ODE becomes an eigen-problem to determine the axial force that causes buckling.

8.3 Integral form

To apply the Galerkin weak form, the governing ODE is multiplied by $v(x)$ and the integral over the length of the beam is set to zero. The highest (4th) derivative term is always present, and needs to be integrated twice by parts to reduce the inter-element continuity requirement. That integral becomes

$$I_E = \int_0^L v \frac{d^2}{dx^2}\left(E(x)I(x)\frac{d^2 v}{dx^2}\right)dx,$$

$$I_E = \left[v\frac{d}{dx}\left(E(x)I(x)\frac{d^2 v}{dx^2}\right)\right]_0^L - \int_0^L \frac{dv}{dx}\frac{d}{dx}\left(E(x)I(x)\frac{d^2 v}{dx^2}\right)dx,$$

$$\begin{aligned} I_E &= \left[v \frac{d}{dx} \left(E(x)I(x) \frac{d^2 v}{dx^2} \right) \right]_0^L - \left[\frac{dv}{dx} \left(E(x)I(x) \frac{d^2 v}{dx^2} \right) \right]_0^L \\ &\quad + \int_0^L \frac{d^2 v}{dx^2} \left(E(x)I(x) \frac{d^2 v}{dx^2} \right) dx. \end{aligned}$$

The integration by parts brings the two possible non-essential boundary conditions, or reactions, into the integral form. Those terms are the transverse shear force, $d(\mathrm{EI}\, d^2v/dx^2)/dx$, and the moment, $\mathrm{EI}\, d^2v/dx^2$. The first NBC term is the product of the displacement and the transverse shear force at the end points, while the second term is the product of the slope (angle) and the point moment at the end points. Both of those products are definitions of mechanical work. Thus, the contributions from the fourth derivative term in the ODE for $N = 0$ are

$$I_E = [vV(x)]_0^L - \left[\frac{dv}{dx} M(x) \right]_0^L + \int_0^L \frac{d^2 v}{dx^2} \left(E(x)I(x) \frac{d^2 v}{dx^2} \right) dx.$$

The second integral term is originally non-symmetric because it also has unbalanced derivative orders. Creating a symmetric form requires integration by parts:

$$I_T = \int_0^L v \left(-T(x) \frac{d^2 v}{dx^2} \right) dx = \left[v \left(-T(x) \frac{dv}{dx} \right) \right]_0^L + \int_0^L \frac{dv}{dx} T(x) \frac{dv}{dx} dx.$$

For small deflections, the slope is approximately equal to its tangent. Thus, the boundary term $T dv/dx \approx T_y$ is the transverse (y-) component of the axial force in the deformed beam. The final integral form contains second derivatives (d^2v/dx^2) as its highest derivative term. The presence of second derivatives in the integral form means that calculus requires the elements to have inter-element continuity of the deflection and the slope, $v(x)$ and $\theta(x) = v'(x)$. Such adjacent regions are said to have a continuity of C^1. The complete governing Galerkin integral form of the beam column equilibrium equation is

$$\begin{aligned} I &= \left[v \left\{ \frac{d}{dx} \left(E(x)I(x) \frac{d^2 v}{dx^2} \right) - T(x) \frac{dv}{dx} \right\} \right]_0^L \\ &\quad - \left[\frac{dv}{dx} \left(E(x)I(x) \frac{d^2 v}{dx^2} \right) \right]_0^L + \int_0^L \frac{d^2 v}{dx^2} \left(E(x)I(x) \frac{d^2 v}{dx^2} \right) dx \end{aligned}$$

$$+\int_0^L \frac{dv}{dx}T(x)\frac{dv}{dx}dx + \int_0^L \frac{dT(x)}{dx}\frac{dv}{dx}dx$$
$$+\int_0^L vkv\,dx - \int_0^L vv_\infty dx - \int_0^L vw(x)dx = 0. \tag{8.1}$$

Note that this integral form is called the weak form because now the approximating basis functions only need a non-vanishing second derivative compared to a fourth derivative in the ODE strong solution. To have at least a piecewise continuous solution, the first derivative needs to be continuous across regional boundaries. In 1D FEA studies, that is easy to achieve using Hermite polynomials. But, that level of continuity is quite difficult to achieve in higher-dimension FEA studies. The guaranteed (default) continuity of splines and NURBS between regions of parametric patches in all dimensions is one of the reasons why IGA offers benefits over FEA.

The first non-essential boundary condition states that when $T \neq 0$, the consistent definition of the transverse shear force is

$$V(x) = \{d(\mathrm{EI}(x)d^2v/dx^2)/dx - T(x)dv/dx\} \equiv V_{\mathrm{EI}} - V_T.$$

Equation (8.1) must be satisfied along with any essential boundary conditions.

8.4 iElement arrays

As before, the symbol $[\boldsymbol{N}^{\boldsymbol{r}}(u)]$ will be used to represent the assumed basis function representation of the unknown within a region, r, and thus can stand for B-Splines, NURBS, or T-splines in a parametric patch region. Then, the transverse deflection of the member is

$$v(u) = [\boldsymbol{N}^{\boldsymbol{r}}(u)]\{\mathbf{AP}_{\boldsymbol{v}}^{\boldsymbol{r}}\},$$

where the values of the transverse displacement, v, at the Control Points of a region are $\{\mathbf{AP}_{\boldsymbol{v}}^{\boldsymbol{r}}\}$. Then, any approximate spatial derivative of those displacements is obtained from the derivatives of those basis functions. The slope of the beam in a region becomes

$$v'(x) = \theta(x) = dv/dx = dv/du\ du/dx = d[\boldsymbol{N}^{\boldsymbol{r}}(u)]/dx\{\mathbf{AP}_{\boldsymbol{v}}^{\boldsymbol{r}}\},$$

and so on up to the third derivative needed to establish the beam design criteria once the generalized analysis control point unknowns, $\{\mathbf{AP}_{\boldsymbol{v}}\}$, have been computed.

When these assumptions are substituted into the equivalent integral forms developed earlier, the unknowns factored out of the integrals, the integrations completed and the resulting regional and system matrices, define the new governing form before the displacement boundary conditions are imposed.

To simplify the notation, let the system unknown displacements be $\{\mathbf{AP}_{\boldsymbol{v}}\} \equiv \boldsymbol{\delta}$, so in matrix notation, the equilibrium equation becomes

$$I = \boldsymbol{\delta}^{\boldsymbol{T}}\boldsymbol{c}_{\mathbf{NBC}} + \boldsymbol{\delta}^{\boldsymbol{T}}\boldsymbol{S}_{\mathbf{EI}}\boldsymbol{\delta} + \boldsymbol{\delta}^{\boldsymbol{T}}\boldsymbol{S}_{\boldsymbol{T}}\boldsymbol{\delta} + \boldsymbol{\delta}^{\boldsymbol{T}}\boldsymbol{S}_{\boldsymbol{K}}\boldsymbol{\delta} - \boldsymbol{\delta}^{\boldsymbol{T}}\boldsymbol{S}_{\boldsymbol{u}}\boldsymbol{\delta} \\ - \boldsymbol{\delta}^{\boldsymbol{T}}\boldsymbol{c}_{\infty} - \boldsymbol{\delta}^{\boldsymbol{T}}\boldsymbol{c}_{\boldsymbol{w}} = 0.$$

This leads to the matrix system

$$[\boldsymbol{S}_{\mathbf{EI}} + \boldsymbol{S}_{\boldsymbol{T}} + \boldsymbol{S}_{\boldsymbol{u}} + \boldsymbol{S}_{\boldsymbol{k}}]\boldsymbol{\delta} = \boldsymbol{c}_{\mathbf{NBC}} + \boldsymbol{c}_{\boldsymbol{P}} + \boldsymbol{c}_{\infty} + \boldsymbol{c}_{\boldsymbol{w}}, \tag{8.2}$$

where the subscript u emphasizes that a variable axial tension leads to an non-symmetric stiffness matrix. The system matrices are assembled from the corresponding element matrices of

$$\boldsymbol{S}^{\boldsymbol{r}}_{\mathbf{EI}} = \int_{L^r} \frac{d^2\boldsymbol{N}(x)}{dx^2}^T \mathrm{EI}^r \frac{d^2\boldsymbol{N}(x)}{dx^2} dx \equiv \int_{L^r} \boldsymbol{B}^e(x)^T\, \mathrm{EI}^r \boldsymbol{B}^r(x) dx,$$

$$\boldsymbol{B}^r(x) \equiv \frac{d^2\boldsymbol{N}(x)}{dx^2}, \quad \boldsymbol{S}^{\boldsymbol{r}}_{\boldsymbol{T}} = \int_{L^r} \frac{d\boldsymbol{N}(x)}{dx}^T T(x)^r \frac{d\boldsymbol{N}(x)}{dx} dx,$$

$$\boldsymbol{S}^{\boldsymbol{r}}_{\boldsymbol{u}} = -\int_{L^r} \boldsymbol{N}(x)^T \frac{dT}{dx}\frac{d\boldsymbol{N}(x)}{dx} dx, \quad \boldsymbol{S}^{r}_{\boldsymbol{k}} = \int_{L^r} \boldsymbol{N}(x)^T k^r \boldsymbol{N}(x) dx,$$

$$\boldsymbol{c_w}^r = \int_{L^r} \boldsymbol{N}(x)^T w^r(x) dx, \quad \boldsymbol{c}^r_{\infty} = \int_{L^r} \boldsymbol{N}(x)^T v^r_{\infty}\, dx. \tag{8.3}$$

In a vast majority of cases (excluding long drill stings), the axial force within a region, $T(x)$, can be taken as a constant, so $dT/dx = 0$, and the non-symmetric contribution, $\boldsymbol{S}^{\boldsymbol{r}}_{\boldsymbol{u}}$, does not appear.

The new matrix, $\boldsymbol{S}^{r}_{\boldsymbol{T}}$, which is dependent on the axial force stress, and thus the axial stress, is usually called a 'geometric stiffness matrix' in the finite element literature. The bending stiffness matrix is $\boldsymbol{S}^{r}_{\mathbf{EI}}$, the foundation stiffness matrix is $\boldsymbol{S}^{e}_{\boldsymbol{k}}$, while the axial force gradient matrix has no common name.

Referring to the two integration by contributions in (8.1), the first row of the reaction/natural boundary condition vector, $\boldsymbol{c}_{\mathbf{NBC}}$, contains the transverse shear $\{V_{\mathrm{EI}}(0) - V_T(0)\}$, the second row contains the point bending moment $\{-M(0)\}$, the next-to-last row contains the transverse shear force $\{-V_{\mathrm{EI}}(L) + V_T(L)\}$, and the last row contains the end point bending moment $\{M(L)\}$.

Any distributed transverse load per unit length is usually defined by the value of that load input at the geometric control points and approximated with a Lagrangian interpolation function or a B-spline basis function, say $\boldsymbol{h}^w(x)$. Assume that $w^r(x) = \boldsymbol{h}^w(x)\boldsymbol{w}^r$. If that approach is used, the resultant iElement load vector changes into an alternate rectangular integral:

$$\begin{aligned}
\boldsymbol{c}_{\boldsymbol{w}}^r &= \int_{L^r} \boldsymbol{N}(x)^T \boldsymbol{h}^w(x)\boldsymbol{w}^r dx \\
&= \left[\int_{L^r} \boldsymbol{N}(x)^T \boldsymbol{h}^w(x) dx\right] \boldsymbol{w}^r \equiv \boldsymbol{R}^r \boldsymbol{w}^r \\
\boldsymbol{R}^e &\equiv \int_{L^r} \boldsymbol{N}(x)^T \boldsymbol{h}^w(x) dx. \qquad (8.4) \\
n_i \times n_n &= (n_i \times 1)(1 \times n_n).
\end{aligned}$$

Temperature changes along the length of the beam do not affect the transverse deflections or generalized forces. However, a temperature change, ΔT, through the thickness, t^e, from *top to bottom* does affect them. It is possible to include those thermal effects in a beam model. Thermal bending is not common, but the result of including it gives another load vector acting on the element:

$$\boldsymbol{c}_{\boldsymbol{\alpha}}^r = \int_{L^r} \boldsymbol{B}^{r^T} \mathrm{EI}^r(x) \frac{\alpha^r \Delta T(x)^r}{t(x)} dx = \frac{\alpha^r \Delta T^r \mathrm{EI}^r}{t^r} \int_{L^r} \boldsymbol{B}^{r^T}(x) dx, \qquad (8.5)$$

where α^r is the coefficient of thermal expansion of the material in the region.

The foundation stiffness matrix, $\boldsymbol{S}_{\boldsymbol{k}}^r$, has the form previously called a generalized mass matrix. If this model is later extended to be time dependent, then the literal mass matrix will be required.

In this case, it is

$$\boldsymbol{m}^r = \int_{L^r} \boldsymbol{N}(x)^T \rho^r A^r \boldsymbol{N}(x) dx, \tag{8.6}$$

where ρ^r is the mass density of the material and A^e is the cross-sectional area of the beam. Of course, the volume of the beam is $A^r L^r$ and its total scalar mass is $m = \rho^r A^r L^r$.

8.5 Derivative discontinuities

Historically, mathematical splines model the smooth physical deformed shapes of thin elastic beams under transverse loads, for a fourth-order ODE. So, it is logical to obtain IGA solutions of fourth-order ODEs using B-splines or NURBS because they are expected to match the solution shape (here deflections) quite accurately. However, an analyst in a field of application should have some expectations of the behavior of the solutions and their post-processing needs. Those post-processing needs almost always include computing various derivatives of the solution.

For example, while some beam designs are governed by their maximum deflection (AP solution value) in high-precision machines, like 3D printers, a vast majority of structural beams are designed by limits to their maximum moment (second partial derivative) and their maximum transverse shear force (third partial derivative) or combinations of the two. For many beam models (fourth-order ODEs), the external source terms (forces) cause discontinuities in the important higher derivatives recovered in post-processing. Such discontinuities are illustrated in Fig. 8.2 where theoretical point sources produce important discontinuities in the shear forces in the beams.

To address the topic of derivative discontinuities and discovery a naïve approach to the leftmost beam in Fig. 8.2 will be discussed. Then, the model will be enhanced to demonstrate the power of the IGA model.

It is known that members like those in Fig. 8.2 are typically cubic polynomials, while varying line loads per unit length are commonly quartic or quintic polynomials. Thus, one can begin the study with a cubic B-spline over a single beam span length of L. The two pined end supports mean that the deflection at those points is zero (the

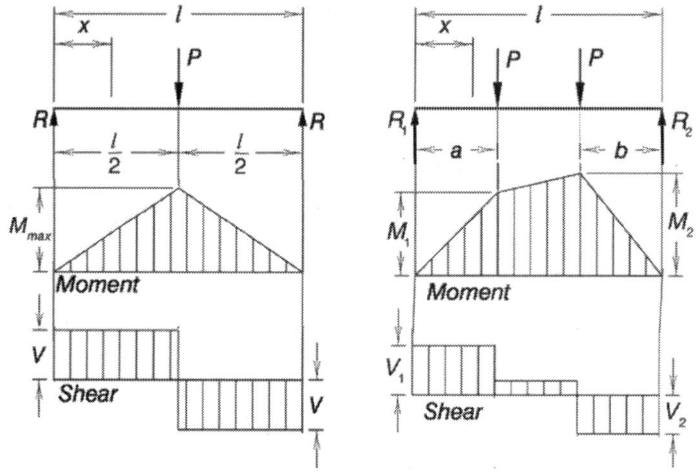

Figure 8.2. Common beam load Cases #5 and #10, from [AISC 1989]

Dirichlet conditions). The bending stiffness EI is taken as a constant. The point load, and the two end reaction forces, is a special form of the load per unit length, $w(x)$, appearing in the ODE:

$$\frac{d^2}{dx^2}\left[\mathrm{EI}(x)\frac{d^2v}{dx^2}\right] - w(x) = 0, \quad v(0) = 0 = v(L).$$

The usual weak form resultant of the load per unit length $\{\boldsymbol{c}^r_{\boldsymbol{w}}\} = \int_{L^r} [\boldsymbol{N}(x)]^T w^r(x)dx$ has the point force represented by the Dirac Delta distribution:

$$w^r(x) = P\ \delta(x - x_P),$$

where $x_P = L/2$ is the location of the point load. The Dirac Delta distribution has properties that

$$\delta(x - x_P) = \begin{cases} 1 & \text{if } x = x_P \\ 0 & \text{otherwise} \end{cases},$$

and for any function of position $F(x)$,

$$\int_{-\infty}^{+\infty} F(x)\delta(x - x_P)dx \equiv F(x_P).$$

Therefore, the distributed load resultant changes to

$$\begin{aligned}\{\boldsymbol{c}_{\boldsymbol{P}}^{r}\} &= \int_{L^r} [\boldsymbol{N}(x)]^T w^r(x) dx \\ &= \int_{L^r} [\boldsymbol{N}(x)]^T P\ \delta(x - x_P) dx = P[\boldsymbol{N}(x_P)]^T, \\ \{\boldsymbol{c}_{\boldsymbol{P}}^{r}\} &= P[\boldsymbol{N}(u_P)]^T \quad \text{with} \quad x_P = [\boldsymbol{N}(u_P)]\{\boldsymbol{x}^{\boldsymbol{r}}\}\end{aligned}$$

In other words, given a point source in physical, x_P, space, the parametric geometry mapping is solved for the corresponding point in parametric space, u_P, and then the analysis basis functions are evaluated there, $[\boldsymbol{N}(u_P)]$, and multiplied by the value of the source to determine the portion of the source to be distributed to each analysis control point:

$$\{\boldsymbol{c}_{\boldsymbol{P}}^{r}\} = P \begin{Bmatrix} N_1(u_P) \\ N_2(u_P) \\ \vdots \end{Bmatrix}.$$

Since the basis functions are a partition of unity, $\sum_i N_i(u_P) \equiv 1$, the sum of all the individual AP loads equals the applied load. This approach smears a point source over all the AP control values connected to an iElement. There is a useful special case where the point source is placed at either end knot. Then, that basis function is unity, all others are zero, and the full point source goes to that AP.

That observation *strongly suggests* that an analyst should begin an IGA study by placing an AP there, or a panel interface, for the most accurate solutions on either side of the point source. This is also true for locations where a load per unit length starts and stops.

8.6 Single knot span cubic model

For the cubic basis functions defined by the analysis knot vector $\mathbb{A}u = \{0, 0, 0, 0, 1, 1, 1, 1\}$, Appendix 1 gives them as

$$N(u) = [-(u-1)^3,\ 3u(u-1)^2, -3u^2(u-1), 2u^3]$$

with $dN(u)/du = [-3(u-1)^2, (9u^2 - 12u + 3), -3u(3u-2), 3u^2]$,

$$d^2N(u)/du^2 = [(6-6u), (18u-12), (6-18u), 6u],$$
$$d^3N(u)/du^3 = [-6, 18, -18, 6].$$

The third derivative equation serves as a reminder that the pth derivative of any degree p polynomial is always constant. For this application, it means that the shear force in an iElement is constant, which is extremely rare in practical applications. From Fig. 8.2, the problem at hand is that the model needs to be able to capture at least multiple constant shear forces. To capture the expected solution(s), one approach could be to introduce another iElement so that the concentrate point load falls on the last (and first) knots of both iElements. That would be a classical FEA approach. A different IGA approach offers to capture the needed behavior and enrich the approximation. The present cubic polynomial is $C^{p-1} = C^2$ continuous, which gives the constant shear force, $V(x)$, along with a continuous bending moment, $M(x)$.

Consider adding one or more interior knots to the knot vector with one at the location of the discontinuous point source, $x = L/2$ and (by inspection) $u = 0.5$. Try the knot vector $\mathbb{A}u = \{0, 0, 0, 0, 0.5, 1, 1, 1, 1\}$ with a new knot and a new analysis control point in the center. The related items of index space, AP locations, and parametric space are sketched in Fig. 8.3.

There are now two knot spans: [0, 0.5) using AP 1–4 and [0.5, 1) using AP 2–5. Appendix 1 gives the basis functions as

$$N(u) = [-(2u-1)^3, 2u(7u^2 - 9u + 3), -2u^2(4u-3)2u^3,$$
$$d^3N(u)/du^3 = [-48, 84, -48, 12].$$

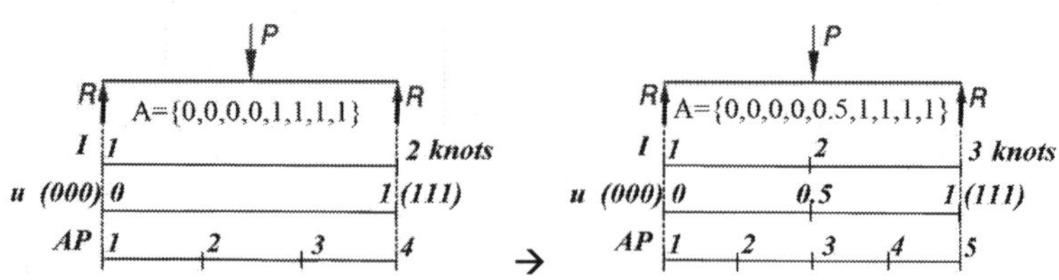

Figure 8.3. Index space (knots), parametric space, and analysis control points

For the first knot span [0, 0.5) and over the second knot span [0.5,1),

$$N(u) = [-(2u-2)(u-1)^2,\ 2(4u-1)(u-1)^2, \ldots$$
$$(-14u^3 + 24u^2 - 12u + 2), (2u-1)^3]$$
$$d^3N(u)/du^3 = [-12, 48, -84, 48].$$

Clearly, the third derivatives now are different constants on both sides of the knot inserted at the point source. Knowledge of the field of application makes this change a more realistic model.

In this simple application, analytic integration is possible instead of the quadrature required when using NURBS as the basis functions. For the first knot span, the bending stiffness matrix

$$\boldsymbol{K}^r_{\mathbf{EI}} = \int_{L^r} \frac{d^2\boldsymbol{N}(x)}{dx^2}^T \mathrm{EI}^r \frac{d^2\boldsymbol{N}(x)}{dx^2} dx,$$

with a linear geometry mapping giving $dx = (L^r/2)du$ and making the usual small beam slope approximation,

$$\frac{d^2(\,)}{dx^2} = \frac{1}{(L^r/2)^2}\frac{d^2(\,)}{du^2}$$

this stiffness matrix becomes

$$\boldsymbol{K}^1_{\mathbf{EI}} = \mathrm{EI}\int_0^{0.5} \frac{4}{L^2}\frac{d^2\boldsymbol{N}(u)^T}{du^2}\frac{4}{L^2}\frac{d^2\boldsymbol{N}(u)}{du^2}(L/2)du$$
$$= \frac{8\mathrm{EI}}{L^3}\int_0^{0.5} \frac{d^2\boldsymbol{N}(u)}{du^2}^T \frac{d^2\boldsymbol{N}(u)}{du^2} du,$$
$$\boldsymbol{K}^1_{\mathbf{EI}} = \frac{6\mathrm{EI}}{L^3}\begin{bmatrix} 16 & -22 & 4 & 2 \\ -22 & 31 & -7 & -2 \\ 4 & -7 & 4 & -1 \\ 2 & -2 & -1 & 1 \end{bmatrix}.$$

The contribution of the point force, P, to the matrix system has changed because it now falls on the third analysis control point. Viewing it as an iElement source, its application point in parametric space is outside the first knot span since $[0, 0.5) < u_P = 0.5$, but does fall into the second knot span $[0.5 = u_P, 1)$. Therefore, $\{\boldsymbol{c}^1_{\boldsymbol{P}}\} = \{\mathbf{0}\}$, and the point source contributions from the second span (with the shown force acting in the negative y-direction) are

$$\{\boldsymbol{c}_{\boldsymbol{P}}^2\} = -P\begin{Bmatrix} N_1(u_P) \\ N_2(u_P) \\ \vdots \end{Bmatrix} = -P\begin{Bmatrix} N_1(0.5) \\ N_2(0.5) \\ \vdots \end{Bmatrix} = \frac{-P}{4}\begin{Bmatrix} 1 \\ 2 \\ 1 \\ 0 \end{Bmatrix}.$$

The second knot span stiffness is

$$\boldsymbol{K}_{\mathbf{EI}}^{\mathbf{2}} = \frac{6\,\mathrm{EI}}{L^3}\begin{bmatrix} 1 & -1 & -2 & 2 \\ -1 & 4 & -7 & 4 \\ -2 & -7 & 31 & -22 \\ 2 & 4 & -22 & 16 \end{bmatrix}.$$

Assembling (scattering) the matrices from the two iElements into the five dofs in the system matrices gives

$$\boldsymbol{K}_{\mathbf{EI}} = \frac{6\,\mathrm{EI}}{L^3}\begin{bmatrix} 16 & -22 & 4 & 2 & 0 \\ -22 & 32 & -8 & -4 & 2 \\ 4 & -8 & 8 & -8 & 4 \\ 2 & -4 & -8 & 32 & -22 \\ 0 & 2 & 4 & -22 & 16 \end{bmatrix}$$

$$= \frac{12\,\mathrm{EI}}{L^3}\begin{bmatrix} 8 & -11 & 2 & 1 & 0 \\ -11 & 16 & -4 & -2 & 1 \\ 2 & -4 & 4 & -4 & 2 \\ 1 & -2 & -4 & 16 & -11 \\ 0 & 1 & 2 & -11 & 8 \end{bmatrix}.$$

Scattering the point source gives $\{\boldsymbol{c}_{\boldsymbol{P}}\} = P[0,\ 1,\ 2,\ 1,\ 0]^T/4$. The equilibrium matrix system

$$[\boldsymbol{K}_{\mathbf{EI}}]\{\boldsymbol{\delta}\} = \{\boldsymbol{c}_{\boldsymbol{P}}\}$$

is singular until modified for the two Dirichlet (essential) conditions of zero displacements at the AP 1 and 5, $\delta_1 = 0 = \delta_5$. After those required modifications, only the displacements at AP 2, 3, and 4 are active. This reduced non-singular system is

$$\frac{12\,EI}{L^3}\begin{bmatrix}16 & -4 & -2\\ -4 & 4 & -4\\ -2 & -4 & 16\end{bmatrix}\begin{Bmatrix}\delta_2\\ \delta_3\\ \delta_4\end{Bmatrix} = \frac{-P}{4}\begin{Bmatrix}1\\ 2\\ 1\end{Bmatrix}.$$

The three interior analysis control point displacement values are

$$\begin{Bmatrix}\delta_2\\ \delta_3\\ \delta_4\end{Bmatrix} = \frac{-\mathrm{PL}^3}{96\,\mathrm{EI}}\begin{Bmatrix}1\\ 3\\ 1\end{Bmatrix}.$$

At this point, all of the analysis point control displacements are downward,

$$\{\boldsymbol{\delta}\} = -\frac{\mathrm{PL}^3[0\ \ 1\ \ 3\ \ 1\ \ 0]^T}{96\,\mathrm{EI}}.$$

But, those are NOT the physical displacements of the beam (except at the end essential boundary condition points). To find the physical displacement at $x = L/2$, substitute $u = 0.5$ in the second span basis functions:

$$\delta(L/2) = [4/8 - 12/4 + 9/2 - 1]\frac{-\mathrm{PL}^3}{96\,\mathrm{EI}}\begin{Bmatrix}0\\ 1\\ 3\\ 1\\ 0\end{Bmatrix} = \frac{-\mathrm{PL}^3}{48\,\mathrm{EI}},$$

which is the exact answer.

The study does not end at this point. A practical use requires one to compute the distribution of the bending moment, $M(x) = \mathrm{EI}d^2\delta(x)/dx^2$, and the transverse shear force, $V(x) = \mathrm{EI}d^3\delta(x)/dx^3$. The slope is usually not important, other than to verify that it is $\ll 1$ radians, as assumed. Before computing the moments, inspection of Fig. 8.3 shows that the correct first and second iElement shears must be half of the point load, $R = P/2$, and acting upward. Gathering the first four analysis control point displacements, and multiplying them by the constant third derivative of the left iElement, listed previously, gives

$$V\left(x < \frac{L}{2}\right) = V\left(u < \frac{1}{2}\right) = \mathrm{EI}\left(\frac{1}{L^r}\right)^3 d^3\boldsymbol{N}(u)/du^3\{\boldsymbol{\delta}\},$$

$$= \frac{8\,\mathrm{EI}}{(L^r)^3}[-48, 84, -48, 12]\frac{-PL^3}{96\,\mathrm{EI}}\begin{Bmatrix} 0 \\ 1 \\ 3 \\ 1 \end{Bmatrix},$$

$$V\left(x < \frac{L}{2}\right) = \frac{-8P}{96}(-48(0) + 84(1) - 48(3) + 12(1)) = \frac{+48\,P}{96}.$$

Likewise, for the second iElement,

$$V\left(x > \frac{L}{2}\right) = \frac{-8P}{96}(-12(1) + 48(3) - 84(1) + 48(0)) = \frac{48\,P}{96}.$$

Referring back to Fig. 8.3, these results make physical sense. Approaching the beam from the left, the air transmits no force. At the leftmost point of the beam, equilibrium requires that the reaction introduce a vertical (+) force $V(x)$ of $+P/2$ into the beam. That shear force in the beam remains constant until another external force is encountered. That happens with the P force at the center which changes the shear force beyond that to $(+P/2 - P = -P/2)$ to the shear force $V(x)$ to the right of center up to the next force, which is the right reaction of $+P/2$. This converts the following shear force, in the air, to $(-P/2 + P/2 = 0)$, as required.

Since the ODE is linear, the solution for multiple load cases would be visualized by superimposing the two or more of the 42 shear and moment diagrams tabulated from common load cases in the American Iron and Steel Construction (AISC) handbook.

8.7 Common beam problems

Historically, common FEA cubic Hermite beam elements are very widely used to establish the moment, $M(x)$, and transverse shear force, $V(x)$, in beam structures by placing a large number of them (say 50–100) within each continuous span so that the many piecewise constant shear force (third derivative) steps reasonably approximate the true smooth distribution of the transverse shear force. The common practice of using cubic polynomials for beams is not recommended. Instead, it is recommended that an IGA or FEA beam study use a quintic (degree 5) polynomial that can yield the exact solution to the most common beam load cases included in the AISC Beam Tables (1989). A quintic Hermite interpolation FEA approach can easily handle any of the common boundary conditions, jumps, and source terms in beam models. Indeed, some FEA solutions can easily be made exact everywhere. IGA quintic basis functions can also yield exact solutions to many beam problems, but they can require more thought in defining the analysis basis functions.

From Appendix 1, the analytic forms of six quintic B-spline basis functions, and their local derivatives, on a single span are defined by the knot vector

$$\begin{aligned}
\mathbb{A}u &= \{0,0,0,0,0,0,1,1,1,1,1,1\}\\
[\boldsymbol{N}(u)] &= [(1-u)^5, 5u(1-u)^4, -10u^2(1-u)^3,\\
&\quad 10u^3(1-u)^2, 5u^4(1-u)u^5],\\
\frac{d[\boldsymbol{N}(u)]}{du} &= [-5(1-u)^4, 5(5u-1)(u-1)^3, -10u(5u-2)(u-1)^2, \ldots\\
&\quad 10u^2(5u^2-8u+3), -5u^3(5u-4), 5u^4]\\
\frac{d^2[\boldsymbol{N}(u)]}{du^2} &= [-20(u-1)^3, 20(5u-2)(u-1)^2,\\
&\quad (-200u^3+360u^2-180u+20), \ldots\\
&\quad + 20u(10u^2-12u+3), -20u^2(5u-3), 20u^3]\\
\frac{d^3[\boldsymbol{N}(u)]}{du^3} &= [-60(u-1)^2, (300u^2-480u+180),\\
&\quad (-600u^2+720u-180), \ldots (600u^2-480u+60),\\
&\quad -60u(5u-2), 60u^2].
\end{aligned}$$

To illustrate the new challenges in IGA methods, a beam with a single physical span will be subject to various boundary and loading conditions using a single quintic B-spline over the full span. The simplified ODE with constant EI and a line load of $w(x)$ is

$$\mathrm{EI}\,\frac{d^4 v}{dx^4} - w(x) = 0, \quad 0 \le x \le L.$$

This ODE represents a typical beam, with constant EI, where the following quantities are considered in engineering studies:

$$\begin{array}{lll} v(x) & = & \text{transverse deflection (L)}, \\ v'(x) & = \theta(x) & \text{slope (L/L)}, \\ \mathrm{EI}\,v''(x) & = M(x) & \text{bending moment (FL)}, \\ \mathrm{EI}\,v'''(x) & = V(x) & \text{transverse shear force (F)}, \\ \mathrm{EI}\,v''''(x) & = w(x) & \text{load per unit length (F/L)}. \end{array}$$

In structural mechanics, member design criteria include determining the largest absolute values in the solution domain of the bending moment, $M(x) = \mathrm{EI}(x)\partial^2 v(x)/\partial x^2$, and the transverse shear force, $V(x) = \mathrm{EI}(x)\partial^3 v(x)/\partial x^3$. To have the moment and shear design criteria varying in a study then the third derivative must be at least linear in x, meaning that the approximating polynomial should be at least degree four to provide useful post-processing data in regions between discontinuities in the data ($K(x)$, $A(x)$, $C(x)$, $Q(x)$, or external values of $M(x)$ or $V(x)$).

8.8 IGA beam models

The example models will begin with the transverse displacement given by $v(u) = [\boldsymbol{N}(u)]\{\mathbf{AP}_{\boldsymbol{v}}\}$ with the local (and here the system) unknowns $\{\mathbf{AP}_{\boldsymbol{v}}\}^T = [v_1, v_2, v_3, v_4, v_5, v_6]$. Since the beam is taken to be straight, a linear position mapping of $x(u) = x_1 + uL$ is used in this approximation so as to render the Jacobian constant, $J = dx/du = L$. This choice leads to a regional stiffness matrix, before imposing the essential boundary conditions, of

$$[\boldsymbol{K}] = \int_L \frac{d^2[\boldsymbol{N}(u)]}{dx^2}^T \mathrm{EI}\,\frac{d^2[\boldsymbol{N}(u)]}{dx^2} dx$$

$$= \mathrm{EI} \int_L \frac{d^2[\boldsymbol{N}(u)]^T}{du^2} \frac{1}{L^2} \frac{d^2[\boldsymbol{N}(u)]}{du^2} \frac{1}{L^2}(L\,du),$$

$$[\boldsymbol{K}] = \frac{\mathrm{EI}}{7L^3} \begin{bmatrix} 400 & -600 & 80 & 60 & 40 & 20 \\ -600 & 1040 & -260 & -160 & -60 & 40 \\ 80 & -260 & 240 & 40 & -160 & 60 \\ 60 & -160 & 40 & 240 & -260 & 80 \\ 40 & -60 & -160 & -260 & 1040 & -600 \\ 20 & 40 & 60 & 80 & -600 & 400 \end{bmatrix},$$

and the resultant force vector is

$$\{\boldsymbol{F}\} = \int_L [\boldsymbol{N}(u)]^T w(x) dx.$$

For a uniform line load of $w(x) = w$, and for a linearly increasing trapezoidal (triangular) load of $w(u) = [(1-u)(u)][w_1 \; w_1]^T$, the resultant vectors are

$$\{\boldsymbol{F}_{\mathbf{unif}}\} = \frac{\mathrm{wL}}{6} \begin{Bmatrix} 1 \\ 1 \\ 1 \\ 1 \\ 1 \\ 1 \end{Bmatrix}, \quad \{\boldsymbol{F}_{\mathbf{tri}}\} = \frac{\mathrm{wL}}{42} \begin{Bmatrix} 1 \\ 2 \\ 3 \\ 4 \\ 5 \\ 6 \end{Bmatrix}, \quad \text{and}$$

$$\{\boldsymbol{F}_{\mathbf{trap}}\} = \frac{L}{42} \begin{bmatrix} 6 & 1 \\ 5 & 2 \\ 4 & 3 \\ 3 & 4 \\ 2 & 5 \\ 1 & 6 \end{bmatrix} \begin{Bmatrix} w_1 \\ w_2 \end{Bmatrix},$$

respectively. Note that the sum of all of the triangular load resultant is wL/2, as expected.

If this model is later to be used in dynamics or vibrations, then for a mass per unit length, ρ, the consistent mass, $[\boldsymbol{M}]$, is

$$[\boldsymbol{M}] = \int_L [\boldsymbol{N}(u)]^T \rho(x) [\boldsymbol{N}(u)] dx.$$

For a constant mass density,

$$[\boldsymbol{M}] = \frac{\rho L}{2772} \begin{bmatrix} 252 & 126 & 56 & 21 & 6 & 1 \\ 126 & 140 & 105 & 60 & 25 & 6 \\ 56 & 105 & 120 & 100 & 60 & 21 \\ 21 & 60 & 100 & 120 & 105 & 56 \\ 6 & 25 & 60 & 105 & 140 & 126 \\ 1 & 6 & 21 & 56 & 126 & 252 \end{bmatrix}.$$

Instead of a consistent mass matrix, as mentioned above, several FEA authors like to use the scaled diagonal to account for the total mass (as finite difference methods do). This form is

$$[\boldsymbol{M}]_{\text{diag}} = \frac{\rho L}{1024} \lceil 252 \quad 140 \quad 120 \quad 120 \quad 140 \quad 252 \rfloor$$

Hughes (2000) has demonstrated better FEA resultants using the simple average of the consistent and diagonal mass matrices. Since this approach may be useful in IGA, that version is also given here:

$$[\boldsymbol{M}]_{\text{ave}} = \frac{\rho L}{354816} \begin{bmatrix} 59787 & 8064 & 3584 & 1344 & 384 & 64 \\ 8064 & 33215 & 6720 & 3840 & 1600 & 384 \\ 3584 & 6720 & 28470 & 6400 & 3840 & 1344 \\ 1344 & 3840 & 6400 & 28470 & 6720 & 3584 \\ 384 & 1600 & 3840 & 6720 & 33215 & 8064 \\ 64 & 384 & 1344 & 3584 & 8064 & 59787 \end{bmatrix}.$$

If the beam is supported by an elastic foundation (in compression or tension) of stiffness k (units F/L^2), then its stiffness matrix is the same as the consistent mass matrix with ρ replaced by k.

As a validation test of this single span beam, consider a non-symmetry beam model shown in Fig. 8.4 (AISC Case 12), known

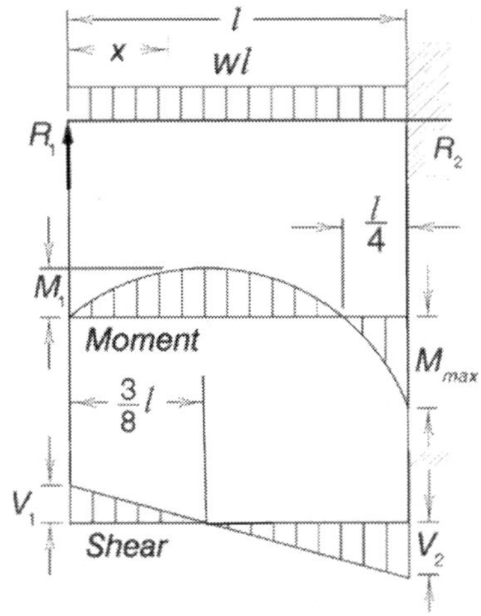

Figure 8.4. Moment and shear force in a propped cantilever with a uniform load

as a propped cantilever beam, where the exact solution is a fifth-degree ($p = 5$) polynomial. At the left of that figure ($x = 0$), there is an essential boundary conditions that the deflection is zero. There is also a natural non-essential condition there of $v''(0) = 0$ because there is no external moment applied, $M(0) = 0$.

The main challenge occurs at the right end, $x = L$, with the two Dirichlet BCs that both the deflection and the slope are zero, $v(L) = 0 = v'(L)$. The B-spline analysis basis function defines only the displacement at the control point at $x = L$, so how can the slope there also be specified? The FEA uses the Hermite family of polynomials because they have physical derivatives (slope, curvature, etc.) at each node that can be used to impose the specified slope. Not having multiple dofs per analysis control point means that this essential condition must be given some thought. One mathematical way to enforce that Dirichlet condition is to write a multipoint constraint (MPC) equation at that point on all AP dofs that are active in the knot span where the condition is to be enforced. But, that is not necessary, yet. Recall the basic property of all B-splines that each end point is tangent to the line connecting to the adjacent control point. To make that line horizontal, one can just set the AP next to the fixed end to zero.

This single iElement has a total of six AP dof values. Three of them are now to be set to zero, so the fixed dofs are [Fixed] = [1, 5, 6] with [Given] = [0, 0, 0] and the remaining dofs are [Free] = [2, 3, 4]. Since there is only one iElement in the structure, the system matrix equilibrium equations are the local matrices:

$$[\boldsymbol{K}]\{\boldsymbol{\delta}\} = \{\boldsymbol{F}_{\mathbf{unif}}\} + \{\boldsymbol{c}_{\mathbf{EBC}}\}.$$

Since there are no externally applied known point forces or couples (yet), the [Free] rows of the external point sources, $\{\boldsymbol{c}_{\mathbf{EBC}}\}$, are all zero and the three [Fixed] reaction values cannot be determined until all AP displacement values are known. The function *Solve_Free_DOF.m* can be used to find the [Fixed] dof given numerical values for all of the coefficients. In this special case, there are only three AP values to find and that can be done analytically. Retaining the [Free] rows and columns of $[\boldsymbol{K}]$ and $\{\boldsymbol{F}_{\mathbf{unif}}\}$ and carrying the (zero) EBC contributions to the right-hand side (RHS) give the reduced (non-singular) equilibrium equations of

$$\frac{\mathrm{EI}}{L^3}\begin{bmatrix} 1040 & -260 & -160 \\ -260 & 240 & 40 \\ -160 & 40 & 240 \end{bmatrix}\begin{Bmatrix} \delta_2 \\ \delta_3 \\ \delta_4 \end{Bmatrix} = \frac{-wL}{6}\begin{Bmatrix} 1 \\ 1 \\ 1 \end{Bmatrix} - \begin{Bmatrix} 0 \\ 0 \\ 0 \end{Bmatrix}.$$

Solving gives the analysis control point displacement values of

$$\begin{Bmatrix} \delta_2 \\ \delta_3 \\ \delta_4 \end{Bmatrix} = \frac{L^3}{\mathrm{EI}}\frac{1}{400}\begin{bmatrix} 4 & 4 & 2 \\ 4 & 16 & 0 \\ 2 & 0 & 13 \end{bmatrix}\frac{-wL}{6}\begin{Bmatrix} 1 \\ 1 \\ 1 \end{Bmatrix} = \frac{\mathrm{wL}^4}{\mathrm{EI}}\frac{-1}{2400}\begin{Bmatrix} 10 \\ 20 \\ 15 \end{Bmatrix}.$$

Again, those are not the physical displacements of the beam; those have to be calculated with the basis functions, $\delta(u) = [\boldsymbol{N}(u)]\{\boldsymbol{\delta}^r\}$.

The reaction forces, in this case, occur at the end points of the beam. Only the first Dirichlet condition was uniquely associated with the left end analysis point. The reaction force there can be found by using just the first of three unused rows of the system equilibrium equations. Substituting the now known displacements into it and solving for the left reaction force gives

$$\frac{\mathrm{EI}}{7L^3}[400\ \ -600\ \ 80\ \ 60\ \ 40\ \ 20]\frac{\mathrm{wL}^4}{\mathrm{EI}}\frac{-1}{2400}\begin{Bmatrix} 0 \\ 10 \\ 20 \\ 15 \\ 0 \\ 0 \end{Bmatrix} = \frac{-\mathrm{wL}}{6}\{1\} + \{R_1\}$$

$$\frac{-\mathrm{wL}}{7(2400)}(0 - 6000 + 1600 + 900 + 0 + 0) = \frac{-\mathrm{wL}}{6} + R_1$$

$$\frac{3500\,\mathrm{wL}}{7(2400)} + \frac{400\,\mathrm{wL}}{2400} = \frac{3\,\mathrm{wL}}{8} = R_1, \text{ which is exact.}$$

The other two Dirichlet conditions produce two off set vertical forces needed to provide the reaction force and reaction couple at the fixed end. It is not clear at this point how to recover them, in part because the other two unused rows produce AP reactions with the units of force.

At this point, the system has become statically determinant. Writing the sum of the vertical forces gives

$$\sum F_y = 0 = \frac{3\,\mathrm{wL}}{8} - wL + R_2 \rightarrow R_2 = \frac{5\,\mathrm{wL}}{8}, \text{ which is exact.}$$

Taking the sum of moments at the fixed end about the z-axis, with CCW positive, gives

$$\sum M_z = 0 = -(L)R_1 + \left(\frac{L}{2}\right)\mathrm{wL} + (0)R_2 + M_2 \rightarrow M_2$$

$$= \frac{-1\,\mathrm{wL}^2}{8}, \text{ which is exact.}$$

Estimates of the moment and shear force reactions are always available from their definitions:

$$M(x) = \mathrm{EI}\,v''(x) = \frac{d^2[\boldsymbol{N}(u)]}{dx^2}\{\boldsymbol{\delta}\}, \quad \text{and}$$

$$V(x) = \mathrm{EI}\,v'''(x) = \frac{d^3[\boldsymbol{N}(u)]}{dx^3}\{\boldsymbol{\delta}\}.$$

Evaluating them at $u = 0$ and $u = 1$ will give estimates of the Neumann conditions there. The moment estimate at $x = L(u = 1)$ is

$$M(L) = \mathrm{EI}\frac{1}{L^2}[0\ \ 0\ \ 0\ \ 20\ \ -40\ \ 20]\frac{-\mathrm{wL}^4}{2400\,\mathrm{EI}}\begin{Bmatrix} 0 \\ 10 \\ 20 \\ 15 \\ 0 \\ 0 \end{Bmatrix}$$

$$= \frac{-\mathrm{wL}^2}{8}, \quad \text{CW which is exact.}$$

Likewise, for the shear force there,

$$V(L) = \mathrm{EI}\frac{1}{L^3}[0\ \ 0\ \ -60\ \ 180\ \ -180\ \ 69]\frac{-\mathrm{wL}^4}{2400\,\mathrm{EI}}\begin{Bmatrix} 0 \\ 10 \\ 20 \\ 15 \\ 0 \\ 0 \end{Bmatrix}$$

$$V(L) = \frac{5\,\mathrm{wL}}{8}, \text{ up, which is exact.}$$

Of course, the last two calculations are very misleading because the exact solution for $v(x)$ is a fifth-degree polynomial which the analysis control points of the $p = 5$ B-spline exactly captured. In general, the AP values will be approximate and the reactions recovered this way are much less accurate than those recovered from the system matrix equilibrium equations. By the way, substituting $u = 0$ into the moment equation gives the exact zero moment there. For a practical study where the AP displacement controls are not exact, the moment there would not be zero. Due to physical insight for the application that implies the other $M(x)$ values probably have the same magnitude error and should be viewed as $M(x) \pm M(0)$.

Before leaving this example, remember the concept of the superposition of linear solutions. The solution obtained here for Fig. 8.4 (AISC Case 12) could also be subjected to other load cases for a single span like Cases (5) and (10) of Fig. 8.2. As noted, at any location, $x(u_P)$, where a discontinuity in the loads occurs, introducing one or more repeated knots into the analysis knot vector can allow the quintic B-spline or NURBS to yield accurate approximations.

8.9 Summary

Physical derivatives of basis functions:

$$\frac{\partial N(u)}{\partial x} = \frac{\partial N(u)}{\partial u}\frac{\partial u}{\partial x},$$

$$\frac{\partial^2 N(u)}{\partial x^2} = \frac{\partial}{\partial x}\left(\frac{\partial N(u)}{\partial u}\frac{\partial u}{\partial x}\right) = \left(\frac{\partial}{\partial x}\frac{\partial N(u)}{\partial u}\right)\frac{\partial u}{\partial x} + \frac{\partial N(u)}{\partial u}\frac{\partial^2 u}{\partial x^2},$$

$$\frac{\partial^2 N(u)}{\partial x^2} = \frac{\partial^2 N(u)}{\partial u^2}\left(\frac{\partial u}{\partial x}\right)^2 + \frac{\partial N(u)}{\partial u}\frac{\partial^2 u}{\partial x^2},$$

$$\text{If } \frac{\partial u}{\partial x} \ll 1 \quad \text{then } \frac{\partial^2 u}{\partial x^2} \approx 0 \text{ and } \frac{\partial^2 N(u)}{\partial x^2} \cong \frac{\partial^2 N(u)}{\partial u^2}\left(\frac{\partial u}{\partial x}\right)^2.$$

Chapter 9

Second-Order Partial Differential Equations

9.1 Introduction

The general form of a second-order partial differential equation (PDE) for a variable, $T(x, y, z)$, in a closed domain, Ω enclosed by a boundary, Γ, is

$$\begin{aligned}&\frac{\partial}{\partial x}\left(a_{11}\frac{\partial T}{\partial x}+a_{12}\frac{\partial T}{\partial y}+a_{13}\frac{\partial T}{\partial z}\right)+\frac{\partial}{\partial y}\left(a_{21}\frac{\partial T}{\partial x}+a_{22}\frac{\partial T}{\partial y}+a_{23}\frac{\partial T}{\partial x}\right)\\&\quad+\frac{\partial}{\partial z}\left(a_{31}\frac{\partial T}{\partial x}+a_{32}\frac{\partial T}{\partial y}+a_{33}\frac{\partial T}{\partial z}\right)\\&\quad+b_1\frac{\partial T}{\partial x}+b_2\frac{\partial T}{\partial y}+b_3\frac{\partial T}{\partial z}+c_1T=f(x,y,z)\end{aligned}$$

The boundary, Γ, is often composed of several different portions where different types of boundary conditions apply. If all of the b_k coefficients are zero, the PDE is known as self-adjoint; otherwise, it is non-self-adjoint (NSA). All of the coefficients can be functions of the spatial coordinates x, y, z. If any of the coefficients depend on the dependent variable, T, then the PDE is nonlinear; otherwise, it is a linear PDE. The dependent variable may be a scalar or a vector with several components.

Currently, it is assumed that the coefficients define an elliptical PDE. This means that some type of boundary condition on $T(x, y, z)$ (and its components) must be given at each point on the boundary. One portion of the boundary for a scalar dependent variable can be subject to a Dirichlet (essential) boundary condition by assigning it specific values as

$$T(x, y, z) = T_D(x, y, z) \quad \text{on boundary portion}\, \Gamma_D.$$

Another, non-overlapping, portion may have a Neumann (non-essential) boundary condition specifying the slope normal to the boundary

$$\frac{\partial T(x, y, z)}{\partial n} = N(x, y, z) \quad \text{on boundary portion}\, \Gamma_N.$$

The remaining portion of the boundary may have a Robin or mixed condition, which couples the solution and normal slope such as

$$\begin{aligned} T(x, y, z) + m(x, y, z)&\frac{\partial T(x, y, z)}{\partial n} \\ &= R(x, y, z) \quad \text{on boundary portion}\, \Gamma_R. \end{aligned}$$

The union of those regions forms the entire boundary of the domain, $\Gamma_D \cup \Gamma_N \cup \Gamma_R = \Gamma$

Probably, the most common second-order PDE is Poisson's equation, and its most common approximation in engineering is probably steady state heat transfer. Allowing for an orthotropic material, the governing PDE for steady linear heat transfer is

$$\kappa_x(x, y)\frac{\partial^2 T}{\partial x^2} + \kappa_y(x, y)\frac{\partial^2 T}{\partial y^2} = Q(x, y),$$

where T is the temperature, κ denotes the material's thermal conductivity, and Q is the rate of internal heat generation per unit volume (like electrical resistance, or curing concrete). A Dirichlet (essential) boundary condition usually specifies the temperature along some portion of the boundary, $T = T_D$ on Γ_D.

The non-essential (Neumann) boundary conditions involve the temperature gradient normal to the boundary, and thus the normal heat flux per unit area:

$$-\kappa_n \frac{\partial T}{\partial n} = q_n \quad \text{on}\, \Gamma_N,$$

which is positive when entering the body. The default (natural) boundary condition is that all surfaces are insulated with $q_n \equiv 0$. This case requires no action in IGA and FEA studies. Another common non-essential boundary condition is the mixed or Robin condition of heat convection on the surface of a heat conducting solid:

$$-\kappa_n \frac{\partial T}{\partial n} = h^b(T - T^b_\infty) \quad \text{on}\, \Gamma_R,$$

where h^b is the surface convection coefficient and T^b_∞ is the known fluid temperature along the boundary segment. Of course, the boundary of the solution domain is the total union of the above-mentioned non-overlapping boundary segments, $\Gamma = \Gamma_D \cup \Gamma_N \cup \Gamma_R$. An edge view of the boundary with those boundary condition options is shown in Fig. 9.1.

It should be noted that in some applications, it is possible to use an exact solution to define only Neumann boundary conditions on the entire boundary of a domain. However, the mathematical solution for that case is then only known to within an arbitrary constant. In other words, the matrix system is singular and an IGA or FEA solution is not obtainable without introducing some analysis trick. The most common trick is specifying a zero Dirichlet value at a single point in the domain. It is usually easy to find a corner of the control net that corresponds to an arbitrary point in the solution domain where a zero solution value can be imposed. Then, the solution values are computed relative to the zero value at the CP with the fake Dirichlet condition. Misleadingly, some papers on the accuracy of IGA solutions have used that trick to show impressive energy norm convergence rates. Such studies are misleading because

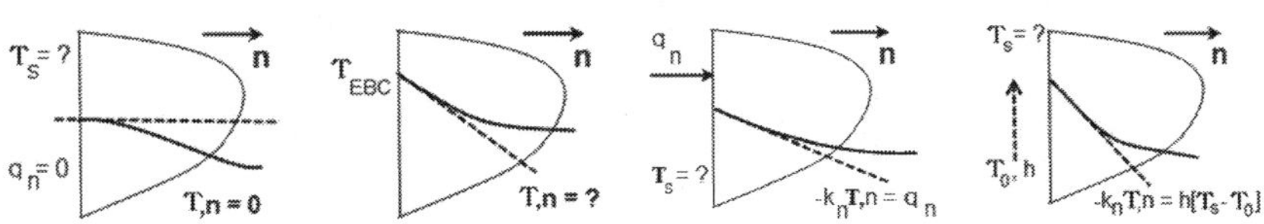

Figure 9.1. Natural, essential, known flux, and convection boundary conditions

they avoid addressing the challenge of generally imposing Dirichlet boundary conditions.

In heat transfer, Fourier's Law defines the orthotropic heat flux vector per unit area, $\boldsymbol{q}$, within the solid in terms of the gradient as

$$\vec{q} = \begin{Bmatrix} q_x \\ q_y \end{Bmatrix} = \boldsymbol{q} = -\left[\boldsymbol{\kappa}\right] \boldsymbol{\nabla}^{\boldsymbol{T}} T(x,y) = -\begin{bmatrix} \kappa_x & 0 \\ 0 & \kappa_y \end{bmatrix} \begin{Bmatrix} \partial T/\partial x \\ \partial T/\partial y \end{Bmatrix}.$$

The normal heat flux per unit area on the boundary of a domain is

$$q_n = \left(-\left[\boldsymbol{\kappa}\right] \boldsymbol{\nabla}^{\mathbf{T}} T(x,y)\right) \cdot \vec{n} = -\kappa_n \partial T/\partial n,$$

where $\vec{\boldsymbol{n}}$ is the outward unit normal vector to the boundary. Its value is often known, and it is usually zero over most of the boundary where the temperature is not specified, that is, wherever a portion of the boundary is insulated. The normal heat flux, q_n, is considered positive when it enters a body. In IGA and FEA studies, all boundaries are automatically insulated ($q_n \equiv 0 \leftrightarrow \partial T/\partial n \equiv 0$) unless otherwise stated and they require no user input. Importantly, if a portion of the boundary is not insulated, then the user must specifically provide the required data.

The amount of power (heat *flow*) entering or leaving a body is the integral of the normal heat flux over the surface:

$$P_n = \int_\Gamma q_n \, dA.$$

9.2 Galerkin Integral Form

Carrying out the Galerkin weighted residual method using Green's Theorem results in the equivalent integral form

$$I = \int_{\Gamma_N} T \left(\kappa_n \frac{\partial T}{\partial n}\right) d\Gamma - \int_\Omega \left[\frac{\partial T}{\partial x}\left(\kappa_x \frac{\partial T}{\partial x}\right) + \frac{\partial T}{\partial y}\left(\kappa_y \frac{\partial T}{\partial y}\right)\right] d\Omega,$$

$$- \int_\Omega T Q d\Omega - \int_{\Gamma_R} T\, h^b T d\Gamma + \int_{\Gamma_R} T h^b T_\infty^b d\Gamma = 0,$$

which satisfies the PDE when the temperature also satisfies the essential boundary conditions.

9.3 Basis Functions Tensor Products

For any reader unfamiliar with the phrase 'tensor product', it will be demonstrated first as arithmetic and then with basis functions. Define two numerical arrays as $\{Nr\} = [r1\ r2\ r3\ r4]'$ and $\{Ns\} = [s1\ s2]'$. To have a basis function array consistent in 1D or higher IGA studies it is desired to assign the analysis control points (APs) numbers, and the associated basis function numbers, first along the r-axis (for $s1 = 0$) from its origin to its end and the end. Then, continue the analysis control point (AP) numbering in the r-direction for the next s-direction value ($s2$), and so on. If the study is 3D, the above-mentioned AP numbers lie on the $t1 = 0$. Then, the numbering continues for the $t2$ plane along the r-direction and s-direction and so on for all t-planes. For the current example, the desired basis function (and AP) ordering is

$$[(r1\,s1),\ (r2\,s1),\ (r3\,s1)(r4\,s1),\ (r1\,s2),\ (r2\,s2),\ (r3\,s2)(r4\,s2)]$$
$$\text{AP}:\ 1,\ 2,\ 3,\ 4,\ 5,\ 6,\ 7,\ 8.$$

In other words, if they were basis functions rather than numbers, there would be four APs (1–4) along the r-axis ($s = 0$) and four more AP (5–8) above them along a parallel line at $s > 0$.

In mathematics literature, the 'tensor product' is denoted as $\{Nr\} \otimes \{Ns\}$. The tensor product of those vectors in matrix notation is

$$\{\boldsymbol{N}r\} \otimes \{\boldsymbol{N}s\} = \{\boldsymbol{N}r\}\{\boldsymbol{N}s\}' = \begin{bmatrix} \mathrm{r1} * \mathrm{s1} & \mathrm{r1} * \mathrm{s2} \\ \mathrm{r2} * \mathrm{s1} & \mathrm{r2} * \mathrm{s2} \\ \mathrm{r3} * \mathrm{s1} & \mathrm{r3} * \mathrm{s2} \\ \mathrm{r4} * \mathrm{s1} & \mathrm{r4} * \mathrm{s2} \end{bmatrix} = [\boldsymbol{N}rs]\,.$$

The rectangular matrix is converted to a row matrix with the Matlab operator, reshape:

$$[\boldsymbol{N}_{\boldsymbol{rs}}] = \text{reshape}\,(\boldsymbol{N}rs, 1, 8) = [r1^*s1,\ r2^*s1, r3^*s1, r4^*s1,$$
$$\times\, r1^*s2, r2^*s2, r3^*s2, r4^*s2].$$

Beginners should think of the above-mentioned numbers as identifying subscripts of IGA basis functions in two parametric directions.

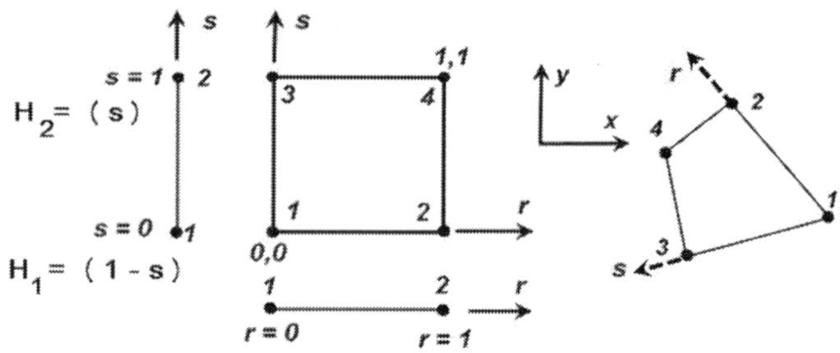

Figure 9.2. Example tensor product of linear IGA and FEA basis functions

In IGA and FEA, the linear basis functions are identical. So, the well-known FEA and IGA bi-linear quadrilateral basis in Fig. 9.2 will be used as the first example where its four two-dimensional basis functions are developed as the tensor product of their one-dimensional basis functions. Figure 9.2 shows that the one-dimensional basis functions in the r-direction are $Nr_1 = (1 - r)$ and $Nr_2 = r$ in the s-direction.

$$[\boldsymbol{N}_{\boldsymbol{r},\boldsymbol{s}}(r, s)] = [\boldsymbol{N}_{\boldsymbol{r}}(r)]^T [\boldsymbol{N}_{\boldsymbol{s}}(s)] = \begin{Bmatrix} (1-r) \\ r \end{Bmatrix} [(1-s) \quad s]$$

$$\begin{aligned} N_{1,1}(r, s) &= N_1(r)N_1(s) = (1-r)(1-s) - 1 - r - s + rs \\ N_{2,1}(r, s) &= N_2(r)N_1(s) = r(1-s) = r - rs \\ N_{1,2}(r, s) &= N_1(r)N_2(s)(1-r)s = s - rs \\ N_{2,2}(r, s) &= N_2(r)N_2(s) = rs, \end{aligned}$$

The four tensor product basis functions in the 2D parametric space of this example domain when reshaped into a row matrix are

$$[\boldsymbol{N}_{\boldsymbol{r},\boldsymbol{s}}(r, s)] = [(1 - r - s + rs)\ (r - rs)\ (s - rs)\ (rs)]\,.$$

These IGA basis functions, and any FEA Lagrange interpolations, have the property of being a 'partition of unity', which means that their algebraic sum at any set of parametric coordinates must be one:

$$\sum_k N_k(r,\ s) \equiv 1, \quad \sum_k \frac{\partial N_k(r,\ s)}{\partial r} \equiv 0, \quad \sum_k \frac{\partial N_k(r,\ s)}{\partial s} \equiv 0.$$

The IGA basis functions also satisfy the 'positivity condition' that they are all non-negative, $N_k(r,\, s) \geq 0$. For this example, it is easily seen that the first two requirements are met.

In the more common IGA literature notation, the above-mentioned parametric surface basis functions are often written as $Ni, j; p, q(r, s)$ where i denotes the number of basis functions of degree p in the parametric r-dimension. Likewise, j denotes the number of basis functions of degree q in the second parametric dimension, s. The basis function $Ni, j; pq$ is the tensor product of each of those terms. A NURBS analysis 'surface' is the weighted sum of all of those basis functions and their respective analysis control mesh point values.

The numeric examples of a tensor product are reproduced as Matlab Script 9.1. Using the Matlab symbolic capabilities, the script outlines how to get 2D basis functions if both analytic forms of the starting basis functions are known.

Script 9.1. Symbolic example of 'basis functions' $p = 1 = q$ tensor product

```
% Tensor product of row basis functions Nr and Ns
%    Analytic bi-linear IGA basis functions
%      s 3*----* 4
%         |    |         N(r) = [(1-r) r]'
%        1*----*2 --> r
syms r s real       % define the symbolic terms
Nr = [(1-r); r]              ; % linear r-basis
Ns = [(1-s); s]              ; % linear s-basis

Nrs = Nr * Ns' ; % rectangular basis functions
% Nrs = [(r - 1)*(s - 1), -s*(r - 1)]
%       [     -r*(s - 1),         r*s]

Nrs = reshape (Nrs, 1, 4)     % row matrix form
% Nrs = [(r-1)*(s-1), -r*(s-1), -s*(r-1), r*s]

simplify( sum (Nrs) )   % ans = 1, check is ok
```

9.4 Single-Span Planar Thermal Analysis

Keep in mind that a NURBS analysis results surface with constant weights becomes a B-spline surface. Thus, using B-spline surface solutions, with low-order polynomials, instead allows more analytic examples of most of the steps required for an IGA study.

As a simple detailed example of solving the 2D Poisson equation, consider a single planar iElement rectangle of lengths of $\Delta x = L_x$ and $\Delta y = L_y$, which is to be modeled as a B-spline surface. Since the IGA and FEA linear basis functions are exactly the same, this example will use a quadratic B-spline in each direction. From Appendix A1, for the open unitary knot vector $\mathbb{K} = \{0\,0\,0\,1\,1\,1\} = \mathbb{A}$, which will define both the geometry and analysis results, the basis functions over the single knot span in each parametric direction are $N(r) = N(s)$ where the single knot span range is $[0,\ 1)$ in each of the two parametric directions:

CP	Basis Functions $N(s)$	Derivatives $dN(s)/ds$
1	$[(s-1)^2]$	$[(2s-2)]$
2	$[2s(1-s)]$	$[(2-4s)]$
3	$[s^2\quad]$	$[2s\quad]$

The physical geometric shape in Fig. 9.3 (rightmost) is defined by the tensor product of these functions, which requires $(3 \times 3) = 9$ geometric control points (CPs) giving their x-, y-coordinates.

The temperature distribution will likewise be defined by nine APs at the same locations and their nine basis functions formed from the

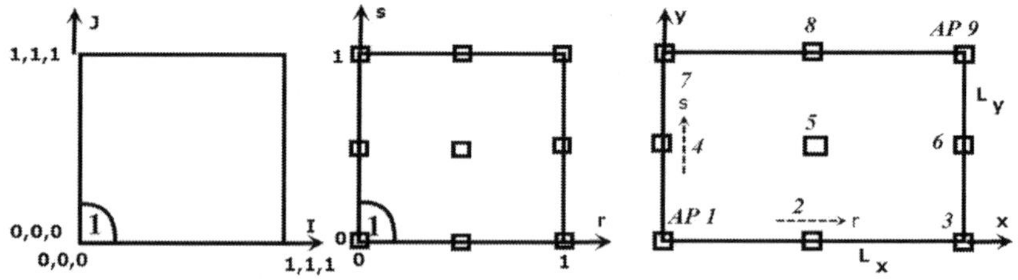

Figure 9.3. Index, parametric, and physical space for $p = 2$, $q = 2$.

tensor product of the 1D equations given earlier:

$$N(r,s) = \left[(r-1)^2\,(s-1)^2\,,\ -2r(r-1)(s-1)^2,\ r^2\,(s-1)^2,\ \ldots\right.$$
$$-2s(r-1)^2(s-1),\ 4rs(r-1)(s-1),\ -2r^2s(s-1),\ldots$$
$$\left. s^2(r-1)^2,\ -2rs^2(r-1), r^2s^2\right]$$

Note that the first three basis functions, at $s = 0$, are supported by the AP numbers one to three. Likewise, the last three basis functions, at $s = 1$, are supported by AP numbers seven through nine. For an application with multiple iElements, some of those APs would also support adjacent iElements.

The physical rectangular shape is shown in Fig. 9.3 along with the nine geometric and APs shown as small boxes. Since there is just one iElement, the connection list for assembly (scatter) is just the list [1:9]. Assuming a uniformly spaced set of geometric CPs, the coordinate array is

$$[\boldsymbol{x}\ \boldsymbol{y}]' = \begin{bmatrix} 0 & \dfrac{L_x}{2} & L_x & 0 & \dfrac{L_x}{2} & L_x & 0 & \dfrac{L_x}{2} & L_x \\ 0 & 0 & 0 & \dfrac{L_y}{2} & \dfrac{L_y}{2} & \dfrac{L_y}{2} & L_y & L_y & L_y \end{bmatrix}.$$

For an orthotropic material, the governing PDE for steady linear heat transfer is

$$\kappa_x(x,\, y)\frac{\partial^2 T}{\partial x^2} + \kappa_y(x,y)\frac{\partial^2 T}{\partial y^2} = Q(x,y),$$

where T is the temperature, κ denotes the material's thermal conductivity, and Q is the rate of internal heat generation per unit volume. Applying the Galerkin weighted residual method and using Green's Identity result in the equivalent integral form

$$I = \int_{\Gamma_N} T(\kappa_n \frac{\partial T}{\partial n})\, d\Gamma - \int_{\Omega} \left[\frac{\partial T}{\partial x}\left(\kappa_x \frac{\partial T}{\partial x}\right) + \frac{\partial T}{\partial y}\left(\kappa_y \frac{\partial T}{\partial y}\right)\right] d\Omega,$$
$$-\int_{\Omega} TQ d\Omega - \int_{\Gamma_R} T\, h^b T\, d\Gamma + \int_{\Gamma_R} T\, h^b T^b_\infty d\Gamma = 0,$$

which satisfies the PDE when the temperature also satisfies the essential (Dirichlet) boundary conditions. The other integrals are weak versions of boundary and/or face convection (Robin) condition

$$-\kappa_n \frac{\partial T}{\partial n} = h^b(T - T_\infty^b) \quad \text{on}\, \Gamma_R,$$

where h^b is the surface convection coefficient and T_∞^b is the known fluid temperature along the boundary segment. The Neumann specified normal heat flux contribution is

$$-\kappa_n \frac{\partial T}{\partial n} = q_n \quad \text{on}\, \Gamma_N.$$

For the patch test example $Q = 0$, $h^b = 0 = T_\infty^b$ and the normal flux, q_n is not specified, but its integral is a thermal reaction term in the system matrix form.

The heat conduction contribution to the scalar integral is

$$I_\kappa = \int_\Omega \left[\frac{\partial T}{\partial x}\left(\kappa_x \frac{\partial T}{\partial x}\right) + \frac{\partial T}{\partial y}\left(\kappa_y \frac{\partial T}{\partial y}\right)\right] d\Omega,$$

which needs to be recast in terms of the basis functions and their physical derivatives. Integrating over all regions,

$$I_\kappa = \sum\nolimits_{R=1}^{R=n} \int_{\Omega(R)} \left[\frac{\partial T(r,s)}{\partial x}\left(\kappa_x \frac{\partial T(r,s)}{\partial x}\right)\right.$$
$$\left. + \frac{\partial T(r,s)}{\partial y}\left(\kappa_y \frac{\partial T(r,s)}{\partial y}\right)\right] d\Omega.$$

Currently, there is only one knot span in each parametric direction, so there is only one B-spline surface region (iElement). In any parametric region, the approximate temperature is

$$T(r,s) = [\boldsymbol{N}(r,s)]\left\{\boldsymbol{\delta}^R\right\},$$
$$1 \times 1 = (1 \times 9)(9 \times 1),$$

with $\{\boldsymbol{\delta}^R\} \subset \{\boldsymbol{\delta}\}$, where the $\{\boldsymbol{\delta}^R\}$ are the AP degrees of freedom for that region and $\{\boldsymbol{\delta}\}$ denotes all of the temperature values at the APs.

Currently, the number of geometry and APs is the same. Thus, the geometry mapping is $\begin{bmatrix} x(r,s) & y(r,s) \end{bmatrix} = [\boldsymbol{N}(r,s)] \begin{bmatrix} \boldsymbol{x}^R & \boldsymbol{y}^R \end{bmatrix}$,

$$1 \times 2 = (1 \times 9)(9 \times 2),$$

and the Jacobian of this geometric mapping is

$$[\boldsymbol{J}(r,\ s)] = \begin{bmatrix} \dfrac{\partial x(r,s)}{\partial r} & \dfrac{\partial y(r,s)}{\partial r} \\ \dfrac{\partial x(r,s)}{\partial s} & \dfrac{\partial y(r,s)}{\partial s} \end{bmatrix} = \begin{bmatrix} \dfrac{\partial \boldsymbol{N}(r,s)}{\partial r} \\ \dfrac{\partial \boldsymbol{N}(r,s)}{\partial s} \end{bmatrix} \begin{bmatrix} \boldsymbol{x}^R & \boldsymbol{y}^R \end{bmatrix}$$

$$2 \times 2 = (2 \times 9)(9 \times 2).$$

However, for this rectangular physical space, the physical coordinates are currently defined as

$$x(r,s) = [\boldsymbol{N}^r(r)] \begin{Bmatrix} 0 \\ 0.5 \\ 1 \end{Bmatrix} L_x, \quad \text{so} \quad \frac{\partial x(r,s)}{\partial s} = 0 \quad \text{and}$$

$$\frac{\partial x(r,s)}{\partial r} = \frac{\partial\,[\boldsymbol{N}^r(r)]}{\partial r} \begin{Bmatrix} 0 \\ 0.5 \\ 1 \end{Bmatrix} L_x = \begin{bmatrix} (2r-2)\ (2-4r)\ 2r \end{bmatrix} \begin{Bmatrix} 0 \\ 0.5 \\ 1 \end{Bmatrix} L_x,$$

$$\frac{\partial x(r,s)}{\partial r} = (0 + (1-2r) + 2r)L_x = L_x.$$

Likewise, in this special example, $\partial y(r,s)/\partial r = 0$ and $\partial y(r,s)/\partial s = L_y$. In other words, this example has been set up for a special case of a parametric to physical space mapping having a constant Jacobian. Thus, its determinant and inverse are also constants.

$$[\mathbb{J}_{r,s:x,y}(r,s)] = \begin{bmatrix} L_x & 0 \\ 0 & L_y \end{bmatrix}, \quad |\mathbb{J}_{r,s:x,y}| = L_x L_y,$$

$$[\mathbb{J}_{r,s:x,y}]^{-1} = \begin{bmatrix} 1/L_x & 0 \\ 0 & 1/L_y \end{bmatrix}$$

As a reality check, recall that the calculus rule for an area of constant thickness, t, is

$$d\Omega = t\,dA \equiv t\,\left|J_{r,s:x,y}\,(r,s)\right|\,dr\,ds.$$

Here, the physical volume is

$$\begin{aligned} V &= \iint_V d\Omega = t \iint_{0,0}^{1,1} \left|\mathbb{J}_{r,s:x,y}\,(r,s)\right| dr\ ds \\ &= t\,L_x L_y \iint_{0,0}^{1,1} dr\ ds = tL_x L_y, \end{aligned}$$

as expected. Usually, this integral is obtained as a by-product of the numerical integration of the conduction matrix as a useful check on the analysis results.

Returning to the chain rule of calculus,

$$\begin{Bmatrix} \dfrac{\partial}{\partial r} \\ \dfrac{\partial}{\partial s} \end{Bmatrix} = \begin{bmatrix} \dfrac{\partial x(r,s)}{\partial r} & \dfrac{\partial y(r,s)}{\partial r} \\ \dfrac{\partial x(r,s)}{\partial s} & \dfrac{\partial y(r,s)}{\partial s} \end{bmatrix} \begin{Bmatrix} \dfrac{\partial}{\partial x} \\ \dfrac{\partial}{\partial y} \end{Bmatrix} = \left[\mathbb{J}_{\boldsymbol{r,s:x,y}}(r,s)\right] \begin{Bmatrix} \dfrac{\partial}{\partial x} \\ \dfrac{\partial}{\partial y} \end{Bmatrix}.$$

The physical derivatives at any point in the parametric space, needed in the PDE, are obtained via the inverse of the geometric Jacobian matrix:

$$\begin{Bmatrix} \dfrac{\partial T(r,s)}{\partial x} \\ \dfrac{\partial T(r,s)}{\partial y} \end{Bmatrix} = \left[\boldsymbol{J}_{\boldsymbol{r,s:x,y}}(r,s)\right]^{-1} \begin{Bmatrix} \dfrac{\partial T(r,s)}{\partial r} \\ \dfrac{\partial T(r,s)}{\partial s} \end{Bmatrix}.$$

In this special case,

$$\begin{Bmatrix} \dfrac{\partial T(r,s)}{\partial x} \\ \dfrac{\partial T(r,s)}{\partial y} \end{Bmatrix} = \begin{bmatrix} 1/L_x & 0 \\ 0 & 1/L_y \end{bmatrix} \begin{Bmatrix} \dfrac{\partial T(r,s)}{\partial r} \\ \dfrac{\partial T(r,s)}{\partial s} \end{Bmatrix} = \begin{Bmatrix} \dfrac{1}{L_x}\dfrac{\partial T(r,s)}{\partial r} \\ \dfrac{1}{L_y}\dfrac{\partial T(r,s)}{\partial s} \end{Bmatrix}.$$

Substituting the B-spline surface approximation of the temperature distribution,

$$\begin{Bmatrix} \dfrac{\partial T(r,s)}{\partial x} \\ \dfrac{\partial T(r,s)}{\partial y} \end{Bmatrix} = \begin{bmatrix} \dfrac{1}{L_x}\dfrac{[\partial \boldsymbol{N}(r,s)]}{\partial r} \\ \dfrac{1}{L_y}\dfrac{[\partial \boldsymbol{N}(r,s)]}{\partial s} \end{bmatrix} \{\boldsymbol{\delta}^R\}$$

$$2 \times 1 = (2 \times 9)(9 \times 1).$$

Here, the 3×3 tensor product matrix has been reordered into the 1×9 row matrix:

$$[\boldsymbol{N}(r,s)] \overleftrightarrow{\text{reshape}} [\boldsymbol{N}(r)]^T [\boldsymbol{N}(s)] = \begin{Bmatrix} (r-1)^2 \\ 2r(1-r) \\ r^2 \end{Bmatrix}$$

$$\times \left[(s-1)^2 \quad 2s(1-s) \quad s^2\right].$$

Or,

$$[\boldsymbol{N}(r,s)] = [N_1^r N_1^s \quad N_2^r N_1^s \quad N_3^r N_1^s \quad N_1^r N_2^s \quad N_2^r N_2^s \quad N_3^r N_2^s \quad N_1^r N_3^s \quad N_2^r N_3^s \quad N_3^r N_3^s]$$

$$AP \quad 1 \quad 2 \quad 3 \quad 4 \quad 5 \quad 6 \quad 7 \quad 8 \quad 9.$$

The parametric partial derivatives likewise form as tensor products:

$$\frac{[\partial \boldsymbol{N}(r,s)]}{\partial r} \leftrightarrow \begin{Bmatrix} (2r-2) \\ (2-4r) \\ 2r \end{Bmatrix} \left[(s-1)^2 \; 2s(1-s) \; s^2\right],$$

$$\frac{[\partial \boldsymbol{N}(r,s)]}{\partial s} \leftrightarrow \begin{Bmatrix} (r-1)^2 \\ 2r(1-r) \\ r^2 \end{Bmatrix} [(2s-2) \quad (2-4s) \quad 2s]$$

Those derivatives can be reshaped into a 2×9 rectangular physical gradient matrix. Following FEA literature notation, multiplying by the inverse Jacobian matrix defines the physical gradient matrix,

$\left[\boldsymbol{B^R}(r,s)\right]$, of the basis functions. For this special case,

$$\left[\boldsymbol{B^R}(r,s)\right] \equiv \begin{bmatrix} \dfrac{[\partial \boldsymbol{N}(r,s)]}{\partial x} \\ \dfrac{[\partial \boldsymbol{N}(r,s)]}{\partial y} \end{bmatrix} = \begin{bmatrix} \dfrac{1}{L_x}\dfrac{[\partial \boldsymbol{N}(r,s)]}{\partial r} \\ \dfrac{1}{L_y}\dfrac{[\partial \boldsymbol{N}(r,s)]}{\partial s} \end{bmatrix}$$

$$2 \times 9 = (2 \times 9).$$

The contribution of a region to the conduction integral rearranges from the scalar form

$$I_\kappa^R = \int_{\Omega(R)} \left[\frac{\partial T(r,s)}{\partial x}\left(\kappa_x \frac{\partial T(r,s)}{\partial x}\right) + \frac{\partial T(r,s)}{\partial y}\left(\kappa_y \frac{\partial T(r,s)}{\partial y}\right)\right] d\Omega$$

to the matrix form

$$I_\kappa^R = t \int_{A(R)} \left[\begin{bmatrix} \dfrac{\partial T}{\partial x} & \dfrac{\partial T}{\partial y} \end{bmatrix} \begin{bmatrix} \kappa_x & 0 \\ 0 & \kappa_y \end{bmatrix} \begin{Bmatrix} \dfrac{\partial T}{\partial x} \\ \dfrac{\partial T}{\partial y} \end{Bmatrix}\right] dA.$$

$$1 \times 1 = (1 \times 2)(2 \times 2)(2 \times 1).$$

Finally, substituting the basis functions' gradient matrix times the unknown AP temperature values gives the matrix form of the conduction contribution

$$I_\kappa^R = t \int_{A(R)} \left(\left[\left[\boldsymbol{B^R}(r,s)\right]\left\{\boldsymbol{\delta}^R\right\}\right]^T \begin{bmatrix} \kappa_x & 0 \\ 0 & \kappa_y \end{bmatrix} \left[\boldsymbol{B^R}(r,s)\right]\left\{\boldsymbol{\delta}^R\right\} \right) dA$$

Bringing the unknown constant analysis control point temperatures, $\{\boldsymbol{\delta}^R\}$, out of the integrals,

$$I_\kappa^R = t \left\{\boldsymbol{\delta}^R\right\}^T \int_{A(R)} \left[\boldsymbol{B^R}(r,s)\right]^T \begin{bmatrix} \kappa_x & 0 \\ 0 & \kappa_y \end{bmatrix} \left[\boldsymbol{B^R}(r,s)\right] dA \left\{\boldsymbol{\delta}^R\right\}$$

$$\equiv \left\{\boldsymbol{\delta}^{R}\right\}^{T}\left[\boldsymbol{K}^{\boldsymbol{R}}\right]\left\{\boldsymbol{\delta}^{R}\right\}$$

$$1 \times 1 = (1 \times 9)(9 \times 2)(2 \times 2)(2 \times 9)(9 \times 1) = (1 \times 9)(9 \times 9)(9 \times 1)$$

defines the IGA conduction matrix for this bi-quadratic region as

$$[\boldsymbol{K}] \equiv t \int_{\mathrm{A}^{(R)}} \begin{bmatrix} \dfrac{1}{L_x} \dfrac{[\partial \boldsymbol{N}(r,s)]}{\partial r} \\ \dfrac{1}{L_y} \dfrac{[\partial \boldsymbol{N}(r,s)]}{\partial s} \end{bmatrix}^{T} \begin{bmatrix} \kappa_x & 0 \\ 0 & \kappa_y \end{bmatrix} \begin{bmatrix} \dfrac{1}{L_x} \dfrac{[\partial \boldsymbol{N}(r,s)]}{\partial r} \\ \dfrac{1}{L_y} \dfrac{[\partial \boldsymbol{N}(r,s)]}{\partial s} \end{bmatrix} d\mathrm{A}.$$

$$(9 \times 9) = (9 \times 2)\,(2 \times 2)(2 \times 9).$$

This physical integral must be evaluated in the parametric space since that is where the basis functions and their derivatives are known. Changing the limits of integration and inserting the Jacobian determinant for the area conversion give the final IGA conduction matrix

$$[\boldsymbol{K}] = t \iint_{00}^{11} [\boldsymbol{B}\,(r,s)]^{T} \begin{bmatrix} \kappa_x & 0 \\ 0 & \kappa_y \end{bmatrix} [\boldsymbol{B}\,(r,s)]\,(L_x L_y\, dr\, ds).$$

In practice, such integrals must be evaluated by numerical integration. However, this simple special case could be evaluated by hand or symbolically using Matlab. This results in a constant thickness conduction matrix of

$$[\boldsymbol{K}] = \frac{t\,\kappa_x}{45}\frac{L_y}{L_x} \begin{bmatrix} 12 & -6 & -6 & 6 & -3 & -3 & 2 & -1 & -1 \\ -6 & 12 & -6 & -3 & 6 & -3 & -1 & 2 & -1 \\ -6 & -6 & 12 & -3 & -3 & 6 & -1 & -1 & 2 \\ 6 & -3 & -3 & 8 & -4 & -4 & 6 & -3 & -3 \\ -3 & 6 & -3 & -4 & 8 & -4 & -4 & -4 & 8 \\ -3 & -3 & 6 & -3 & 6 & -3 & -3 & -3 & 6 \\ 2 & -1 & -1 & 6 & -3 & -3 & 12 & -6 & -6 \\ -1 & 2 & -1 & -3 & 6 & -3 & -6 & 12 & -6 \\ -1 & -1 & 2 & -3 & -3 & 6 & -6 & -6 & 12 \end{bmatrix}$$

$$+\frac{t\,\kappa_y}{45}\frac{L_x}{L_y}\begin{bmatrix} 12 & 6 & 2 & -6 & -3 & -1 & -6 & -3 & -1 \\ 6 & 8 & 6 & -3 & -4 & -3 & -3 & -4 & -3 \\ 2 & 6 & 12 & -1 & -3 & -6 & -1 & -3 & -6 \\ -6 & -3 & -1 & 12 & 6 & 2 & -6 & -3 & -1 \\ -3 & -4 & -3 & 6 & 8 & 6 & -3 & -4 & -3 \\ -1 & -3 & -6 & 2 & 6 & 12 & -1 & -3 & -6 \\ -6 & -3 & -1 & -6 & -3 & -1 & 12 & 6 & 2 \\ -3 & -4 & -3 & -3 & -4 & -3 & 6 & 8 & 6 \\ -1 & -3 & -6 & -1 & -3 & -6 & 2 & 6 & 12 \end{bmatrix}$$

Any IGA conduction matrix (or stiffness matrix) must be symmetric, and the sum of each row and each column must be zero. These rules provide a quick check of the typed matrix. Also, each knot span conduction (or stiffness) matrix is singular. So is the assembly of all regional conduction matrices until the system matrix has essential boundary conditions enforced.

Simplify this example further by making the material isotropic with $\kappa = \kappa_x = \kappa_y$ and making the rectangle square, so $L = L_x = L_y$, so the summed region conduction matrix becomes

$$[\boldsymbol{K}] = \frac{t\,\kappa}{45}\begin{bmatrix} 24 & 0 & -4 & 0 & -6 & -4 & -4 & -4 & -2 \\ 0 & 20 & 0 & -6 & 2 & -6 & 0 & -6 & -4 \\ -4 & 0 & 24 & -4 & -6 & 0 & -2 & -4 & -4 \\ 0 & -6 & -4 & 20 & 2 & -2 & 0 & -6 & -4 \\ -6 & 2 & -6 & 2 & 16 & 2 & -6 & 2 & -6 \\ -4 & -6 & 0 & -2 & 2 & 20 & -4 & -6 & 0 \\ -4 & -4 & -2 & 0 & -6 & -4 & 24 & 0 & -4 \\ -4 & -2 & -4 & -6 & 2 & -6 & 0 & 20 & 0 \\ -2 & -4 & -4 & -4 & -6 & 0 & -4 & 0 & 24 \end{bmatrix}$$

Since this single knot span (iElement) model has one interior AP, this formulation can be checked with a numerical 'patch test'.

Beginners should also calculate the physical size of each iElement and the assembled size of the full domain as a useful by-product of the above-mentioned matrix operation. From calculus,

$$d\Omega = t\,dA = t\,\left|J_{r,s:x,y}\right|\,drds,$$

and for any iElement,

$$t\,A = t\int_A dA = t\iint_{0,0}^{1,1} \begin{bmatrix} \partial N(r,s)/dr \\ \partial N(r,s)/ds \end{bmatrix} dr\,ds\,\begin{bmatrix} \boldsymbol{x}^R & \boldsymbol{y}^R \end{bmatrix}$$

Here, there is one

$$A = \left| \tfrac{1}{3} \begin{bmatrix} -1 & 0 & 1 & -1 & 0 & 1 & -1 & 0 & 1 \\ -1 & -1 & -1 & 0 & 0 & 0 & 1 & 1 & 1 \end{bmatrix} \begin{bmatrix} \boldsymbol{x}^R & \boldsymbol{y}^R \end{bmatrix} \right| = L_x L_y.$$

$$1 \times 1 = \det|(2 \times 9)(9 \times 2)| \;= 1 \times 1.$$

Usually, this by-product is calculated in the numerical integration loop of NURBS to form the conduction matrix.

All of the above-mentioned steps were obtained by the symbolic Matlab script in Script 9.2.

The output from the script is listed in Figure 9.4. In practice, the conduction and source matrices must be obtained by using numerical integration for two reasons. The user defines the control point x–y coordinates, rendering the Jacobian matrix a variable instead of a constant. NURBS are the ratio of polynomials so their integration is much more difficult, even for made-up geometries with a constant Jacobian matrix.

Script 9.2. Symbolic derivation of conduction and source matrices for $p = 2 = q$

```
% ================= Script_K_p2_q2_rect.m ===================
% Symbolic derivation of a conduction & source matrices
% for a one span orthotropic rectangular (1 iElement)
%           Ar = {0,0,0,1,1,1} = As bi-quadratic
% manual steps for constant Jacobian, not via quadratures

syms r s Lx Ly kx ky Q real                    % symbolic variables
CPx = [0 0.5 1]*Lx ; CPy = [0 0.5 1]*Ly      ; % rectangle only
```

```
%     r- and s-direction basis functions from Appendix 1
R = [(r-1)^2, 2*r*(1-r), r^2]; nb_r = 3             ; % r-basis
S = [(s-1)^2, 2*s*(1-s), s^2]; nb_s = 3             ; % s-basis
dR = diff (R,r) ; dS = diff (S,s)     % parametric derivatives
%      Verify CP coordinates give CONSTANT Jacobian
J = [dR*CPx', 0; 0, dS*CPy']; J = simplify(J)% rectangle only
det_J = det (J),    inv_J = inv(J)                   ; % always
RS = R' * S                     ; % tensor product basis functions
N = reshape (RS, 1, nb_r*nb_s) ; % basis functions row matrix
Nsum= sum(sum(N)) ; Nsum=simplify(Nsum)   ; % check, sum == 1?
disp('Sum of basis functions = '), disp(Nsum)      ; % check !
%
%    Analytically integrate [N] for a constant source, Q
Nir = int(N,r,0,1) ; Nirs = int(Nir,s,0,1)*det_J ;% wrt r- s-
c_Q = simplify (Nirs) * Q             % resultant source vector
Qsum = sum(c_Q); disp('Heat generation total = '), disp(Qsum)

%    Get parametric & derivatives of [N(r,s)]       (B matrix)
Ndrs = [diff(N,r) ; diff(N,s)]                ; % r-, s-derivatives
B = inv (J) * Ndrs ; B = simplify (B)    ; % x-, y-derivatives

%   Integrate for kx conduction for a constant Jacobian
Just_x = [kx  0; 0  0]                        ; % just kx conduction
BtKB    = B' * Just_x*B ;     BtKB = simplify (BtKB); % product
I_BtKB = int(BtKB,r,0,1); I_BtKB = simplify(I_BtKB); % wrt r-
I_BtKB = int(I_BtKB,s,0,1) * det_J                   ; % wrt s-
K_x     = simplify(I_BtKB)          % convert differential area
%   Integrate for ky conduction for a constant Jacobian
Just_y = [0  0; 0  ky]                        ; % just ky conduction
BtKB    = B' * Just_y * B ; BtKB = simplify (BtKB) ; % product
I_BtKB =int(BtKB,r,0,1); I_BtKB = simplify(I_BtKB) ; % wrt r-
I_BtKB = int(I_BtKB,s,0,1) * det_J                   ; % wrt s-
K_y     = simplify(I_BtKB)          % convert differential area
% K     = Kx + Ky      conduction matrices complete [K]{T}={c_Q}
% end Script_K_p2_q2_rect.m ==================================
```

Running a patch test is useful in validating that a user knows how to properly use a black box IGA code, FEA code, and/or to validate a personal IGA implementation before undertaking one of the many more complicated validation problems. For analysts, running a patch test can ensure the correctness of any implementation.

```
%           function [] = NAFEMS_test_92_p2q2.m
% From Script_K_p2_q2_rect.m for a rectangular iElement
%    with CP located to create a CONSTANT Jacobian
%      Quarter symmetry with heat generation, Q
%              symmetry      Ly = delta_y/2
%          7 *------*8------9
%     T=0    |   Q          |   Q =(1e7 W/C m^3)
%          4 *       *5     6
%            | (1)          |  Symmetry
%          1 *------*2------3  Lx = delta_x/2
%                T=0
%   with manual assembly into system matrices

clear all                                      ; % be safe
Lx = 0.8/2, Ly = 0.6/2   ; % physical lengths % thick = 1
kx = 52; ky = 52; Q = 1e6                 ; % thermal data
Fixed = [1, 2, 3, 4, 7]          ; % AP with Dirichlet BC
Given = [0, 0, 0, 0, 0]           ; % Dirichlet BC values
Free  = [5, 6, 8, 9]          ; % temperatures to compute

%       Basis functions tensor product (show this)
%N =   [(r-1)^2*(s-1)^2, -2*s*(r-1)^2*(s-1),  s^2*(r-1)^2]
%[-2*r*(r-1)*(s-1)^2,  4*r*s*(r-1)*(s-1), -2*r*s^2*(r-1)]
%[        r^2*(s-1)^2,      -2*r^2*s*(s-1),         r^2*s^2]

%   Basis functions as a row matrix, for result plotting
%N =[(r-1)^2*(s-1)^2, -2*r*(r-1)*(s-1)^2, r^2*(s-1)^2,...
%-2*s*(r-1)^2*(s-1), 4*r*s*(r-1)*(s-1),-2*r^2*s*(s-1),...
%    s^2*(r-1)^2, -2*r*s^2*(r-1), r^2*s^2]
%                    Sum of basis functions = 1, checked!

%                     Heat source resultant
c_Q = [1, 1, 1, 1, 1, 1, 1, 1, 1]*Q*Lx*Ly/9;% Q resultant
%                Heat generation total = Lx*Ly*Q
c = c_Q'  ; % manual assembly (scatter) to system sources
%   Fill the kx conduction part.   (effect of x-gradient)
K_x = [12, -6, -6,  6, -3, -3,  2, -1, -1;
       -6, 12, -6, -3,  6, -3, -1,  2, -1;
       -6, -6, 12, -3, -3,  6, -1, -1,  2;
        6, -3, -3,  8, -4, -4,  6, -3, -3;
       -3,  6, -3, -4,  8, -4, -3,  6, -3;
       -3, -3,  6, -4, -4,  8, -3, -3,  6;
        2, -1, -1,  6, -3, -3, 12, -6, -6;
       -1,  2, -1, -3,  6, -3, -6, 12, -6;
       -1, -1,  2, -3, -3,  6, -6, -6, 12]*kx*Ly/Lx/45

%   Fill the ky conduction part.   (effect of y-gradient)
K_y = [12,  6,  2, -6, -3, -1, -6, -3, -1;
        6,  8,  6, -3, -4, -3, -3, -4, -3;
```

Figure 9.4. Output from the symbolic derivations of Script 9.2.

```
          2,  6, 12, -1, -3, -6, -1, -3, -6;
         -6, -3, -1, 12,  6,  2, -6, -3, -1;
         -3, -4, -3,  6,  8,  6, -3, -4, -3;
         -1, -3, -6,  2,  6, 12, -1, -3, -6;
         -6, -3, -1, -6, -3, -1, 12,  6,  2;
         -3, -4, -3, -3, -4, -3,  6,  8,  6;
         -1, -3, -6, -1, -3, -6,  2,  6, 12]*ky*Lx/Ly/45

%     Manually assemble into 9x9 system conduction
K = K_x + K_y              ; % matrices complete [K]{T}={c}

%    Compute temperatures, recover thermal reactions
[T,R]= Solve_Free_DOF (K, c, Free, Fixed, Given) % answer
%    Graph x(r) and true T(r) on top symmetry line
% set s = 1, loop over k r-values, get real x & T:
% x(k) = N(r(k),1) * CP_x; T_top(k) = N(k) * T       etc.
% end NAFEMS_test_92_p2q2 ================================
```

Figure 9.4. (*Continued*)

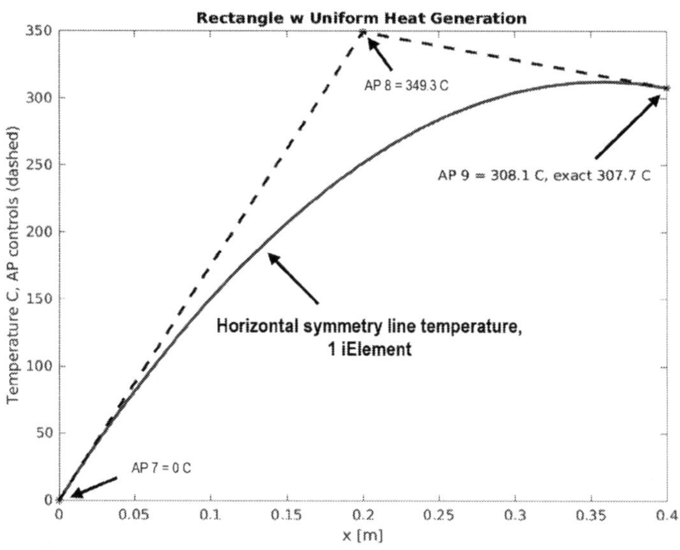

Figure 9.5. Approximate temperature on quarter-symmetry line, one iElement.

9.5 Patch Tests

Originally, the concept of a patch test was based on engineering judgment that, as the number of sub-regions (IGA iElements and/or FEA elements) of an analysis domain approaches infinity, the size of each sub-region approaches zero and the solution gradient in each sub-region should become a constant (Irons and Razzaque, 1972). Thus, all valid analysis formulations must be capable of producing a solution with arbitrary constant gradients.

Subsequently, the use of a 'patch test' has been proven to be a mathematically valid convergence test as well as a numerical validation of an IGA or an FEA implementation. Consider a patch (or sub-assembly) of an analysis region containing at least one internal analysis control point (or an FEA node). Let the analysis problem be formulated by an integral statement with the highest derivatives of order n. Assume an arbitrary 2D polynomial function, $P(x, y)$, whose nth order derivatives are constant. Use this function to prescribe the essential (Dirichlet) boundary conditions on the dependent variable on all boundary APs (or nodes) of the patch. Solve for the internal analysis control point value(s) of the dependent variable, as well as its nth order derivatives in each analysis sub-domain. To be a convergent and correctly implement the formulation:

(1) The internal control point values *must* agree exactly with the assumed function, $P(x, y)$, evaluated at each and every internal analysis control point.
(2) The calculated n-th order derivatives *must* exactly agree with the constant derivative values defined by the assumed function.
(3) These results *must* be independent of the shape of the sub-region(s).

The patch test is very important from the engineering point of view since it can be executed numerically and gives a numerical check of the entire program used to formulate the patch test and the recovery of the solution gradients.

As an example, assume that the weak integral form contains the first derivatives with respect to x and y (as does the Poisson equation). Then, the assumed solution can be any complete linear polynomial with three arbitrary coefficients, say $P(x, y) = 1+3x-4y$, which has the constant derivatives of $\partial P(x, y)/\partial x = 3$ and $\partial P(x, y)/\partial y = -4$ everywhere in the interior domain.

The required Dirichlet boundary condition values for all CPs on the external boundary are found by simple substituting their x- and y-coordinates into the assumed form. When those essential boundary conditions are imposed, then only internal DOFs remain as unknowns in the matrix system. When the system is solved, the internal DOFs become available for validation. At each internal analysis control point, substituting its x- and y-coordinates into the assumed form gives a value that must exactly match the value computed from the matrix system. If those results satisfy the first requirement of the test, then post-processing is used to validate the presence of the constant gradient components at any points in each sub-region. Usually, that is done at the numerical integration points since the gradient of the basis functions have previously been evaluated there while forming the conduction (or stiffness) matrix. If the second step of the test is also met, then change the shape of the patch to some other arbitrary shape and run the test at least a second time.

9.6 Numerical Thermal Patch Test

Next, a numerical patch test of the above-mentioned formulation will be run on a square of length $L = L_x = L_y = 8\,\text{cm}$ having a homogeneous material with a heat conduction coefficient of $\kappa = 3\,W/\text{cm}^\circ\text{C}$, and no internal heat generation, $Q = 0$. The geometric control point data become

$$[\boldsymbol{x}\ \boldsymbol{y}]' = \begin{bmatrix} 0 & 4 & 8 & 0 & 4 & 8 & 0 & 4 & 8 \\ 0 & 0 & 0 & 4 & 4 & 4 & 8 & 8 & 8 \end{bmatrix} \text{cm}.$$

The specified (fixed) Dirichlet conditions will be enforced on the eight exterior APs and the internal (fifth AP) temperature will be computed from thermal equilibrium. The AP degrees of freedom (dof) are separated into two vector subscripts:

$$[\text{Fixed}] = [1 \quad 2 \quad 3 \quad 4 \quad 6 \quad 7 \quad 8 \quad 9], \quad \text{and} \quad [\text{Free}] = [5].$$

The assumed exact temperature for the patch test is $T(x, y) = 3 + 2x + 4y$. The APs fall exactly on the perimeter of the shape, and the EBC will be applied to those points and will vary between them.

Substituting the above-mentioned eight boundary coordinates gives the following values to be enforced as Dirichlet conditions:

$$[\text{Given}] = \begin{bmatrix} 3 & 11 & 19 & 19 & 35 & 35 & 43 & 51 \end{bmatrix} {}^\circ\text{C}.$$

Likewise, substituting the remaining interior (fifth) AP coordinates of (4, 4) cm means the computed internal temperature will have to be $T(4, 4) = T_5 = 27^\circ$ in order to pass this patch test. The details for obtaining this patch test solution are given in Script 9.3 and the post processing steps are given in Script 9.4.

For this example, the system contains only a single parametric region, and the system conduction matrix is just from this one region. The system thermal matrix equilibrium equations become

$$[\boldsymbol{K}]\{\boldsymbol{\delta}\} = \{\boldsymbol{c_Q} = \boldsymbol{0}\} + \{\boldsymbol{c_q}\} \equiv \{\boldsymbol{P}\}.$$

They are singular until modified to include at least one essential (Dirichlet) boundary condition. Here, $\{\boldsymbol{c_q}\,(\text{Fixed})\}$ (defined in the following) contains the resultant external input boundary heat flow values at every AP (here also zero) and/or the reaction heat flow values necessary to maintain any Dirichlet condition(s).

Utilizing the algorithm *Solve_Free_DOF.m* with arguments $[T, R]$ = Solve_Free_DOF (K, P, Free, Fixed, Given) returns all of the AP temperature values, $T \to \{\boldsymbol{\delta}\}$, and all reactions, $R \to \{\boldsymbol{R}\}$, at analysis point degrees of freedom having a given Dirichlet condition. The use of the reactions, R, is optional and depends on the field of application. For heat transfer, the reactions are the thermal equilibrium heat flows at the APs. The complete patch test operations, for a unit thickness, are in Script 9.3. The post-solution calculations for this patch test are given in Script 9.4.

Script 9.3. Typical patch test model for bi-quadratic B-splines

```
%                conduction_patch_test.m
% Bi-quadratic conduction on a square, p = 2 = q
k = 3; L = 8;  % W/cm C, cm specific region data
xy_AP=[0,0   ; L/2,0   ; L,0   ; ...  % AP 1,2,3
       0,L/2 ; L/2,L/2 ; L,L/2 ; ...  % AP 4,5,6
       0,L   ; L/2,L   ; L,L   ];     % AP 7,8.9
% Generic bi-quadratic conduction matrix for sq
K =[24,  0, -4,  0, -6, -4, -4, -4, -2;
     0, 20,  0, -6,  2, -6, -4, -2, -4;
    -4,  0, 24, -4, -6,  0, -2, -4, -4;
     0, -6, -4, 20,  2, -2,  0, -6, -4;
    -6,  2, -6,  2, 16,  2, -6,  2, -6;
    -4, -6,  0, -2,  2, 20, -4, -6,  0;
    -4, -4, -2,  0, -6, -4, 24,  0, -4;
    -4, -2, -4, -6,  2, -6,  0, 20,  0;
    -2, -4, -4, -4, -6,  0, -4,  0, 24]*k/45 ;
```

```
% Define the dof to be imposed and computed
Fixed = [1,2,3,4,6,7,8,9]      ; % all boundary APs
Free  = [5]                    ; % all interior APs
n_bc  = max(size(Fixed))          ; % number of EBC
n_d   = max(size(K))      ; % number of system eqs
n_f   = n_d - n_bc      ; % number of interior AP
P     = zeros(n_d,1) ;    % set all AP input heat
% set given EBC, for patch T = 3 + 2*x + 4*y
for j = 1:n_bc% loop over all AP EBC ===>> ===>>
  n = Fixed(j)    ;       % get dof number for EBC
  x = xy_AP(n, 1);           % get AP x-coordinate
  y = xy_AP(n, 2);           % get AP x-coordinate
  %  Assign assumed test value T=3+2*x+4*y
  Given(j) = 3 + 2*x + 4*y        ; % this dof EBC
end; % for j EBC   <<=== <<=== <<=== <<=== <<===
disp('Fixed dof'); disp(Fixed)     % echo EBC dof
disp('Values    '); disp(Given)  % echo EBC value

% Compute temperatures,recover thermal reactions
[T,R]= Solve_Free_DOF (K, P, Free, Fixed, Given)
%   Compare against patch test requirement
n = Free(1); x = xy_AP(n, 1); y = xy_AP(n, 2);
disp('dof '),disp(n),disp(' is '), disp(T(n))
disp('required is '), ans = 3+2*x+4*y; disp(ans)
disp('Reactions'),disp(R)% heat flow at Fixed AP
disp('Sum '),disp(sum(R))       % net heat in/out
```

Script 9.4. Post-processing the bi-quadratic IGA solution result

```
%                     patchTest_post.m,
%Patch test post-processing for test T(x,y) = 3 + 2x + 4y
%       Temperatures output from the patch test
T = [3     11     19     19     27     35     35     43     51] ;
Lx = 8; Ly = 8; r = 1/4; s = 1/2;           % Specific data
%
% Recompute basis functions & derivatives (used to get K)
Nr      = [(r-1)^2, 2*r*(1-r), r^2];    % r basis functions
dNrdr   = [(2*r -2), (2 - 4*r), 2*r]; % r derivative of Nr
Ns      = [(s-1)^2, 2*s*(1-s), s^2];    % s basis functions
dNsds   = [(2*s -2), (2 - 4*s), 2*s]; % s derivative of Ns
```

```
Nrs    = Nr' * Ns;% N(r,s) tensor product basis functions
dNrsdr = dNrdr' * Ns ;         % dNrs/dr as tensor product
dNrsds = Nr' * dNsds ;         % dNrs/ds as tensor product
%
%             Convert to row matrix form
Nrs    = reshape(Nrs,1,9) ; % N(r,s) reshaped as a vector
dNrsdr = reshape(dNrsdr,1,9);          % dNrs/dr as vector
dNrsds = reshape(dNrsds,1,9);          % dNrs/ds as vector
%
%            Convert to physical derivatives
dNrs_dx = dNrsdr / Lx ;       % for diagonal Jaconian  only
dNrs_dy = dNrsds / Ly ;       % for diagonal Jaconian  only
%
%   Multiply by solution to obtain gradient components
dT_dx = dNrs_dx*T'          % Running gives {2
dT_dy = dNrs_dy*T'          %                 {4. as expected
```

As expected the computed temperature values from the above-mentioned algorithm are the vector

$$\{\boldsymbol{\delta}\}' = \{\boldsymbol{\delta}^R\}' = [3 \quad 11 \quad 19 \quad 19 \quad \mathbf{27} \quad 35 \quad 35 \quad 43 \quad 51]\,^\circ\mathrm{C},$$

where the previously unknown internal (fifth) AP temperature value is shown in bold and exactly matches the required first step of the test.

Now that all of the AP temperature values are known, the next step is to check the temperature gradient at random points (usually the quadrature points) inside each IGA sub-region or to graph them over each sub-domain. Here, there is only one IGA region, so $\{\boldsymbol{\delta}^R\} = \{\boldsymbol{\delta}\}$, and the temperature gradient at any point (for this special Jacobian) is

$$\begin{Bmatrix} \dfrac{\partial T(r,s)}{\partial x} \\ \dfrac{\partial T(r,s)}{\partial y} \end{Bmatrix} = \left[\boldsymbol{B^R}(r,s)\right]\{\boldsymbol{\delta}^R\} \equiv \begin{bmatrix} \dfrac{[\partial \boldsymbol{N}(r,s)]}{\partial x} \\ \dfrac{[\partial \boldsymbol{N}(r,s)]}{\partial y} \end{bmatrix}$$

$$\{\boldsymbol{\delta}^R\} = \begin{bmatrix} \dfrac{1}{L_x}\dfrac{[\partial \boldsymbol{N}(r,s)]}{\partial r} \\ \dfrac{1}{L_y}\dfrac{[\partial \boldsymbol{N}(r,s)]}{\partial s} \end{bmatrix}\{\boldsymbol{\delta}^R\}.$$

Randomly selecting $r = 1/4$, $s = 1/2$ and using the typical steps in Script 9.4 give

$$\begin{Bmatrix} \dfrac{\partial T(1/4,\ 1/2)}{\partial x} \\ \dfrac{\partial T(1/4,\ 1/2)}{\partial y} \end{Bmatrix} = \begin{bmatrix} \dfrac{1}{L_x} \dfrac{[\partial \boldsymbol{N}(1/4,\ 1/2)]}{\partial r} \\ \dfrac{1}{L_y} \dfrac{[\partial \boldsymbol{N}(1/4,\ 1/2)]}{\partial s} \end{bmatrix} \{\boldsymbol{\delta}^R\},$$

$$\begin{Bmatrix} \dfrac{\partial T}{\partial x} \\ \dfrac{\partial T}{\partial y} \end{Bmatrix} = \frac{1}{384} \begin{bmatrix} -18 & 12 & 6 & -36 & 24 & 12 \\ -27 & -18 & -3 & 0 & 0 & 0 \end{bmatrix.$$

$$\times \begin{matrix} -18 & 12 & 6 \\ 27 & 18 & 3 \end{bmatrix} \{\boldsymbol{\delta}^R\} = \begin{Bmatrix} 2 \\ 4 \end{Bmatrix} {}^\circ\mathrm{C/cm}$$

as required to pass the second step of the patch step, which essentially validates the derivation of the conduction matrix and all of the programming steps.

As another spot check is that the sum of each row in any IGA $\left[\boldsymbol{B}^{\boldsymbol{R}}(r, s)\right]$ matrix must be zero, as they are above.

It is generally useful to have the components of the physical gradient vector stored as AP values as well. This can be easily done by adding a loop over each parametric sub-region and gathering the AP parametric coordinates

$$[\boldsymbol{r}\ \boldsymbol{s}]' = \begin{bmatrix} 0 & 0.5 & 1 & 0 & 0.5 & 1 & 0 & 0.5 & 1 \\ 0 & 0 & 0 & 0.5 & 0.5 & 0.5 & 1 & 1 & 1 \end{bmatrix}.$$

Gather the region's temperatures, evaluate the B-matrix at each AP, multiply that B-matrix by the temperatures, and append the gradient components to the temperature values at the AP. Unlike FEA results, the IGA solutions generally have at least C^1 continuous temperatures, so an AP does not need to have gradients from adjacent sub-regions averaged. That is, each AP gradient only needs to be calculated once.

Returning to the eight Dirichlet reactions at the [Fixed] analysis points, their values are

$$\{\boldsymbol{R}\}' = \begin{bmatrix}-48 & -32 & -16 & -16 & 16 & 16 & 32 & 48\end{bmatrix} W.$$

Without heat generation, Robin (convection) conditions, or non-zero Neumann conditions (normal heat flux), the (positive) amount of heat flowing in must equal the (negative) values flowing out of the boundaries. That is, the above-mentioned example reaction values must sum to zero, and they do.

For completeness here, remember that the system thermal reactions come from the original weak form integral of the normal heat flux

$$I_N = t\int_{\Gamma_N} (T_\Gamma)\left(\kappa_n \frac{\partial T}{\partial n}\right) d\Gamma = t\int_{\Gamma_N} (T_\Gamma)(-q_n)\, d\Gamma$$

$$= -t\int_{\Gamma_N} (T_\Gamma)(q_n)\, d\Gamma,$$

where q_n is the heat flux per unit area entering the boundary. When the temperature (T_Γ) is specified as a Dirichlet condition, the corresponding normal heat flow can be recovered. Of course, the boundary of the domain is the non-overlapping union of individual boundary sub-regions,

$$\Gamma = \sum_b \Gamma^b, \text{ so } \{\boldsymbol{c_q}\} = \sum_b \left\{\boldsymbol{c}_q^b\right\}.$$

In this rectangular analysis geometry, the four boundary segments are lines of either constant r- or constant s-values. Consider the temperature distribution for the three analysis points along the boundary of $s = 0$. The general interpolation of

$$[\boldsymbol{N}(r, s = 0)] = \begin{Bmatrix}(r-1)^2 \\ 2r(1-r) \\ r^2\end{Bmatrix} \begin{bmatrix}(s-1)^2 & 2s(1-s) & s^2\end{bmatrix}$$

$$\rightarrow \begin{Bmatrix}(r-1)^2 \\ 2r(1-r) \\ r^2\end{Bmatrix} \begin{bmatrix}1 & 0 & 0\end{bmatrix} \equiv \left[\boldsymbol{N^b}(r)\right]$$

reduces to only three non-zero entries on that quadratic boundary and the others. A typical applied external heat flux per unit area or Dirichlet reaction resultant on a boundary segment is

$$I_N = t\int_{\Gamma_N} (T_\Gamma)(\kappa_n \frac{\partial T}{\partial n})d\Gamma = t\int_{\Gamma_N} (T_\Gamma)(-q_n)d\Gamma = -t\int_{\Gamma_N} (T_\Gamma)(q_n)\, d\Gamma,$$

$$I_N = -t\int_{\Gamma_N} \left[\left[\boldsymbol{N^b}(r)\right]\left\{\boldsymbol{\delta}^b\right\}\right]^T q_n(r)\, d\Gamma = -\left\{\boldsymbol{\delta}^b\right\}^T t$$
$$\times \int_{\Gamma_N} \left[\boldsymbol{N^b}(r)\right]^T q_n(r)\, d\Gamma = -\left\{\boldsymbol{\delta}^b\right\}^T \left\{\boldsymbol{c}_q^b\right\},$$

where the normal heat flux source vector is defined as

$$\left\{\boldsymbol{c}_q^b\right\} = t\int_{\Gamma_N} \left[\boldsymbol{N^b}(r)\right]^T q_n(r)\, d\Gamma.$$

Assume the input heat flux to be a constant and the boundary be $\Gamma(r) = \boldsymbol{x}_1^b + r\,L_x$. Then, the physical differential line length is $d\Gamma = L_x\, dr$, leading to the source integral being completely defined parametrically:

$$\left\{\boldsymbol{c}_q^b\right\} = q_n\, t\int_{r=0}^{r=1} \left[\boldsymbol{N^b}(r)\right]^T (L_x\, dr) = q_n\, t\, L_x\int_{r=0}^{r=1} \left[\boldsymbol{N^b}(r)\right]^T dr.$$

Here,

$$\left\{\boldsymbol{c}_q^b\right\} = q_n\, t\, L_x\int_{r=0}^{r=1} \begin{Bmatrix} (r-1)^2 \\ 2r(1-r) \\ r^2 \end{Bmatrix} dr = \frac{q_n\, t\, L_x}{3}\begin{Bmatrix} 1 \\ 1 \\ 1 \end{Bmatrix}.$$

Since $t\,L_x$ is the surface area for one boundary edge of the rectangle and q_n is the heat flux per unit area entering the boundary, the total heat flux entering, $q_n t L_x$, is distributed equally to the three APs on the edge.

Even when using low-order basis functions, an actually IGA application uses numerical integration to evaluate the integrals in parametric space. For example, the internal heat source resultant contribution was earlier listed as

$$I_q = -t \int_\Omega T(x, y)\, Q(x, y) d\Omega = -t \int_\Omega T\,(r, s)\, Q(r, s)\, d\Omega,$$

$$I_q = -t \int_\Omega \left[\left[\boldsymbol{N^R}(r, s) \right] \left\{ \boldsymbol{\delta}^R \right\} \right]^T Q(r, s)\, d\Omega = \left\{ \boldsymbol{\delta}^R \right\}^T t$$

$$\times \int_\Omega \left[\boldsymbol{N^R}(r, s) \right]^T Q(r, s)\, d\Omega \equiv \left\{ \boldsymbol{\delta}^R \right\}^T \left\{ \boldsymbol{c}_Q^R \right\}.$$

If the rate of heat generation per unit volume, Q, is constant, then the resultant source vector is just that rate times the integral of the basis functions

$$\left\{ \boldsymbol{c}_Q^R \right\} = Q\, t \int_\Omega \left[\boldsymbol{N^R}(r, s) \right]^T d\Omega,\ 0 \le rs \le 1,$$

$$\left\{ \boldsymbol{c}_Q^R \right\} = Q\, t \iint_{00}^{11} \left[\boldsymbol{N^R}(r, s) \right]^T (L_x L_y\, dr\, ds)$$

$$= \frac{Q\, t\, L_x L_y}{9} \begin{bmatrix} 1 & 1 & 1 & 1 & 1 & 1 & 1 & 1 & 1 \end{bmatrix}.$$

Observe that the total heat generated in the region, QtL_xL_y, is equally distributed to each of the sub-regions APs.

Remember, to get the differential volume and physical derivatives, $\left[\boldsymbol{B^R}(r, s) \right]$, of the basis functions, it is always necessary to utilize the Jacobian matrix from the unit coordinate space to physical space. That is,

$$d\Omega = t\, dA \equiv t\, \left| \mathbb{J}_{r,s:x,y}\,(r, s) \right| (dr\, ds)$$

and

$$\left[\boldsymbol{B^R}(r, s) \right] \equiv \begin{bmatrix} \dfrac{[\partial \boldsymbol{N}(r, s)]}{\partial x} \\[2ex] \dfrac{[\partial \boldsymbol{N}(r, s)]}{\partial y} \end{bmatrix} = \left[\boldsymbol{J}_{r,s:x,y}(r, s) \right]^{-1} \begin{bmatrix} \dfrac{[\partial \boldsymbol{N}(r, s)]}{\partial r} \\[2ex] \dfrac{[\partial \boldsymbol{N}(r, s)]}{\partial s} \end{bmatrix}.$$

To evaluate this IGA integral, and all others, by numerical integration means mapping from where the quadrature data are tabulated to the dimensionless basis function space. There are at least three such dimensionless coordinate tabulations of the same quadrature data: baracentric coordinates, natural coordinates, and unit coordinates. Either of the first two choices will require calculation of a second Jacobian, and its determinant, say $[\boldsymbol{J}_{a,b,r,s}(a,b)]$. This is a trivial task, but in most practical IGA applications, it must be done millions of times. To avoid any unnecessary calculations in IGA studies, simply stay in the unit coordinates when doing the numerical integration. This just requires the one-time-ever step of tabulating the Gauss Quadrature data in unit coordinates. The tabulated 1D Gauss unit coordinate data are accessed via function *qp_rule_unit_Gauss.m*, and for convenience the original data are in function *qp_rule_Gauss.m*.

The mapping from the 1×1 unit coordinate square to the 2×2 natural coordinate Gauss quadrature square has a constant Jacobian matrix with the determinant of four, $|J_{a,b:r,s}\,(a,b)| = 4$. The vast majority of IGA literature uses natural coordinates for integration and requires the second Jacobian term in the numerical integration equations. Herein, since the single knot span covers [0, 1), the second Jacobian is not necessary, but the symbol $|J_{r,s:r,s}| \leftrightarrow |J_{r,s:r,s}\,(r,s)| \equiv 1$ will be included in equations herein as a reminder that most IGA literature shows a second Jacobian to correctly carry out numerical integration.

Evaluating the source term by numerical integration for an arbitrary shape converts it to a summation:

$$\{\boldsymbol{c}_Q^R\} = Q\,t \iint_{00}^{11} \left[\boldsymbol{N}^{\boldsymbol{R}}(r,s)\right]^T (|\mathbb{J}_{r,s:x,y}\,(r,s)|\,dr\,ds),$$

$$\{\boldsymbol{c}_Q^R\} = Q\,t\,|\mathbb{J}_{r,s:r,s}| \sum_{q=1}^{n_q} \left[\boldsymbol{N}^{\boldsymbol{R}}(r_q,s_q)\right]^T |\mathbb{J}_{r,s:x,y}(r_q,s_q)|\,w_q,$$

where the r_q, s_q are tabulated locations each having an associated weight of w_q. In the prior special example, the determinant, $|J_{r,s:x,y}| = L_x L_y$, was the physical area of the rectangle.

IGA numerical integration uses the tensor product of the first-direction quadrature data with the second-direction quadrature data.

To determine the number of quadrature points in each direction, it is necessary to calculate the polynomial degree of the integrand in each parametric example. The basis functions are quadratic in the r-direction (degree 2). In the prior example, it was constant (degree 0). Thus, the total polynomial degree in the r-direction is $(2 + 0) = 2 = D_r$. The same degrees exist in the s-direction. Gauss proved that the number of points required in the r-direction to exactly integrate the polynomial is

$$D_r \leq (2nr_q - 1) \quad \text{or} \quad 2 \leq (2nr_q - 1) \Rightarrow nr_q = 2.$$

Here, the s-polynomial contributes the same polynomial degree and also requires $ns_q = 2$. The tensor product rule requires a total number of $n_q = nr_q * ns_q = 4$ terms in the summation. That is often called a 2-by-2 rule. In general, those two numbers differ in each direction. When the Jacobian matrix is not constant, usually one extra point is added in each direction for 'badly' distorted regions. NURBS and T-splines are rational polynomials which are not exactly integrated by Gaussian quadratures. For them, the number of points may be raised or a more economical method like collocation may be used.

9.7 Validation Options

Soon after the development of FEA technology, it was applied to components involved with general engineering, national defense, and public safety. Various engineering societies including the American Society of Mechanical Engineers (ASME), government agencies like the US Nuclear Regulatory Commission (USNRC), and international organizations including the International Association for the Engineering Analysis Community (NAFEMS) developed benchmark and verification simulation tests to compare FEA results with known analytical and/or experimental results (available from 1950s). In the USA, a commercial FEA system cannot be sold to the public until it has satisfied all such problems within its range of capabilities. There are hundreds of such validation models. Some FEA codes, like ANSYS, IDEAS, and NASTRAN, publish large validation manuals (now online) with their supporting data to verify the accuracy of their software.

NAFEMS also certifies engineers as being competent in selected fields and specific FEA codes by requiring that they independently solve the verification problems starting with only the given problem statements. In Europe, there are some engineering jobs where you must have the proper NAFEMS certification(s) to be employed.

9.8 Example IGA Validation Study

This study will be compared to the NAFEMS heat transfer problem number 9.2. It is for a rectangular region having a constant rate of heat generation and a constant temperature on the outer perimeter. As shown in Fig. 9.6, the solution has quarter-symmetry, so this IGA study will only use the lower left quarter of the interior space. The Dirichlet (essential) boundary conditions apply to the analysis control points on the outer perimeter (AP numbers 1–4, 5, 9, 13). The internal symmetry lines have the Neumann (natural) boundary condition of zero normal slopes, at AP numbers 8, 12, 14, 15, and 16.

The analytic solution is known, and was compared to the temperature and heat flux results from a 5,000 dof finite element study displayed in Fig. 9.7. The IGA implementation for this application with four bi-quadratic iElements is listed in Script 9.5, which illustrates using the analytic basis functions.

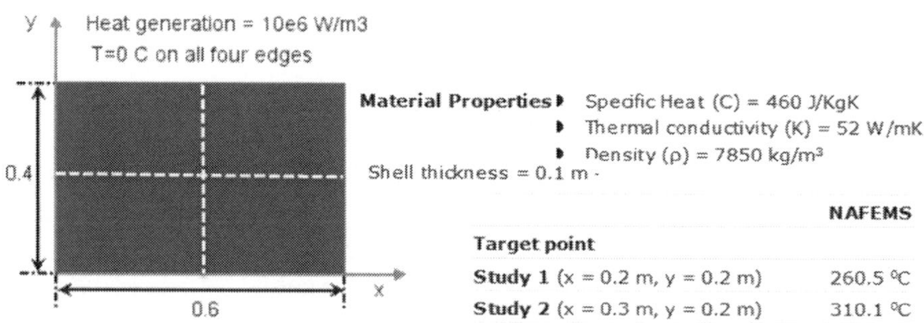

Figure 9.6. Symmetric region with internal heat generation

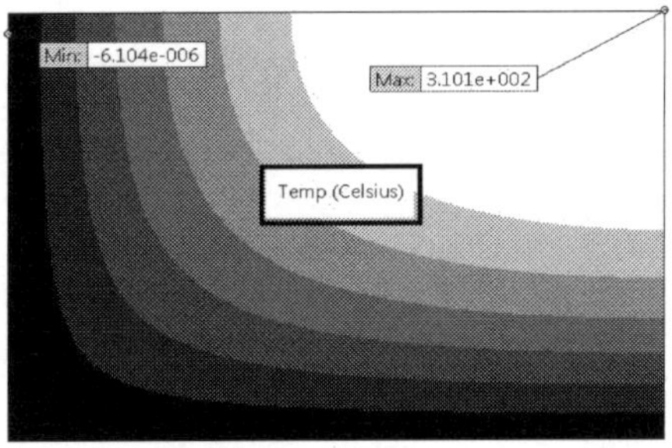

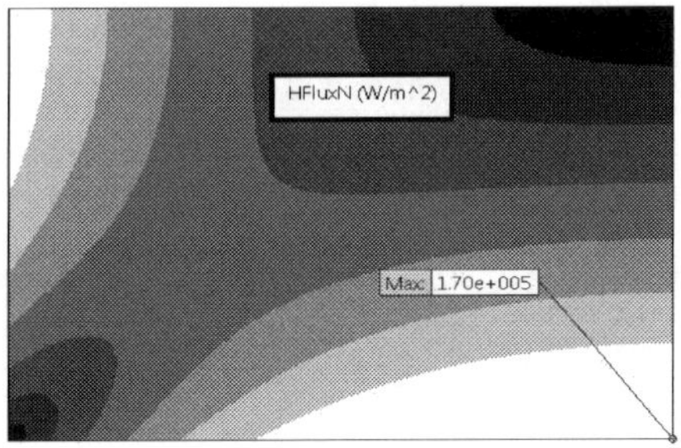

Figure 9.7. Temperature and heat flux from 5,000 dof FEA solution

Script 9.5. Four iElement numerically integrated solution to NAFEMS 9.2

```
%                          NAFEMS_92_example.m
%  NAFEMS_9.2 rectangle w uniform heat generation, 1/4 symmetry
diary NAFEMS_92.txt                               % save the results
%                        Symmetric       Ly
%          13 o------o14--+--o15-----o16       o = control point with
%             |iE3         |iE4        |       temperatures (16 dof)
%           9 o      o10  |  o11     o12
%    T = 0    |-----------+----------|   Symmetric
%           5 o      o6   |  o7      o8
%             |iE1         |iE2        |    Q constant generation
%           1 o------o2---+--o3------o4   Lx
%                       T = 0
clear all             % Here using bi-variant basis functions, p=q=2
%        Problem data and geometry, V = [0 0 0 0.5 1 1 1]
Q = 1e6; k = 52; K = [k  0; 0  k]; Lx = 0.3; Ly=  0.2;     % data
n_g = 1; n_q = 4; n_r = 2; n_s = 2 ; % dof/AP, qp, B rows, spaces

%         Selected AP locations (also try 0 1/3 2/3 1)
AP_list=[1 2 3 5 6 7 9 10 11; 2 3 4 6 7 8 10 11 12; ...      % AP
         5 6 7 9 10 11 13 14 15; 6 7 8 10 11 12 14 15 16];  % pts
AP_x = [0, 1/4, 3/4, 1, 0, 1/4, 3/4, 1, ...          % x-coordinates
        0, 1/4, 3/4, 1, 0, 1/4, 3/4, 1]*Lx      ; % rectangle only
AP_y = [0, 0, 0, 0, 1/4, 1/4, 1/4, 1/4, ...          % y-coordinates
        3/4, 3/4, 3/4, 3/4,1, 1, 1,  1]*Ly      ; % rectangle only
```

```
%       Count number of iElements and AP per iElement
n_e = size (AP_list, 1); n_n = size (AP_list, 2)    ; % constants
n_cp = max(max(AP_list)); n_d = n_cp*n_g; n_i = n_n*n_g;
SS = zeros(n_d,n_d); cS = zeros(n_d,1)         ; % system matrices

%       Boundary conditions for quarter symmetry (try none)
Fixed = [1 2 3 4 5 9 13]; Given = [0 0 0 0 0 0 0] ; % EBC, values
Free  = [6 7 8 10 11 12 14 15 16]              ; % manually picked
AP_xy(:, 1) = AP_x; AP_xy(:, 2) = AP_y    ; % reshape coordinates

%        Knot vector data in parametric and index spaces
range_U=[0 0.5; 0.5 1] ; range_V=[0 0.5; 0.5 1]   ; % param space
range_I=[0 0.5; 0.5 1] ; range_J=[0 0.5; 0.5 1]   ; % index space
span_U = range_U(:,2)- range_U(:,1)            ; % parametric length
span_V = range_V(:,2)- range_V(:,1)            ; % parametric length
Index=[1 1; 2 1; 1 2; 2 2]             ; % I,J for e in index space

%     Open units needed for post-processing operations
el_unit = fopen ('el_store.bin', 'w');% open for iElement writing
qp_unit = fopen ('qp_store.bin', 'w')     ; % open for GQ writing
[r_q, s_q, w_q] = qp_rule_unit_quad (n_q)   % get quadrature data

%             Form matrices, assembly system of equations
Area = 0                    ; % Initialize domain area (for checking)

  for e= 1:n_e;% Loop over iElements (knot spans) ===>> ===>> ===>>
  Area_e = 0                   ; % initialize area for this iElement
  dof    = AP_list (e, :)    ; % Gather AP dof # for this iElement
  xy_e   = AP_xy(dof, :)    ; % Gather their x- and y- coordinates
  I  = Index (e ,1); J = Index (e ,2); % iElement column & row #s
  cQ = zeros (n_i, 1) ; Se = zeros (n_i, n_i)    ; % allocate the
  B  = zeros (n_r, n_i)                          ; % local arrays

  for qp = 1:n_q ;% loop over Gauss points    ---> ---> ---> --->
    %  Compute quadrature pt coordinates in parameter space
    R = range_I(I,1) + span_U(I)*r_q(qp); % location in knot span
    S = range_J(J,1) + span_V(J)*s_q(qp); % location in knot span
    J2 = span_U(I)*span_V(J) / 1     ; % for unit rule Gauss [0,1]

    % Analytically evaluate basis functions and first derivative
    if ( I == 1 )    % use numerical function here
      Nr=[(2*R - 1)^2, -2*R*(3*R - 2), 2*R^2]   ; % first I-span
      dNr_dr = [(8*R - 4), (4 - 12*R),4*R]    ; % dNr/dr dNr/ds=0
    else                                         ; % second I-span
      Nr=[(2*R - 2)*(R - 1), (-6*R^2 + 8*R - 2),(2*R - 1)^2] ;
```

```
      dNr_dr=[(4*R - 4),(8 - 12*R), (8*R - 4)]; % dNr/dr dNr/ds=0
    end                                ; % if which r-span to use
    if ( J == 1 )
      Ns=[(2*S - 1)^2, -2*S*(3*S - 2), 2*S^2]     ; % first J-span
      dNs_ds = [(8*S - 4), (4 - 12*S),4*S]    ; % dNr/ds, dNr/dr=0
    else                                            ; % second J-sp
      Ns=[(2*S - 2)*(S - 1), (-6*S^2 + 8*S - 2),(2*S - 1)^2] ;
      dNs_ds=[(4*S - 4),(8 - 12*S), (8*S - 4)] ;
    end                                ; % if which r-span to use

    %  Tensor products of basis functions & partial derivatives
    Nrs      = Nr' * Ns      ; Nrs      = reshape (Nrs,      1, 9);
    dNrs_dr = dNr_dr' * Ns ; dNrs_dr = reshape (dNrs_dr, 1, 9);
    dNrs_ds = Nr' * dNs_ds ; dNrs_ds = reshape (dNrs_ds, 1, 9);
    Grad_rs = [dNrs_dr; dNrs_ds]          ; % Local Gradient matrix

    %    Find the physical coordinates, Jacobian, determinant
    xy_q    = Nrs * xy_e ; Jac = Grad_rs * xy_e;% at quadrature pt
    det_J   = det(Jac) ; inv_J = inv(Jac)      ; % inverse Jacobian
    Area_e = Area_e + J2 * det_J * w_q(qp; % iElement area update
    B_q     = inv_J * Grad_rs        ; % Get physical gradient matrix

    %      Update the iElement heat source Int Q N' dArea
    cQ = cQ + Q * Nrs'   * J2 * det_J * w_q(qp)   ; % generate heat
    Se = Se + B_q' * K * B_q *J2 * det_J * w_q(qp) ; % conduction
    %    Save data at quadrature point for post-processing
    fwrite (qp_unit, xy_q, 'double'); % save physical coordinates
    fwrite (qp_unit, K,     'double')              ; % save material
    fwrite (qp_unit, B_q,   'double')     ; % save physical gradient
    fwrite (qp_unit, Nrs,   'double'); % save basis functions here

  end ; % for qp loop over quadrature points  <--- <--- <--- <---
  Area = Area + Area_e     ; % update total domain area as a check

  %     Scatter iElement cQ into c & Se into S, system arrays
  cS(dof) = cS(dof)+cQ(:); SS(dof, dof) = SS(dof, dof)+Se(:,:);

  % Save element arrays for post-processing T and heat flux
  fwrite (el_unit, xy_e, 'double')           ; % save AP coordinates
  fwrite (el_unit, K,     'double')              ; % save properties
  fwrite (el_unit, Se,    'double')              ; % save conduction
end  ; % for e loop over iElements  <<=== <<=== <<=== <<=== <<===

fprintf ('Area of domain is %g \n', Area)     ; % checking results
fprintf ('Expected value is %g \n', Lx*Ly)    ; % checking results
fprintf ('Sum of heat generation is %g \n', sum(cS)) ; % checking
fprintf ('Expected total value is %g \n', Lx*Ly*Q)    ; % checking
```

```
% Now solve the temperatures and Dirichlet (thermal) reactions
[T, DR] = Solve_Free_DOF(SS, cS, Free, Fixed, Given);%solve for T
disp('Expecting center T = 309.7540 computed is '), disp(T(n_d))

% Post-process for temperatures, heat flux mag. and vectors
fprintf ('iElement Post-processing: \n')
fclose (qp_unit); fclose (el_unit); % now available below to read

%     Open (inefficient) text files for plotting or reading
qp_unit = fopen ('qp_store.bin', 'r')         ; % open for reading
el_unit = fopen ('el_store.bin', 'r')         ; % open for reading
q_id    = fopen ('iga_qp_xyz_fluxes.txt', 'w') ; % heat flux etc.
% file iga_qp_xyz_fluxes.txt header line is n_e, n_cp, n_i, n_q

fprintf(q_id,'%i %i %i %i %i %i \n', n_e, n_cp, n_i, n_q, 0, 0);
for j = 1:n_e; % loop over iElements   ====>> ====>> ====>> ====>>
  dof = AP_list (e, :)       ; % Gather AP dof # for this iElement
  xy_e = AP_xy(dof, :)      ; % Gather their x- and y- coordinates
  T_e = T (dof)           ; % Gather temperatures for this iElement

  %     Recover previously calculated iElement arrays
  [xy_e] = fread (el_unit, [n_n, n_s], 'double');% AP coordinates
  [K]    = fread (el_unit, [n_r, n_r], 'double')   ; % properties
  [Se]   = fread (el_unit, [n_i, n_i], 'double')   ; % conduction

  for qp = 1:n_q ;% loop over Gauss points     ---> ---> ---> --->
    %     Recover previous calculations at this quadrature point
    [xy_q] = fread (qp_unit, [1, n_s],    'double');%physical cord
    [K]    = fread (qp_unit, [n_r, n_r], 'double')   ; % material
    [B_q]  = fread (qp_unit, [n_r, n_i], 'double')    ; % grad op
    [Nrs]  = fread (qp_unit, [1, n_i],    'double')   ; % BF at pt
    %         Get temperature and its gradient at point
    T_q      = Nrs * T_e ; Grad_q = B_q * T_e ;
    Vector_q = - K * Grad_q    ; % heat flux vector, Fourier's Law
    Flux     = sqrt(Vector_q(1)^2+Vector_q(2)^2);% flux magnitude

    %             List quadrature point results
    fprintf('\n')                                     ; % blank line
    fprintf('In iElement =           %i          \n',j)
    fprintf('Q_Pt,X-Y Coordinates %i, %g %g \n',qp,xy_q (1:2))
    fprintf('Q_Pt,Temperature      %i, %g    \n',qp,T_q)
    fprintf('Q_Pt,Gradient Vector %i, %g %g \n',qp,Grad_q (1:2))
    fprintf('Q_Pt,Heat Flux        %i, %g    \n',qp,Flux)
    fprintf('Q_Pt,Flux Vector      %i, %g %g \n',qp,Vector_q(1:2))
```

```
    % save to iga_qp_xyz_fluxes.txt for contours, quiver, etc.
    fprintf(q_id,'%g %g, %g, ', xy_q(1,1), xy_q(1,2), T_q)
    fprintf(q_id,'%g, %g %g ', Flux, Vector_q(1,1),Vector_q(2,1))
    fprintf(q_id,'\n')                                  ; % line feed
  end  ; % for qp quadrature points <--- <--- <--- <--- <--- <---
end  ; % for each j iElement in mesh  <<==== <<==== <<==== <<====
fclose(q_id)                        ; % close the flux data plot file
% end NAFEMS_92_example ========================================
```

Script 9.5 with the 4 iElements and 16 dof was run to check the maximum temperature against the analytic solution, at the top right corner (at AP number 16), and against the 5,000 dof FEA study. That temperature difference was less than 1%. The maximum heat flux value occurs at the lower right corner at AP number 4. The quadrature point closest to that control point has a heat flux magnitude was within 8% of the analytic maximum heat flux value.

To enhance the IGA Poisson solver model, input arguments were added to specify the number of iElements in each direction, along with a shared polynomial degree. In each direction, the number of iElements of the given degree was used to build a governing uniform knot vector which then defines the basis functions.

Each of the two knot vectors was created with the simple steps in *uniform_knot_vector.m*, in Script 9.6. That enhancement required replacing the analytic basis functions in Script 9.5 with numerical values of the basis functions from the mesh knot vectors evaluated at the quadrature point. That step was preceded by taking the tensor product of the $(p + 1)$ non-zero segments of each of the basis functions. Of course, rather than using a uniform knot vector from Script 9.5 a user could define two non-uniform mesh knot vectors manually before calling the Poisson solver.

The physical mesh AP coordinates were scaled by the Greville abscissa of each of the two knot vectors, since that is known to cause a constant Jacobian for each iElement and reduces the required number if integration points. For a given analysis knot vector of degree p with n control points the Greville abscissa are defined as averaged points from the knot vector:

$$G_I = (r_{I+1} + r_{I+2} + \cdots + r_{I+p})\,/p,\ I = 1 \cdots n$$

For example, V $= \{0\,0\,0\,1\,2\,3\,4\,4\,4\}$ gives $n = 6$ so $G = \{0\,1\,3\,5\,7\,8\}\,/8$ relative locations.

Script 9.6. Convert the number of iElements into a knot vector

```
function [V_r] = uniform_knot_vector (n_e, deg) % ==========
% Create open uniform knot vector with internal knots 1:n_e,
% preceeded with deg+1 zeros, and appended with deg+1 values
% of n_e. For example, arguments (5,2) give the uniform knot
% vector V_r = [0 0 0 1 2 3 4 5 5 5]
%
% n_e = number of iElements in r direction
% deg = polynomial order, default is cubic,3
% V_r = returned open uniform knot vector
%
if ( nargin == 1 ); deg = 3; end          ; % if deg defaults
open = n_e + 1 + 2*deg                   % size of output vector
V_r = zeros (1, open)                     ; % allocate memory
q = deg + 1                                ; % polynomial order
V_r (1:q) = 0                            % initial (deg+1) zeros
V_r (q+1:q+n_e-1) = (1:n_e-1)              ; % interior knots
V_r (q+n_e:2*q+n_e-1) = n_e)              ; % pad (deg+1) n_e
% end uniform_knot_vector ===================================
```

The enhanced version of Script 9.5 was re-run with a relatively crude 5×5 quadratic mesh with 49 dof. Figure 9.8 shows the Matlab *surf* AP temperature plot for the 49 dof model along with the Matlab *contourf* color filled contour plot of those temperatures. Those contours should be compared to the 5,000 dof FEA temperature contours in Fig. 9.8, with slightly different color bars. The actual IGA temperatures in the domain must be plotted using the product of the shown AP values times their basis functions. That process is much slower, but gives smoother and more accurate physical results.

The accuracy of the heat flux vectors and their maximum values is usually more important than the temperature accuracy. On the left in Fig. 9.9, the IGA heat flux vectors at the quadrature points are shown, and the right portion shows the averaged FEA nodal heat flux vectors, and they agree very well. The extra post-processing steps for producing these fast IGA plots follow after the last line given in Script 9.5, and are listed in Script 9.7.

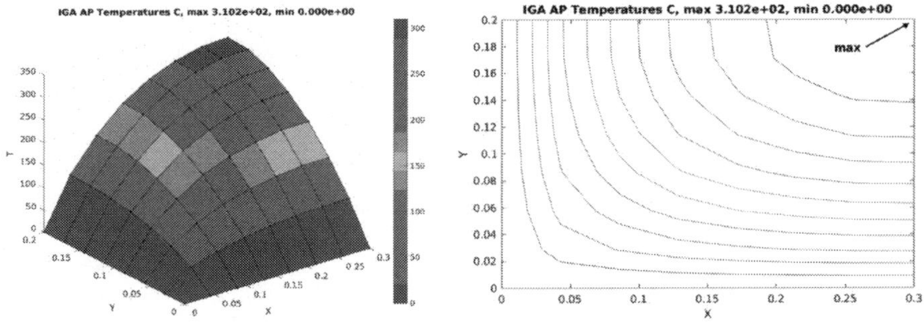

Figure 9.8. Matlab plots of the 49 dof IGA AP temperatures and their contours

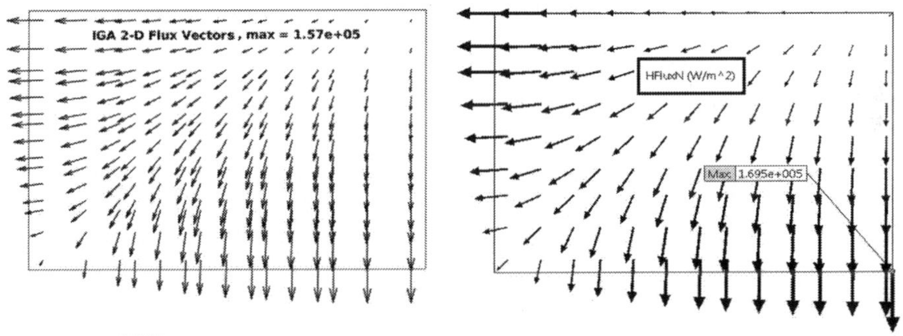

Figure 9.9. Heat flux vectors from 49 dof IGA (left) and 5,000 dof FEA studies

Script 9.7 Additional commands for plotting Poisson equation AP results

```
fclose(q_id)         ; % close the flux data file for plots, etc.
if ( vector == 1 )    ; % plot q_id flux at all quadrature pts
iga_quiver_qp_flux (1.5,1,1)    ; % inputs (scale, inc_g, sign)
end                              ; % if show qp heat flux vectors
if ( color == 1 )                ; % prepare plots, with colormap
colormap(jet(15))            ; % override the ugly OpenGL colors
AP_x = reshape (AP_x, [x_CP, y_CP])    ; % for Matlab functions
AP_y = reshape (AP_y, [x_CP, y_CP])    ; % for Matlab functions
T = reshape (T', [x_CP, y_CP])         ; % for Matlab functions
%                         AP result surface
clf                                 ; % clear the graphics frame
plot3(AP_x, AP_y, T,'k.')            ; % FORCE 3d axes to appear
```

```
hold on                                    ; % keep the following
grid on                          ; % grid the two background walls
surf (AP_x, AP_y, T)              ; % AP temperature color surface
colorbar                     ; % add a color bar to show the scale
xlabel('X'); ylabel('Y'); zlabel('T')              ; % axis labels
title (['IGA AP Temperatures C, max ', ...                % title
    num2str(max_T,'%8.3e'), ', min ', num2str(min_T,'%8.3e')])
print ('-dpng', ['AP_values_surface'])  ; % name,save the plot
v_text = ['Created AP_values_surface.png'] ; % record the name
fprintf (1,'%s', v_text) ; fprintf (1, ' \n')     ; % note name
%                           color fill AP contours
clf ;                                  % clear the graphics frame
contourf (AP_x, AP_y, T, 8); % fill colors between AP contours
hold on                                    ; % keep the following
colorbar                     ; % add a color bar to show the scale
axis('equal'); xlabel('X'); ylabel('Y')            ; % axis labels
title (['IGA AP Temperatures C, max ', ...                % title
    num2str(max_T,'%8.3e'), ', min ', num2str(min_T,'%8.3e')])
print ('-dpng', ['contour_AP_values'])        ; % save the plot
v_text = ['Created contour_AP_values.png']  ; % save plot name
fprintf (1,'%s', v_text) ; fprintf (1, ' \n' )    ; % note name
end                                          ; % if color plots
```

9.9 Discussion of Validation Studies

Consider a thermal problem including specified temperatures, insulated (symmetric) boundaries, edge convection (NAFEMS thermal analysis test 9, 1986), and two knot spans, which is shown in Fig. 9.10. The dashed arrows in the corner are indication that the geometry is a

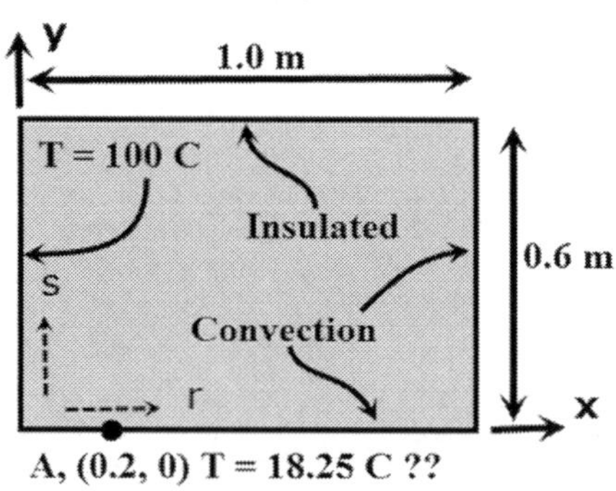

Figure 9.10. Thermal validation problem 1

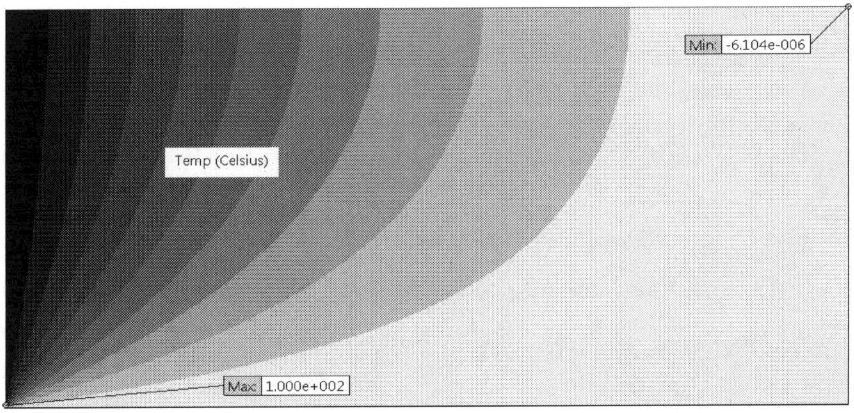

Figure 9.11. FEA validation test case temperatures from an FEA study

one-to-one mapping from the parametric space to the physical space. The rectangular plate with $\Delta x = 0.6\,\mathrm{m}$ by $\Delta y = 1\,\mathrm{m}$ and $t = 0.01\,\mathrm{m}$ thick is subjected to a Dirichlet condition uniform temperature of $100°\mathrm{C}$ on the $y = 0$ edge, is insulated or symmetric along the $x = 0$ edge (a natural condition with $q_n \equiv 0$), and has Neumann convection at the $x = 1\,\mathrm{m}$ edge and the $y = 0\,\mathrm{m}$ edge. The heat transfer coefficient of convection on those edges is $h^b = 750\,W/\mathrm{m}^{2}°\mathrm{C}$ with the surrounding fluid temperature being $T_\infty^b = 0°\mathrm{C}$. The thermal conductivity of the solid is $k = 52\,W/\mathrm{m}°\mathrm{C}$. The temperature computed at point A ($x = 0.2$ m, $y = 0x4$ m) should be $T_A = 18.25°\mathrm{C}$ according to NAFEMS. The analytical solution is known via a doubly infinite series. A fine mesh FEA solution for the temperature is shown in Fig. 9.11. A graph of the temperature along $y = 0$ shows a rapid increase in the temperature approaching the origin. In the neighborhood of the origin, a zoom of the distribution of the heat flux vectors, length scaled to its magnitude, confirms the presence of rapidly changing heat flux vectors, also meaning large temperature gradient vectors.

There are many such classical heat transfer problems used to validate numerical simulations. Such 'exact' analytic solutions are usually obtained by doubly infinite series. Such problems often have one constant temperature given as a Dirichlet condition on one edge with a different constant temperature prescribed on another edge.

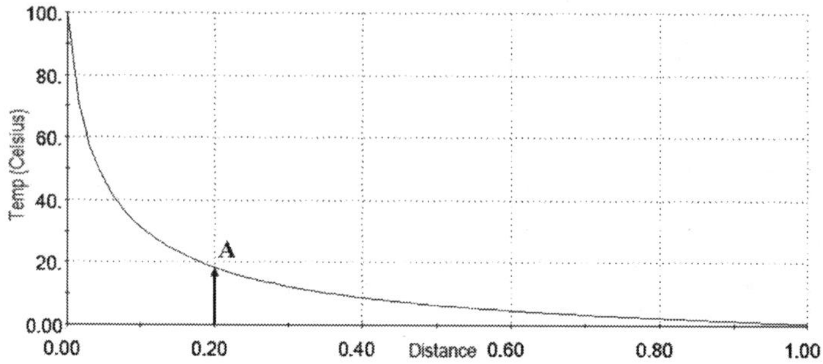

Figure 9.12. FEA validation temperature graph along $y = 0$

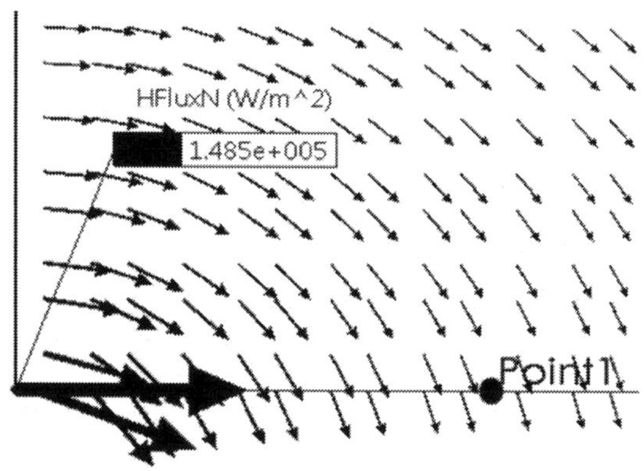

Figure 9.13. FEA heat flux magnitudes and heat flux vectors near point A

The question then arises as to what is the proper IGA Dirichlet condition at a corner where such lines meet? Is it the larger, the smaller, or the average of the two constants? The practical answer is the value last appearing in a data file.

Such field analysis problems have a mathematical singularity at such intersection points, meaning that in theory there is an infinite solution gradient at such locations. Many terms in an infinite series can come close to matching that theoretical prediction. Any IGA simulation will not recover an infinite gradient, nor should any user attempt to do so. Very good numerical results near a singularity can be obtained in IGA studies with a few layers of refinements near such a point. Practically, such refinements require the second IGA concept of using T-spline basis functions for refinements. As will be discussed later, T-spline basis functions can be extracted into another form which still retains the exact geometry and can be easily interfaced into unstructured finite element software. The set of extracted polynomials herein is denoted as an iGe and called an extract as a reminder of how different it is from classical finite elements. The IGA literature simply calls it an element or Bezier extraction element.

When an IGA study clearly needs many, say 20 to 40, iElements (knot spans) in each parametric direction, the approach of using analytically generated basis functions lacks the simplicity of using strictly numerical evaluations of the basis functions and their derivatives. There is no restriction on the length of the knot vector input to the function *symbolic_N_bases.m* to create the basis functions for many knot spans. For example, if a user wanted eight knot spans in a parametric direction, the analysis knot vector $\mathbb{A} = \{0\ 0\ 0\ 1\ 2\ 3\ 4\ 5\ 6\ 7\ 8\ 8\ 8\}$ could be used. It would require using the Matlab 'switch' structure to pick the knot span (iElement) in which a particular quadrature point falls. The much more common procedure is to use the same knot vector, plus another, as input into the function *num_basis_rect.m* (given later) to automatically numerically evaluate the 2D basis functions using the Cox-de Boor recurrence algorithm.

Returning to test problem in Fig. 9.10, note that it requires one to include convection sources along two of the four edges to the system matrix equations for thermal equilibrium. Each contribution requires two line integrals. The square matrix term is

$$\left[\boldsymbol{M}_h^b\right] = \int_{L^b} \left[\boldsymbol{N}(r)^b\right]^T h^b \left[\boldsymbol{N}(r)^b\right] dL(r),$$

and the column source vector is

$$\left\{\boldsymbol{c}_h^b\right\} = \int_{L^b} \left[\boldsymbol{N}(r)^b\right]^T h^b T_\infty^b dL(r).$$

Those line integrals are not included in the symbolic 2D conduction matrix evaluations in Script 9.6, which is considered next, but those evaluations can be easily extended to carry out the simpler 1D matrix generations.

9.10 Choice of NAFEMS Basis Functions

A practical solution of the Poisson problem in Fig. 9.10 by IGA will involve the use of NURBS or T-splines of at least degree $p = 3$, and in practice all of the analysis stages must be evaluated numerically. For educational benefits and to clarify the individuals steps involved, this study will initially utilize low-degree B-splines. Experience suggests that the temperature changes more quickly in the x-direction, and a corresponding quadratic analysis knot vector of $\mathbb{A}^{(r)} = \{0,\ 0,\ 0,\ 1/2,\ 1,\ 1,\ 1\}$ is selected in the associated r-direction. In the y-direction, the even simpler analysis knot vector is chosen as the linear $\mathbb{A}^{(s)} = \{0,\ 0,\ 1,\ 1\}$ in the associated s-direction. The Bezier basis functions in the x-direction have four CPs. In unit coordinates (r), the non-zero valued basis functions are listed in Appendix A1 as

Degree = 2, Number of CP = 4, Number of spans = 2 on [0, 0.5), [0.5, 1).

Iff CP x-coordinates = $[0\ 1/4\ 3/4\ 1]L_x$ then $|\mathbb{J}_{r:L}| = L_x/2$ for both spans.

Count	Basis Functions $N(r)$		Derivatives: $dN(r)/dr$	
1	$[(1-2r)^2,$	$0]$	$[8r-4,$	$0]$
2	$[2r(2-3r),$	$2(1-r)^2]$	$[4-12r,$	$4r-4]$
3	$[2r^2,$	$-6r^2+8r-2]$	$[4r,$	$8-12r]$
4	$[0,$	$(1-2r)^2]$	$[0,$	$8r-4]$
iE	1	2	1	2

The cited control x-coordinates were selected to make the determinant of the geometry mapping Jacobian constant, $|J_{r:L}| = L_x/2$, in both r-knot spans. This is rarely possible in practical applications.

The linear Bezier basis functions in the second direction have two CPs. In unit coordinates, the s-basis functions from Appendix A1 are

Degree = 1, Number of CP = 2, Number of spans = 1 on [0, 1).

For CP x-coordinates = $[0,\ 1]\ L_y$ the Jacobian is constant, $|\mathbb{J}_{s:L}| = L_y$.

Count	Basis Functions $N(s)$	Derivatives $dN(s)/ds$
1	$[(1-s)]$	$[-1]$
2	$[s]$	$[+1]$
iE	1&2	1&2

where the control point y-coordinates also make the determinant of the geometry mapping Jacobian a constant, $|\mathbb{J}_{s:L}| = L_y$. To obtain the conduction matrices for both the non-zero sub-regions (iElements), the parametric derivatives of the above-mentioned basis functions will also be utilized.

A new important feature is that, since there are now two knot spans (two iElements), additional data are required for the list of AP CPs governing each knot span. The array, [AP_list], has one row per iElement giving the AP numbers governing that knot span. Those AP numbers are shown in Fig. 9.10 and listed in Table 9.1.

In a practical IGA study, the following example would have to be done with NURBS and numerical integration. But, if the weights of all the NURBS are constant, then they degenerate to B-spline basis functions as used in the following Matlab symbolic derivation of the conduction matrix for each knot span in Script 9.8. The resulting conduction matrices are listed in Script 9.9.

Table 9.1. System AP numbers controlling each knot span (iElement)

Knot span	**AP connection list**
1	1, 2, 3, 5, 6, 7
2	2, 3, 4, 6, 7, 8

Script 9.8. Partial symbolic derivation of knot span conduction matrices

```
% ================ Scripted_K_p2_q1_rect.m ==================
%    Symbolic derivation of a conduction & source matrices
%          for an isotropic rectangular iElement
clear all                                          % play safe
syms r s Lx Ly k L Q real                  % symbolic variables
diary scripted_K_2.txt                          % keep a record
syms r s Lx Ly k L Q real                  % symbolic variables
```

```
CP_x = [0, 1/4, 3/4, 1, 0, 1/4, 3/4, 1]*Lx       ; % rectangle
CP_y = [0, 0,   0,   0, 1, 1,   1,   1]*Ly       ; % rectangle
AP_list = [1, 2, 3, 5, 6, 7; 2, 3, 4, 6, 7, 8]  % DOF numbers
r- and s-direction basis functions
R1 = [(1-2*r)^2, 2*r*(2-3*r), 2*r^2]; nb_r = 3  ; % r [0,0.5)
R2 = [2*(1-r)^2, 8*r-6*r^2-2, (1-2*r)^2]        ; % r [0.5,1)
S = [(1-s), s]              ; nb_s = 2           ; % s [0, 1)
%
R1S = R1' * S              ; % tensor product [0,0.5) x [0, 1)
R2S = R2' * S              ; % tensor product [0.5,1) x [0, 1)
N1 = reshape(R1S, 1, nb_r*nb_s)          ; % basis row matrix 1
N2 = reshape(R2S, 1, nb_r*nb_s)          ; % basis row matrix 2
N1sum= sum(sum(N1)) ; N1sum=simplify(N1sum)     ; % sum == 1?
N2sum= sum(sum(N2)) ; N2sum=simplify(N2sum)     ; % sum == 1?
disp('Sum of basis functions set 1 = '), disp(N1sum); % check
disp('Sum of basis functions set 2 = '), disp(N2sum); % check
%
%        Get parametric derivatives of [N(r,s)]
N1drs = [diff(N1,r) ; diff(N1,s)]       ; % r-, s-derivatives 1
N2drs = [diff(N2,r) ; diff(N2,s)]       ; % r-, s-derivatives 2

%        Get Jacobian matrix for each span
Size(N1drs)
J1 = N1drs * [CP_x; CP_y]
J2 = N2 drs * [CP_x; CP_y]
det_J1 = det(J1), det_J2 = det(J2) % span physical area

%          Get physical derivatives, [B(r,s)]
B1 = inv(J1)*N1drs; B1 = simplify (B1) ; % x-, y-derivatives 1
B2 = inv(J2)*N2drs; B2 = simplify (B2) ; % x-, y-derivatives 2

%       Integrate [N] for a constant source, Q
N1ir = int(N1,r,0,0.5) ; N1irs = int(N1ir,s,0,1); % wrt r- & s-
N2ir = int(N2,r,0.5,1) ; N2irs = int(N2ir,s,0,1); % wrt r- & s-

c_Q1 = simplify (N1irs)*Q*det_J1    ; % resultant source vector
Qsum = sum(c_Q1); disp('Heat generation 1 = '), disp(Qsum)

c_Q2 = simplify (N2irs)*Q*det_J2    ; % resultant source vector
Qsum = sum(c_Q2); disp('Heat generation 2 = '), disp(Qsum)
% Integrand for isotropic conduction matrix, iElement 1 x
Mat = [k  0; 0  x]                    ; % Get just x contribution
BtKB1 = B1' * Mat* B1 ; BtKB = simplify (BtKB1) ; % product 1
%         Integrate [0,0.5)x[0,1), for constant J1
```

```
I_BtKB = int(BtKB,r,0,0.5); I_BtKB=simplify(I_BtKB); % wrt r-
I_BtKB = int(I_BtKB,s,0,1); I_BtKB=simplify(I_BtKB); % wrt s-
K_1x = I_BtKB * det_J1 ; % convert differential area span 1
disp('K_1x sum ='), disp(simplify(sum(sum(K_1x)))) ; % check

% *** New limits of integration for second knot span***
BtKB2 = B2' * Mat* B2 ; BtKB = simplify (BtKB2) ; % product 2
%          Integrate [0,0.5)x[0,1), for constant J2
I_BtKB = int(BtKB,r,0.5, 1); I_BtKB=simplify(I_BtKB); % wrt r-
I_BtKB = int(I_BtKB,s,0,1); I_BtKB=simplify(I_BtKB); % wrt s-
K_1x = I_BtKB * det_J2 ; % convert differential area span 2
disp('K_x2 sum ='), disp(simplify(sum(sum(K_1x)))) ; % check
```

Earlier, when discussing geometric properties, it was shown that the physical length of the single s-knot span is L_y. Likewise, the physical length of each of the two r-knot spans is $L_x/2$. So, if each non-zero IGA knot span sub-region is thought of as element-like, then they would form an FEA 'mesh' with a dividing line at $r = 0 \rightarrow x = L_x/2$.

Before going on, the matrices created so far will be checked without convection by running a high school physics test (here called HSPT); here, two opposite edges have two constant temperatures prescribed and the other two edges are insulated (or lines of symmetry). Those boundary conditions convert the study to a 1D model calculated with a 2D code. The exact answer for the temperature is just a linear variation between the two specified Dirichlet conditions. The script for computing the temperature is listed in Script 9.9 and the post-processing of those results for the heat flow thermal reactions at the AP with the EBC is listed in Script 9.10. The script also includes the calculation of the temperature gradient vector at the center of the domain (in the second knot span), and confirms an accurate simulation. The data are often needed at multiple points for plotting the heat flux vectors over the domain.

It is suggested that the reader continue with this simplified example by adding the three convection line integrals and comparing the approximate temperatures with the expected values at point A in Fig. 9.14. This can be done using the r-direction basis functions and the temperature values at supporting analysis control point numbers 1, 2, 3, and 4.

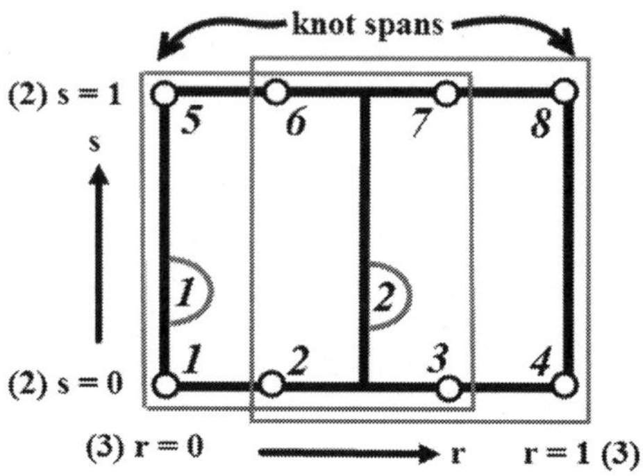

Figure 9.14. **Eight User-defined Control Points for $p = 3$, $q = 1$**

Script 9.9. Assembly and solution for two iElements' conduction check

```
%                    Check_T_PDE_HSPT.m
% High School Physics Test (HSPT) of two assembled
%    iElement conduction matrices from two spans
%    Ar = {0 0 0 0.5 1 1 1} & As = {0 0 0 1 1 1}
%       Specific data, W?(m C), m, m
k = 52 ; Lx = 1 ; Ly = 0.6 ; Q = 0        ; % NAFEMS test
% AP connection lists for iElement 1 and 2 for assembly
%     5 *-*6----|-----*7-*8  two iElements (e)
%       |  : (1)|  (2):  |   CP at r=[0 1/4 3/4 1]Lx
%     1 *-*2----|-----*3-*4  CP left to right then up
diary ReCheck.txt
CP_x = [0, 1/4, 3/4, 1, 0, 1/4, 3/4, 1]*Lx; % rectangle
CP_y = [0, 0,   0,   0, 1, 1,   1,   1]*Ly; % rectangle
[Lists] = [1,2,3,5,6,7; 2,3,4,6,7,8]  ; % left to right
n_d = max (max (Lists) )  ; % for 1 DOF per AP, n_g = 1
%      Allocate final system matrices sizes
K = zeros(n_d,n_d);c_Q = zeros(n_d,1);T = zeros(1,n_d);
% Combine iElement 1 x- and y-gradient matrices
c_Q1 = [2, 3, 1, 2, 3, 1]*Lx*Ly*Q/24    % sum Lx*Ly*Q/2
K_1x = [ 8, -6, -2,  4, -3, -1;
        -6,  6,  0, -3,  3,  0;
        -2,  0,  2, -1,  0,  1;
         4, -3, -1,  8, -6, -2;
        -3,  3,  0, -6,  6,  0;
        -1,  0,  1, -2,  0,  2]*k*Ly/Lx/9 ;
```

```
K_1y = [ 12,   7,  1, -12,  -7, -1]
       [  7,  17,  6,  -7, -17, -6]
       [  1,   6,  3,  -1,  -6, -3]
       [-12,  -7, -1,  12,   7,  1]
       [ -7, -17, -6,   7,  17,  6]
       [ -1,  -6, -3,   1,   6,  3]*k*Lx/Ly/120 ;
K_1  = K_1x + K_1y ; % conduction matrix for iElement 1

c_Q2 = [1, 3, 2, 1, 3, 2]*Lx*Ly*Q/24     % sum Lx*Ly*Q/2
K_2x = [ 2,  0, -2,  1,  0, -1;
         0,  6, -6,  0,  3, -3;
        -2, -6,  8, -1, -3,  4;
         1,  0, -1,  2,  0, -2;
         0,  3, -3,  0,  6, -6;
        -1, -3,  4, -2, -6,  8]*k*Ly/Lx/9 ;

K_2y = [ 3,   6,   1, -3,  -6,  -1;
         6,  17,   7, -6, -17,  -7;
         1,   7,  12, -1,  -7, -12;
        -3,  -6,  -1,  3,   6,   1;
        -6, -17,  -7,  6,  17,   7;
        -1,  -7, -12,  1,   7,  12] *k*Lx/Ly/120 ;

K_2  = K_2x + K_2y ; % conduction matrix for iElement 2
%  Assemble (scatter) the source vector, last dof first
dof = Lists(2,:); c_Q(dof) = c_Q2'          ; % iElement 2
dof = Lists(1,:); c_Q(dof) = c_Q(dof)+c_Q1'   ; % iEl 1
%     Assemble (scatter) the total source vector
dof = Lists(2,:); K(dof,dof) = K_2          ; % iElement 2
dof = Lists(1,:); K(dof,dof) = K(dof,dof) + K_1;% iEl 1

% Add HSPT EBC: r = 0, 100 C AP 1&5, r = 1 0 C AP 4&8
[Fixed] = [1, 4, 5, 8]; [Given] = [100 0 100 0] ; % EBC
%B % [Free] = get_free_from_fixed (n_d, Fixed)   algorithm
[Free]  = [2, 3, 6, 7]        ; % Just cite them manually
% Solve for the HSPT temperatures and thermal reactions
[T, R] = Solve_Free_DOF (K, c_Q, Free, Fixed, Given)
```

Script 9.10. Conduction post-processing for reactions and center gradient

```
%                  -----------------------
%           *** post-processing options ***
%                  -----------------------
% Evaluate temperature at point (r,s)=(0.2,0) in span 1
r = 0.2 ; s = 0. ; dof = Lists(1,:);
[Nrs_1] = [-(2*r-1)^2*(s-1), 2*r*(3*r-2)*(s-1), ...
           -2*r^2*(s-1), s*(2*r-1)^2, ...
           -2*r*s*(3*r-2), 2*r^2*s] ; % basis functions
T_pt = [Nrs_1] * T(dof)     ; % extract edge temperature
% Evaluate gradient at right edge of span 1, r=s=0.5
r = 0.5 ; s = 0.5                    ; % edg: of knot span 1
B_1x=[-(16*r-8)*(s-1),(24*r-8)*(s-1),-8*r*(s-1), ...
      2*s*(8*r-4),-8*s*(3*r-1),8*r*s]/Lx       ; % dN/dx
B_1y=[-(2*r - 1)^2, 2*r*(3*r - 2), -2*r^2, ...
       (2*r - 1)^2, -2*r*(3*r - 2), 2*r^2]/Ly ; % dN/dy
B_1 = [B_1x ; B_1y]     ; % heat_flux-temperature matrix
Grad_1xy = B_1 *  T(dof)            % x,y gradient vector
disp('Exited CHK_T_PDE_sec_p2_q1_HSPT_M')
%                      Running gives
% Fixed = 1     4     5     8
% Given = 100   0   100     0          % given EBC values
% Free =  2     3     6     7
% T = [100. 75. 25. 0. 100. 75. 25. 0.]'        % correct
% R = 3,120 [ 1 -1 1 -1]         % heat flows for Given T
% Grad_1xy = -100.0 -0.000  % gradient vector at center
```

The prior special cases of rectangular physical domains with constant Jacobians are intended to be educational. However, in IGA, any practical analysis means that matrix integrations are carried out numerically. A practical way to do that is to evaluate the 1D r-basis functions at a tabulated 1D quadrature point, do the same for the 1D s-direction basis functions, and numerically compute the tensor product of the 1D arrays to obtain the 2D basis functions and their derivatives at the 2D quadrature point. Usually, there are many (say 20–100) knot spans in each direction of an IGA region with the same degree odd-order basis functions (usually cubic or quintic) used in each parametric direction. The easy way to describe such regions is to create, in each direction, a uniform open knot vector. If the r-direction was to have five iElements, $n_{kr} = 5$, of degree two, the r-knot vector would contain 1, 2, 3, 4, 5, and would be supplied as $Ar = \{0, 0, 0, 1, 2, 3, 4, 5, 5, 5\}$, etc., in the other parametric directions.

The quadrature points locations in an iElement (r_q, s_q) are known. At first glance, it seems like a daunting task to determine in which knot span, in each direction, such a point falls. But, given an analysis knot vector, the Cox-de Boor recurrence relations automatically generate the numerical values of the basis functions, and their parametric derivatives, at the supplied quadrature point. Script 9.11 shows a simple script that does just that by evaluating the 1D recurrence relations in the r- and s-directions for a 2D application. If the study is 3D, just one more function call is needed to evaluate the basis functions.

Script 9.11. Using 1D basis functions for 2D tensor products

```
function[Nr, dNr, Ns, dNs]=num_basis_rect_line ...
                            (r, s, n_r, n_s, deg)
%  This forms basis functions and derivatives at
%  parametric point (r,s) for polynomials of the
%  same degree in r & s by calling the 1D version
%      for r, and calling a second time for s.
%
% r    = coord. of 1st parametric direction
% s    = coord. of 2nd parametric direction
% n_r = end of sequence 1,2,3, ..., n_r
% n_s = end of sequence 1,2,3, ..., n_s
% deg  = polynomial order
%
% Nr  = basis functions r direction
% dNr = derivatives of basis functions r direction
% Ns  = basis functions s direction
% dNs = derivatives of basis functions s direction
%
% Note: num_basis_line will accept a 4-th argument
```

```
%       of knot vector Vr=[0,0,..,1,2,...,n_r,n_r]
%       which speeds the function (do same here?)
[Nr, dNr] = num_basis_line(r,n_r,deg)
%  Basis functions Ns and derivatives
[Ns, dNs] = num_basis_line(s,n_s,deg)
% end num_basis_rect_line ========================
```

In the script, its first use would be to accept the coordinates of a quadrature point and return the values of the r-basis function and the s-basis functions to combine into a tensor product row matrix at the quadrature point. A warning here is that the 1D quadrature point weights must be multiplied together when using such 2D integrations.

The 1D functions cited in the above-mentioned script have a fourth optional argument which is the corresponding knot vector instead on the maximum knot number. The 1D function actually uses the maximum knot number to compute the knot vector and pass it along to the recurrence relations. For frequent calls to the above-mentioned function, it would be more efficient to include the two corresponding knot vectors. Post-processing contouring functions call the above-mentioned script many more times than the numerical integration of the basis functions.

9.11 IGA Steps Summary

The algorithmic steps to perform an IGA study of a 2D scalar field problems are listed in the following:

(1) Allocate the system conduction and convection matrices and heat source vectors for $n_g = 1$ degrees of freedom per analysis control point in the system. If all iElements will use the same quadrature point configuration, gather those coordinates and weights. If using products of 1D rules, remember that their weights are multiplied together.

If the domain is for a rectangular patch, specify the number of iElements in the r- and s-parametric directions, say n_{kr} and n_{ks}, as they are needed in the function that numerically evaluates the basis functions at quadrature points.

(2) Loop over the iElements.

Allocate and clear the matrices needed for all iElement calculations, such as the iElement conduction and convection matrices, heat source vectors, material matrix, basis functions and their local and physical derivatives, iElement AP connections, and iElement AP physical coordinates.

Gather the 2D quadrature rule non-dimensional coordinates and their weights. There are several possible configurations; remember that the sum of the (products of) quadrature weights will define the measure of the quadrature space, $M_q = \sum_q w_q$, which along with the non-dimension measure of the knot span, $\mathcal{M}_{rs} = (r_{I+1} - r_I) \times (s_{J+1} - s_J)$, defines the first (and only constant) Jacobian scalar $\mathbb{J}_{rs:q} = \mathcal{M}_{rs}/\mathcal{M}_q$. Herein, the default Gaussian quadrature data are evaluated in unit coordinates so $\mathcal{M}_q \equiv 1$, but most IGA and FEA studies follow Gauss and use natural coordinates and then $\mathcal{M}_q \equiv 2$ in 1D, 4 in 2D, and 8 in 3D.

Determine the parametric B-spline or NURBS coordinates for each knot span, $[r_I, r_{I+1}) \times [s_J s_J)$ using range_r and range_s.

Gather the AP connectivity list for the current iElement, their physical coordinates, and the iElement material properties data.

(3) Loop over the quadrature points and their weights, say $(u_q,\ v_q)$ and w_q.

Map the quadrature point location into the B-spline or NURBS knot span $(r_q,\ s_q)$, as well as to the physical location, $(x_q,\ y_q)$. Evaluate the scalar Jacobian matrix of the mapping, $J_{rs:q} = M_{rs}/M_q$.

Compute the numerical values of both 1D basis functions and their derivative at the quadrature point using script *num_basis_rect.m*. Take the numerical tensor product to create the row matrix $[\boldsymbol{N}(r, s)]$ of the 2D basis functions in the knot span (iElement) at location $(r_q,\ s_q)$.

The product of the first parametric derivatives of the basis functions with respect to r- and s-directions, say the rectangular matrix $[\boldsymbol{dN}(r, s)]$, with the two physical spatial coordinates of the iElement's APs in the rectangular matrix, say $\left[\boldsymbol{XY}^{iE}\right]$, forms the 2×2 square Jacobian matrix $[\mathbb{J}_{\boldsymbol{xy:q}}]$. Numerically compute its scalar determinant $|\,\mathbb{J}_{xy:q}\,|$ and inverse $[\mathbb{J}_{\boldsymbol{xy:q}}]^{-1}$.

Compute the two rows of physical gradients of the basis functions at $(r_q,\, s_q)$ by multiplying the above-mentioned local gradient, $[\boldsymbol{dN}(r,s)]$, by the inverse of the second Jacobian matrix, say $[\boldsymbol{B}(r,s)] = [\mathbb{J}_{\boldsymbol{xy:q}}]^{-1}\, [\boldsymbol{dN}(r,s)]$. The resulting matrix has the x- and y-partial derivatives of the basis functions.

Add this quadrature point's increment to the iElement conduction matrix formed by the triple matrix product of the gradient matrix and the material matrix, followed by the two Jacobian determinant, and the weight of the quadrature point:

$$\left[\boldsymbol{K}^{iE}\right] = \left[\boldsymbol{K}^{iE}\right] + [\boldsymbol{B}(r_q,\, s_q)]^T\, [\boldsymbol{k}]\, [\boldsymbol{B}(r_q,\, s_q)]\, \mathbb{J}_{rs:q}\, |\, \mathbb{J}_{xy:q}|\, w_q.$$

If there is an internal rate of heat generation per unit volume, Q, integrate for its resultant source vector:

$$\left\{\boldsymbol{F}_B^{iE}\right\} = \left\{\boldsymbol{F}_B^{iE}\right\} + [\boldsymbol{N}(r_q,\, s_q)]^T\, Q(r_q,\, s_q)\, \mathbb{J}_{rs:q}\, |\, \mathbb{J}_{xy:q}\, |\, w_q.$$

Save most arrays evaluated at the quadrature point to a file for later use in post-processing and plotting. Save at least $(x_q,\, y_q)$, $[\boldsymbol{N}(r_q,\, s_q)]$, $[\boldsymbol{B}(r_q,\, s_q)]$, and, if variable, save the material matrix, $[\boldsymbol{k}(r_q,\, s_q)]$.

If the convection is present, form its square matrix and source vector:

$$\left[\boldsymbol{M}^{iE}\right] = \left[\boldsymbol{M}^{iE}\right] + [\boldsymbol{N}(r_q,\, s_q)]^T\, h^{iE}\, [\boldsymbol{N}(r_q,\, s_q)]\; \mathbb{J}_{rs:q}\, |\mathbb{J}_{xy:q}|\, w_q,$$

$$\left\{\boldsymbol{F}_B^{iE}\right\} = \left\{\boldsymbol{F}_B^{iE}\right\} + [\boldsymbol{N}(r_q,\, s_q)]^T\, h^{iE}\, T_\infty\, \mathbb{J}_{rs:q}\, |\mathbb{J}_{xy:q}|\, w_q.$$

Some field problems require the domain integral of the primary variable as well as its gradient. The scalar result from post-processing is a loop over the iElements and its quadrature points, summing the AP values times the vector

$$[\boldsymbol{N}(r_q,\, s_q)]^T\; \mathbb{J}_{rs:q}\, |\mathbb{J}_{xy:q}|\, w_q.$$

The vector has already been computed, so include it in the list of items to save to the binary file for quadrature arrays.

End loop over the quadrature points.

Save the completed iElement matrices to binary storage for possible post-processing. The material matrix, $[\boldsymbol{k}(r_q,\, s_q)]$, is always needed for evaluating the heat flux vector (and its magnitude) at the quadrature points.

Gather the degree-of-freedom numbers at each analysis control point supporting the basis functions and scatter (assembly) all of the rows and columns of the iElement area integrals into the larger system matrices:

$$[\boldsymbol{K}] = [\boldsymbol{K}] + \left[\boldsymbol{K}^{iE}\right], \; [\boldsymbol{M}] = [\boldsymbol{M}] + \left[\boldsymbol{M}^{iE}\right], \; \{\boldsymbol{F}_B\} = \{\boldsymbol{F}_B\} + \left\{\boldsymbol{F}_B^{iE}\right\}.$$

End loop over all the elements.

(4) Here, the area was assumed to have a unit thickness. Even if the thickness is not unity, it multiplies all iElement matrices as a constant and simply cancels out when assembled.

(5) Usually, there are non-zero Neumann or Robin conditions on portions of the edges bounding the area. They require a similar 1D quadrature calculation to be used for their source resultants. Then, the Jacobian matrix $[\mathbb{J}_{\boldsymbol{xy}:\boldsymbol{q}}]$ is a scalar, 1×1, involving a square root of differential line lengths, as discussed elsewhere.

(6) Also remember that the planar concept includes the common axisymmetric domains that follow exactly the same steps, but have an additional step in defining the differential volume per radian as $dV = R(r_q, s_q)\, dA$. Note that the usual constant π term is not present. This is because it factors out all of the matrices after assembly and there is no need to include its presence millions of times just to cancel it at the end.

9.12 Review

Unlike here, some of the IGA literature used a different concept of the integer Index Space where the first and last padded entries of the analysis knot vector are included around the perimeter of their index space. An example of that choice is given in Fig. 9.15. That figure can be interpreted as representing all of the knot spans of the knot vector in each parametric direction, independent of their relative lengths. The difference in that figure and the one used herein is that the current version uses the unique numbers in the knot vector. The choice of using the unique entries in the knot vector makes it easier to identify the regions of non-zero basis functions in the parametric space.

Following the approach of others, the interpretation of Fig. 9.15 is that each non-zero knot span in a knot vector will then define one

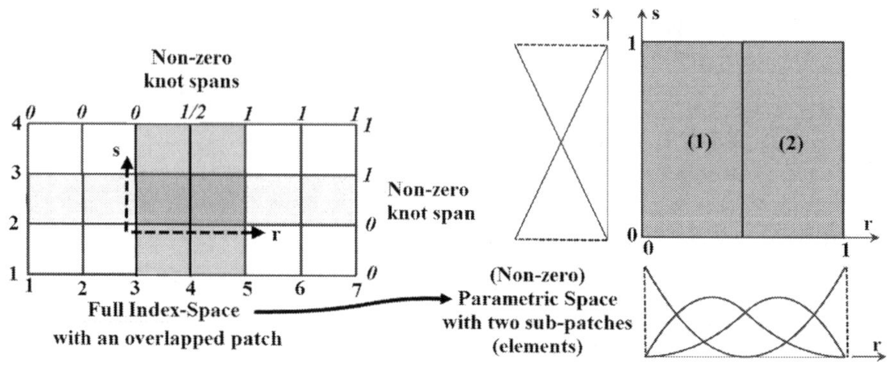

Figure 9.15. Integer index space and the sub-patches in parameter space

sub-range (or iElement) along its parametric coordinate direction. In IGA, many of the basis functions are zero. The focus in the Index Space is which knot span contains a non-zero function and where it overlaps with any other non-zero basis functions in the other parametric direction(s). The shaded knot spans in the r-direction contain two non-zero knot spans, while the s-direction contains only one shaded span, leading to a 'patch' of overlapping non-zero basis functions containing a total of $2 \times 1 = 2$ sub-ranges (or iElements).

The overlapping basis function support regions (if any) in the Index Space define the patch's Parameter Space, which in this example contains only the two overlapping non-zero knot spans (or iElements). The Parameter Space is also depicted on the right in Fig. 9.15, along a sketch of the magnitudes of the non-zero quadratic basis functions (below the r-axis). Likewise, the magnitudes of the other non-zero linear basis functions are sketched beside the s-parametric direction.

In Fig. 9.15, the r-dimension basis functions in each knot span are connected to six of the eight APs in the Control Mesh. In Fig. 9.16, the gray boxes enclose the six APs of the respective knot span. Note that the chosen AP numbering is not unique and must be defined by the user and used consistently during an IGA study.

Also, since this is a thermal analysis, there is only one unknown temperature per analysis control point. Therefore, in this example there will be eight degrees of freedom (DOFs) in the governing matrix system. If this was a plane stress IGA study with two displacements per AP, then the governing matrix system would have 16 DOFs. The APs' 'connection list' of each knot span provides the data for

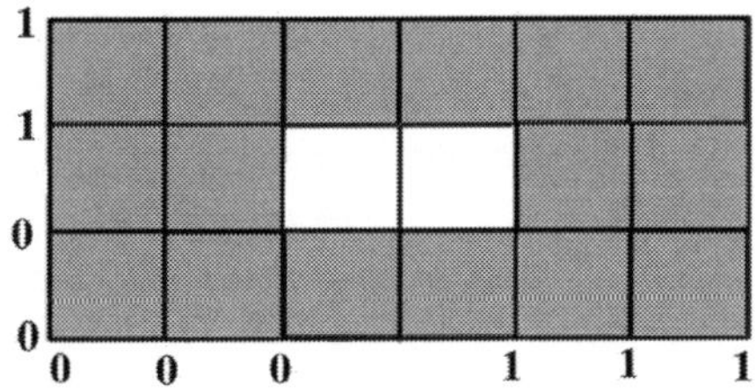

Figure 9.16. Repeated end knots always cause p zero end basis functions

assembling (scattering) the various knot span IGA matrices into the larger system matrices.

Remember, IGA literature discusses Index Spaces for NURBS, but actually uses Open Non-Uniform Rational B-Splines which interpolate the two end points and are tangent to the first and last control legs. To do that, ONURBS have to repeat the first and last knot values in the knot vector $(p+1)$ times. This means that any Index Space has an outer perimeter, in each parametric direction, with p cells where the basis functions are zero. In other words, in the above-mentioned Index Space in the parametric r-direction, there are $p = 2$ vertical columns of zero basis function cells at each end. Likewise, in the s-direction, there is $p = 1$ horizontal row of zero basis function cells at its (top and bottom) ends.

9.13 Summary

Cox-de Boor recursive relation for B-splines basis functions of degree p:

$$p = 0: \quad N_{j,0}(u) = \begin{cases} 1 & \text{if } u_j \le u \le u_{j+1} \\ 0 & \text{otherwise} \end{cases},$$

$$p = 1, 2, 3, \ldots \quad N_{j,p} = \frac{u - u_j}{u_{j+p} - u_j} N_{j,p-1}(u)$$

$$+ \frac{u_{j+1} - u}{u_{j+p+1} - u_{j+1}} N_{j+1,p-1}(u),$$

$$x(u)/\partial u = [\boldsymbol{N}(u)/\partial u] \{\boldsymbol{x}_{CP}\}, \quad \sum_j dN_j(u)/du = 0.$$

Recursive relation for derivatives of B-splines of degree p:

$$N'_{j,p} = \frac{p}{u_{j+p} - u_j} N_{j,p-1}(u) - \frac{p}{u_{j+p+1} - u_{j+1}} N_{j+1,p-1}(u),$$

$$\frac{d^k}{du^k} N'_{j,p} = \frac{p}{u_{j+p} - u_j} \left(\frac{d^{k-1}}{du^{k-1}} N'_{j,p-1}(u) \right) \frac{p}{u_{j+p+1} - u_{j+1}}$$

$$\times \left(\frac{d^{k-1}}{du^{k-1}} N_{j+1,p-1}(u) \right).$$

Greville abscissa: For a knot vector of degree p with n control points, the Greville abscissas defined in the u-parametric space are:

$$G_I = \frac{1}{p} (u_{I+1} + u_{I+2} + \cdots + u_{I+p}), \quad I = 1, \ldots, n.$$

Chapter 10

Elasticity

10.1 Introduction

The subject of elasticity deals with the stress analysis applications where the component returns to its original size and shape when the loads act on the component. The IGA method has proven to be a very practical process for any elasticity analysis. A vast majority of finite element elasticity models utilize the displacements as the primary unknowns. A general solid analysis requires three displacement components that lead to six stress components. An elasticity analysis involves four major topics: the displacement components, the six strain components obtained from the gradients of the displacement components (known in solid mechanics as the strain–displacement relation), and the experimental material property knowledge that relates the strain components to the stress components. It is more common to state the material stress–strain law, which is the inverse of the experimental relation. After the displacements are known, they are post-processed to give the strains and stresses. Once the strains and stresses are known, they must be compared to establish the material failure criterion.

All stress analyses involve a solid object. Historically, many mathematical simplifications have been developed to model common special shapes in order to avoid actual solid stress analysis. Those mathematical approximations include bending of curved three-dimensional shells, axisymmetric solids, two-dimensional plane stress (where a plane's normal stress is zero), two-dimensional plane strain (where the normal strain is zero), bending of two-dimensional flat plates, two-dimensional torsion of non-circular shafts, one-dimensional torsion of circular shafts, bending of one-dimensional beams, and one-dimensional axial bars. All of those special cases have a different set of governing differential equations. However, they can all be solved by a single IGA approach, with the same matrix symbols, by using different types of basis functions for each class. A vast majority of the matrix operations are the same with only the sizes of the matrices changing.

Today, it is practical to conduct a three-dimensional stress analysis, but it is not always the most accurate or the most efficient approach. For example, in certain cases, thousands of solid elements approximating a straight beam can be less accurate than tens of beam-bending elements. Whatever element type is used in an analysis, all IGA study results should be independently validated by some handbook solution and/or some simplified mathematical model in an attempt to set upper and lower bounds for the answer of the actual problem being solved.

The governing differential and integral formulations for the equilibrium of three-dimensional solids were known for about two hundred years before the advent of digital computers. Now, IGA and FEA make it practical for novice engineers to solve such systems.

The governing differential equations for elasticity problems can be converted to corresponding integral forms by using the Method of Weighted Residuals, and in particular Galerkin's method. However, the variational calculus of Euler and the Principle of Virtual Work can give the equivalent integrals as can the work-energy approach known as the Theorem of Minimum Total Potential Energy. Those methods, and the Galerkin method, give identical forms of the element matrices.

The simpler Theorem of MTPE is used here for the IGA formulation of the displacement based stress analysis of elastic bodies. The total potential energy, Π, is defined as the strain energy (potential energy), U, stored internal to a body minus the mechanical work, W, done by externally applied sources:

$$\Pi = U - W. \tag{10.1}$$

The theorem states the following: The displacement field that satisfies the essential displacement boundary conditions and renders stationary the total potential energy is the unique displacement field that corresponds to the state of static equilibrium. Stationary points are minimum (stable) points, maximum (unstable) points, or inflection (neutral) points. For elastic solids, it can be shown that the stationary state is the stable minimum state, thus herein it is referred to as the minimum principle. Therefore, if $\boldsymbol{\delta}$ represents all of the displacement components, then the equations of static equilibrium are

$$\frac{\partial \Pi}{\partial \boldsymbol{\delta}} = \mathbf{0}.$$

This gives one equation per displacement component that must be satisfied for the elastic body to be in equilibrium:

$$\frac{\partial \Pi}{\partial \delta_k} = 0_k, \quad \text{for } k = 1, 2, \ldots, n_d.$$

The displacements must also satisfy the essential boundary conditions (EBCs), and the above-mentioned equations also yield the reactions needed for system equilibrium, developed at the EBC locations.

10.2 Mechanical work

The mechanical work due to the external force is defined as the product of the force and the displacement component in the direction of the force. In general, they are not parallel, and the work is the scalar (dot) product of the force vector and the displacement vector. Beginning with undergraduate courses in statics, most engineers are familiar with writing force vector components with x, y, z subscripts:

$$\overrightarrow{\boldsymbol{F}} \leftrightarrow \boldsymbol{F} = \begin{Bmatrix} F_x(x, y, z) \\ F_y(x, y, z) \\ F_z(x, y, z) \end{Bmatrix}.$$

In statics, it is taught that several classes of problems are more easily solved using work-energy methods. Structural analysis using IGA is one such class. Of course, Newton's Laws can still be used to verify that the applied external forces are equal and opposite to the reaction forces. If not, then the user made a mistake in specifying the external loads.

In the theory of elasticity, the three displacement components are written as (u, v, w), but in the more recent numerical literature, they are often written as $(\delta_x, \delta_y, \delta_z)$. The former approach is used here to simplify most equations:

$$\vec{\boldsymbol{u}} \leftrightarrow \boldsymbol{u} = \begin{Bmatrix} u(x,y,z) \\ v(x,y,z) \\ w(x,y,z) \end{Bmatrix} = \boldsymbol{\delta} = \begin{Bmatrix} \delta_x(x,y,z) \\ \delta_y(x,y,z) \\ \delta_z(x,y,z) \end{Bmatrix}.$$

Thus, the scalar mechanical work of a point force undergoing a displacement is

$$W = \vec{\boldsymbol{u}} \cdot \vec{\boldsymbol{F}} = \boldsymbol{u}^T \boldsymbol{F} = [u\ v\ w] \begin{Bmatrix} F_x \\ F_y \\ F_z \end{Bmatrix} = uF_x + vF_y + wF_z.$$

Since work is a scalar quantity, or a 1×1 matrix, it is equal to its transpose, $W = W^T$, and the matrix algebra rule is that the transpose of a product of matrices is the product of their transposes written in reverse order: $(\boldsymbol{A}\ \boldsymbol{B}\ \boldsymbol{C})^T = (\boldsymbol{C})^T(\boldsymbol{B})^T(\boldsymbol{A})^T$; it is also valid to write the mechanical work as $W = \boldsymbol{F}^T\boldsymbol{u}$.

10.2.1 *Surface forces*

It is necessary to consider additional types of external forces acting on the elastic body. A viscous fluid, and other sources like electromagnetic surface eddy currents, can apply an external force per unit area to solid surfaces. These are called 'surface tractions', and they are denoted in matrix forms by $\vec{\boldsymbol{T}} \leftrightarrow \boldsymbol{T}(x,y,z) = [T_x\ T_y\ T_z]^T$. The differential force at a point on the surface area dS becomes $\overrightarrow{\boldsymbol{dF}} = \vec{\boldsymbol{T}} dS$. When that point on the surface is subjected to a displacement, the differential amount of work done by the surface traction vector is

$$dW_T = \vec{\boldsymbol{\delta}} \cdot \vec{\boldsymbol{T}} dS = \boldsymbol{\delta}^T \boldsymbol{T} dS.$$

Thus, the total work done by all surface tractions is

$$W_T = \int_S \boldsymbol{\delta}^T \boldsymbol{T} dS.$$

10.2.2 *Volumetric forces*

Gravity causes volumetric forces (weight) that are easily included in an IGA study. Likewise, external translational acceleration, rotational velocities, or rotational accelerations, and other sources like the effects of external magnetic fields acting on electric currents flowing through a body, will cause forces per unit volume to occur inside an object. These are called 'body forces', and they are denoted in vector and matrix forms by

$$\vec{\boldsymbol{f}} \leftrightarrow \boldsymbol{f} = [f_x \; f_y \; f_z]^T,$$

respectively, and the differential force acting in a differential volume is $\overrightarrow{\boldsymbol{dF}} = \vec{\boldsymbol{f}} dV$. When the differential volume is subjected to a displacement, the scalar work done by any body force vector is

$$dW_f = \vec{\boldsymbol{\delta}} \cdot \vec{f}\, dV = \boldsymbol{\delta}^T \boldsymbol{f} dV,$$

so the total work done by body forces is

$$W_f = \int_V \boldsymbol{\delta}^T \boldsymbol{f} dV.$$

Combining all of the possible external forces, their mechanical work done on a body is

$$W = \int_V \boldsymbol{\delta}^T \boldsymbol{f} dV + \int_S \boldsymbol{\delta}^T \boldsymbol{T} dS + \sum_{k=1}^{k=n_P} \boldsymbol{\delta}_k^T \boldsymbol{F}_k.$$

It must be noted here that modern physics is making significant progress in creating super-conducting materials and expanding the applications where such materials can be used in the future. The electrical current vector, $\vec{\boldsymbol{J}}$, flowing in a super-conducting material lying within an artificial magnetic field, $\vec{\boldsymbol{B}}$, can cause an extremely large electromagnetic body force, $\vec{\boldsymbol{f}} = \vec{\boldsymbol{J}} \times \vec{\boldsymbol{B}}$. For currents in household wiring and the Earth's magnetic field, such forces are millions of times smaller than foreseeable industrial applications.

Similar to what was seen in the heat transfer examples, each mechanical work source will lead to an iElement resultant force vector. Specifically, for body forces,

$$\{\boldsymbol{c}_f^e\} \equiv \int_{V^e} [\boldsymbol{N}^T(x, y, z)]\{\boldsymbol{f}(x, y, z)\} dV$$

and for boundary regions with non-zero tractions, Γ^b

$$\{\boldsymbol{c}_T^b\} \equiv \int_{\Gamma^b} [\boldsymbol{N}^T(x, y, z)]\{T(x, y, z)\}dS.$$

10.3 Strain energy

The energy part of work-energy in stress analysis is the strain energy stored in an elastic body by the interactions of the strains and stresses acting at every point in a body. Strain energy is also called potential energy in statics because it has the potential for doing work. Recall the elastic springs or 'linear springs' usually introduced in a physics or statics course. The spring, in Fig. 10.1, has a stiffness of k. The unrestrained end has an axial force, F, that causes an axial displacement, u. When the external force reaches the equilibrium displacement, u, the strain (potential) energy of the linear spring is

$$U = \frac{1}{2}ku^2 = \frac{1}{2}uku.$$

Looking Ahead: It should not be surprising that when the IGA matrix equations of equilibrium are established, the strain energy stored in the entire linear elastic body is half the triple product of all of the displacements, $\boldsymbol{\delta}$, with the stiffness matrix, $\boldsymbol{K}$, of the body:

$$U = \frac{1}{2}\boldsymbol{\delta}^T\boldsymbol{K}\boldsymbol{\delta}.$$

This form also suggests that the strain energy relationships to follow must be expressed in terms of the displacements at a point. Since the stresses are experimental material properties times the strains, the key is how the mechanical strains are related to the displacements. That relationship is widely called the 'Strain–Displacement' relation. It is actually a strain–displacement–gradients relation.

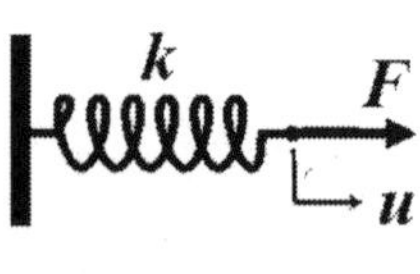

Figure 10.1. A linear spring

There are three normal stresses and three shear stresses, and a corresponding set of strains. Most IGA elasticity studies use a matrix formulation that represents the strain and stress components as column vectors. This is called the Voigt representation and contrasts with the alternate tensor representation. In matrix notation, the material strain and stress components are usually ordered as

$$\boldsymbol{\varepsilon}^T = [\varepsilon_x \ \varepsilon_y \ \varepsilon_z \ \gamma_{xy} \ \gamma_{xz} \ \gamma_{yz}]$$

and

$$\boldsymbol{\sigma}^T = [\sigma_x \ \sigma_y \ \sigma_z \ \tau_{xy} \ \tau_{xz} \ \tau_{yz}].$$

In the theory of elasticity, it is shown that the differential amount of strain energy per unit volume, also called the 'strain energy density' (SED), is half the scalar product of the six strains with the six stresses:

$$\frac{dU}{dV} = \frac{1}{2}\boldsymbol{\sigma}^T\boldsymbol{\varepsilon} = \frac{1}{2}[\boldsymbol{\sigma}_x \ \sigma_y \ \sigma_z \ \tau_{xy} \ \tau_{xz} \ \tau_{yz}]\begin{Bmatrix} \varepsilon_x \\ \varepsilon_y \\ \varepsilon_z \\ \gamma_{xy} \\ \gamma_{xz} \\ \gamma_{yz} \end{Bmatrix} = \frac{1}{2}\varepsilon^T\boldsymbol{\sigma}.$$

Writing this product out in detail gives the differential strain energy of

$$dU = \frac{1}{2}(\sigma_x\varepsilon_x + \sigma_y\varepsilon_y + \sigma_z\varepsilon_z + \tau_{xy}\gamma_{xy} + \tau_{xz}\gamma_{xz} + \tau_{yz}\gamma_{yz})dV = \frac{1}{2}\boldsymbol{\sigma}^T\boldsymbol{\varepsilon}dV,$$

and the total strain energy is obtained from its volume integral:

$$U = \frac{1}{2}\int_V \boldsymbol{\sigma}^T\boldsymbol{\varepsilon}dV = \frac{1}{2}\int_V \boldsymbol{\varepsilon}^T\boldsymbol{\sigma}dV.$$

But, the mechanical stresses are defined in Hooke's Law as a square symmetric matrix, $\boldsymbol{E}$, of experimental material constants multiplying the strains: $\boldsymbol{\sigma} = \boldsymbol{E}\boldsymbol{\varepsilon}$. The material matrix will be considered later, but the focus now remains on the strains.

10.4 Yield criteria

In many metals, the value of the SED predicts if the material has yielded and is no longer linear. If the material-specific yield criterion is exceeded, the theory of plasticity has to be used to construct an IGA model and that is not considered here. More importantly, the whole purpose of a linear stress analysis is to verify that the material has not exceeded its specific yield criterion and that the displacements are not too large to violate the 'small deflection theory' used in elasticity.

The representation of the SED at the onset of material yield can be rewritten as a scalar quantity having the *units* of stress even though it is not a physical stress. The SED measure is called the von Mises Effective Stress (to emphasize it is not an actual stress), but it is frequently just lazily called the von Mises stress. Its definition in terms of the stress components, and the principle stress components, is given in the Summary. For classic metals, the Effective Stress is the default yield criterion. However, for a vast number of materials, like bone, wood, and 3D printed exotic materials, this criterion is not appropriate. Frequently, the Burzynski criterion (see Summary) is a much more general one and it degenerates to the von Mises criterion when a material's yield stresses in tension and compression are identical.

10.5 Displacement gradients

At any point, there are three displacement vector components, $[u\ v\ w]^T$, and the gradient vector contains three spatial derivatives, $[\partial/\partial x\ \partial/\partial y\ \partial/\partial z]^T$, so there are a total of nine displacement gradients. Three of them cause 'normal strains', three cause 'shear strains', and three cause 'rigid body motions' and are of no interest from the viewpoint of elasticity theory. Those distinctions are illustrated for the 2D plane stress and plane strain approximations in Fig. 10.2.

The first group of displacement gradients represents a displacement change of the faces of a differential area (or volume) in directions normal to their original positions (left part of the figure).

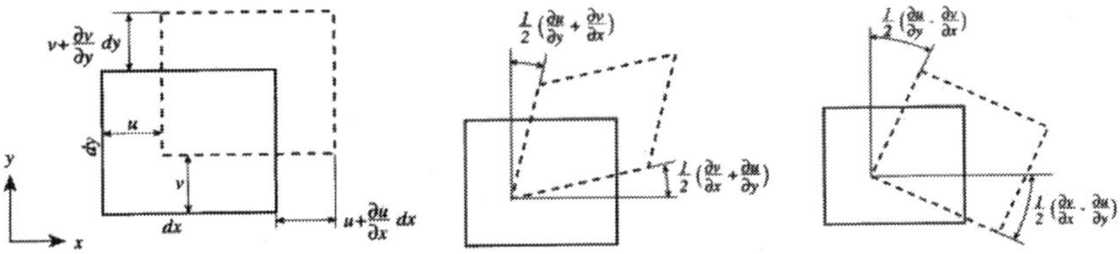

Figure 10.2. Decomposition of the planar displacement gradient components into normal, shear, and unimportant combinations

Those diagonal components are known as the 'normal strains':

$$\varepsilon_x = \frac{\partial u}{\partial x}, \quad \varepsilon_y = \frac{\partial v}{\partial y}, \quad \varepsilon_z = \frac{\partial w}{\partial z}.$$

The second group changes the shape (middle of Fig. 10.2) of a differential area (or volume) by distorting the rectangle into a parallelogram, without changing its area:

$$\gamma_{xy} = \frac{\partial u}{\partial y} + \frac{\partial v}{\partial x} = \gamma_{yx}, \quad \gamma_{xz} = \frac{\partial u}{\partial z} + \frac{\partial w}{\partial x} = \gamma_{zx},$$

$$\gamma_{yz} = \frac{\partial v}{\partial z} + \frac{\partial w}{\partial y} = \gamma_{zy}.$$

In this form, these are called the 'engineering shear strains', because they omit the factor of 1/2 shown in the figure and are the common IGA form and the form common in most undergraduate classes on stress analysis. When that factor is included, then the strain matrix, like the stress matrix, can be recognized as a second-order tensor which has several useful mathematical properties not related to stress analysis. In this case, they are called the strain tensor.

When a single source program is used to solve multiple classes of stress analysis, then it is not unusual to find the positions of the last normal stress, σ_z, and the first shear stress, τ_{xy}, interchanged. This allows for 2D approximations to include an axisymmetric solid formed by revolving a 2D area around an axis. Since the area no longer has a constant thickness, there is an additional 'hoop strain',

$$\varepsilon_\theta = (L' - L)/L = (2\pi(r + u) - 2\pi r)/2\pi r = u/r,$$

and 'hoop stress' (in Fig. 10.3), which needs to be included in the SED and the final stiffness of the part. This alternate stress ordering

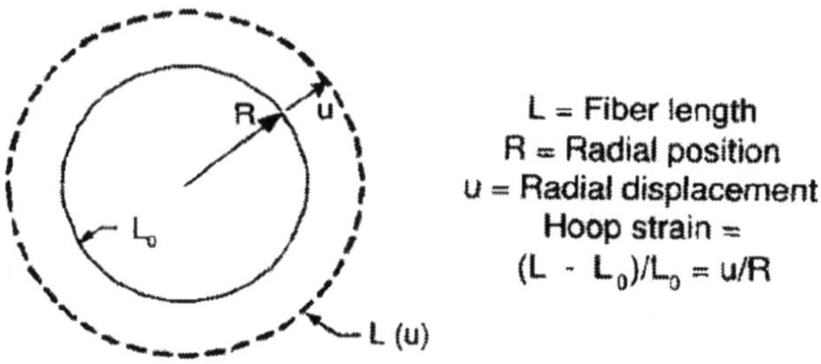

Figure 10.3. Hoop strain caused by a radial displacement

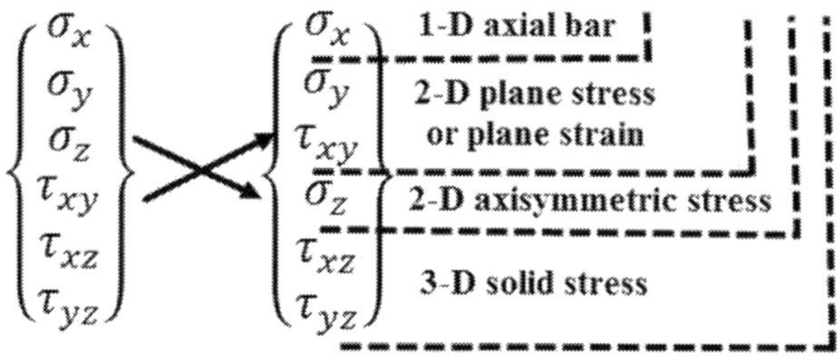

Figure 10.4. Stress (and strain) components order in IGA stress analysis

is used here in the example Matlab scripts for both the stresses and the strains. The application-dependent ordering of the stress components is shown in Fig. 10.4.

10.6 Applying basis functions to displacements and strains

The previous chapters dealt with interpolating primary unknowns (like temperature), which are scalar quantities requiring only a single unknown dof per analysis control point, $n_g = 1$. Now, those approaches must be extended to interpolate the three displacement vector components, $n_g = 3$, at every AP which governs the elasticity formulations. The x-component of displacement was earlier used for axial bars and one-dimensional basis functions.

In the previous chapter on field problems (heat transfer), the tensor product of the two basis functions was reshaped into a single basis function vector, $[\boldsymbol{Nrst}(r, s, t)]$, to evaluate the matrices needed for that scalar analysis. The same basis function vector concept is used here, after adding the basis functions in the third t-direction, by applying them to all three scalar components of the displacement vector within each iElement (3D knot span):

$$\begin{aligned} u(x, y, z) &= [\boldsymbol{Nrst}(r, s, t)]\{\boldsymbol{u}^{\mathbf{e}}\}, \\ v(x, y, z) &= [\boldsymbol{Nrst}(r, s, t)]\{\boldsymbol{v}^{\mathbf{e}}\}, \\ w(x, y, z) &= [\boldsymbol{Nrst}(r, s, t)]\{\boldsymbol{w}^{\mathbf{e}}\}. \end{aligned} \tag{10.2}$$

The analysis requires placing a displacement vector at each analysis control point connected to the iElement to define its degrees of freedom (dof). The ordering used here is number u, v, w values at the first AP, then those at the second AP, and so on. The iElement displacement dof are

$$\boldsymbol{\delta}^{\mathbf{e}^T} = [u_1, v_1, w_1, u_2, \ldots, \ldots, u_{n_n}, v_{n_n}, w_{n_n}]^e. \tag{10.3}$$

Likewise, numbering all analysis control points in the system gives the vector of system unknowns:

$$\boldsymbol{\delta}^T = [u_1, v_1, w_1, u_2, v_2, \ldots, \ldots, u_{n_m}, v_{n_m}, w_{n_m}]. \tag{10.4}$$

It is common in the IGA and FEA literature to write the interpolation of the displacement vector at a point in an element using the symbol $[\boldsymbol{N}(r, s, t)]$, namely, (for $n_g = 3$),

$$\begin{gathered} \left\{ \begin{matrix} u(x, y, z) \\ v(x, y, z) \\ w(x, y, z) \end{matrix} \right\} = \{\boldsymbol{u}(x, y, z)\} \equiv [\boldsymbol{N}(r, s, t)]\{\boldsymbol{\delta}^{\mathbf{e}}\}. \\ n_g \times 1 = (n_g \times n_i)(n_i \times 1). \end{gathered} \tag{10.5}$$

Here, $n_g = 3$ is the number of generalized unknowns (here, displacements) per AP, so the number of independent iElement dofs is $n_i = n_g n_n$ for an iElement with n_n AP per iElement. The first three columns of the rectangular basis functions matrix $[\boldsymbol{N}(r, s, t)]$ use only the three displacements of the iElement's first AP. Likewise, the kth AP is only associated with three columns, ending with column $3k$.

In other words, the first row of (10.5) uses only x-displacement components:

$$\begin{aligned} u(x, y, z) &= Nrst_1(r, s, t)u_1^e + 0 + 0 + Nrst_2(r, s, t)u_2^e + 0 + 0 + \cdots \\ &\quad + Nrst_{n_n}(r, s, t)\, u_{n_n}^e + 0 + 0, \end{aligned}$$

and two-thirds of its columns are zero. Therefore, a typical partition of the rectangular $[\boldsymbol{N}(r, s, t)]$ at each AP contains a scalar basis function, $Nrst_k(r, s, t)$, and zeros in each row. In this three-dimensional case,

$$\begin{aligned} [\boldsymbol{N}(r, s, t)] &= [\boldsymbol{N}_1 | \boldsymbol{N}_2 | \cdots | \boldsymbol{N}_{n_n}], \\ [\boldsymbol{N}_k(r, s, t)] &= \begin{bmatrix} Nrst_k(r, s, t) & 0 & 0 \\ 0 & Nrst_k(r, s, t) & 0 \\ 0 & 0 & Nrst_k(r, s, t) \end{bmatrix}, \\ [\boldsymbol{N}_k(r, s, t)] &= Nrst_k(r, s, t)[\boldsymbol{I}], \end{aligned}$$

where $[\boldsymbol{I}]$ is a 3×3 identity matrix. Therefore, the rectangular basis function array, $[\boldsymbol{N}(r, s, t)]$, can be partitioned into, and programmed as, a sequence of $n_g \times n_g$ sparse matrices associated with each AP dof connected to an iElement.

Now, finally, having all displacements in an iElement defined by IGA basis functions, $\boldsymbol{u}(x, y, z) \equiv [\boldsymbol{N}(r, s, t)]\{\boldsymbol{\delta}^e\}$, it is possible to relate the mechanical strains to the basis functions and the unknown displacements by simply substituting the six useful displacement gradients into the strain–displacement–(gradient) relationships.

10.7 The strain–displacement matrix

Next, the six strains within an iElement, $\boldsymbol{\varepsilon}$, are now rewritten in terms of an iElement's displacements, $\boldsymbol{\delta}^e$. Referring back to the basis functions form of the displacement components in (10.2),

$$\begin{aligned} u(x, y, z) &= [\boldsymbol{Nrst}(r, s, t)]\{\boldsymbol{u}^e\}, \\ v(x, y, z) &= [\boldsymbol{Nrst}(r, s, t)]\{\boldsymbol{v}^e\}, \\ w(x, y, z) &= [\boldsymbol{Nrst}(r, s, t)]\{\boldsymbol{w}^e\}, \end{aligned}$$

the theory of elasticity definition of the x-direction normal strain, $\varepsilon_x = \partial u/\partial x$, is converted to the IGA form as

$$\varepsilon_x = \frac{\partial u}{\partial x} = \frac{\partial[\boldsymbol{Nrst}(r,s,t)]}{\partial x}\{\boldsymbol{u^e}\} \equiv [\boldsymbol{Nrst}]_{,\boldsymbol{x}}\{\boldsymbol{u^e}\}.$$

In the other two normal directions, their normal strains, in the above-mentioned shorthand notation, are

$$\varepsilon_y = \frac{\partial v}{\partial y} = [\boldsymbol{Nrst}]_{,\boldsymbol{y}}\{\boldsymbol{v^e}\}, \quad \varepsilon_z = \frac{\partial w}{\partial z} = [\boldsymbol{Nrst}]_{,\boldsymbol{z}}\{\boldsymbol{w^e}\cdot\}.$$

Likewise, in the shorthand notation for the basis function derivatives, the IGA forms are

$$\gamma_{xy} = \frac{\partial u}{\partial y} + \frac{\partial v}{\partial x} = [\boldsymbol{N}rst]_{,\boldsymbol{y}}\{\boldsymbol{u^e}\} + [\boldsymbol{Nrst}]_{,\boldsymbol{x}}\{\boldsymbol{v^e}\},$$
$$\gamma_{xz} = \frac{\partial u}{\partial z} + \frac{\partial w}{\partial x} = [\boldsymbol{Nrst}]_{,\boldsymbol{z}}\{\boldsymbol{u^e}\} + [\boldsymbol{Nrst}]_{,\boldsymbol{x}}\{\boldsymbol{w^e}\},$$
$$\gamma_{yz} = \frac{\partial v}{\partial z} + \frac{\partial w}{\partial y} = [\boldsymbol{Nrst}]_{,\boldsymbol{z}}\{\boldsymbol{v^e}\} + [\boldsymbol{Nrst}]_{,\boldsymbol{y}}\{\boldsymbol{w^e}\}.$$

Following the IGA literature, these six strain–displacement forms can be rearranged into the matrix format

$$\begin{Bmatrix} \varepsilon_x \\ \varepsilon_y \\ \varepsilon_z \\ \gamma_{xy} \\ \gamma_{xz} \\ \gamma_{yz} \end{Bmatrix}^e = \{\boldsymbol{\varepsilon^e}\} = [\boldsymbol{B^e}(r,s,t)]\{\boldsymbol{\delta^e}\}.$$
$$(n_r \times 1) = (n_r \times n_i)(n_i \times 1).$$

The strain–displacement array, $[\boldsymbol{B^e}(r,s,t)]$, has six rows and $n_i = n_n n_g$ columns that correspond to the displacement values at the iElement's analysis control points, $\{\boldsymbol{\delta^e}\}$. They are NOT the physical displacements (except at the two end knots of an open knot vector).

Like the vector interpolations, the rectangular $[\boldsymbol{B}^{\boldsymbol{e}}(r,s,t)]$ strain–displacement matrix can be partitioned into, and programmed as, rectangular AP contributions

$$[\boldsymbol{B}^{\boldsymbol{e}}(r,s,t)] = [\boldsymbol{B}_1^{\boldsymbol{e}}|\boldsymbol{B}_2^{\boldsymbol{e}}|\cdots|\boldsymbol{B}_{n_n}^{\boldsymbol{e}}],$$

with a typical AP partition (with $n_r = 6$ rows of strains and $n_g = 3$ displacement component columns) being

$$[\boldsymbol{B}_k^e(r,s,t)] = \begin{bmatrix} N_{k,x} & 0 & 0 \\ 0 & N_{k,y} & 0 \\ 0 & 0 & N_{k,z} \\ N_{k,y} & N_{k,x} & 0 \\ N_{k,z} & 0 & N_{k,x} \\ 0 & N_{k,z} & N_{k,y} \end{bmatrix}, \quad N_{k,x} = \partial Nrst_k/\partial x \text{ etc.}$$

$$(n_r \times n_g). \tag{10.6}$$

All five options for the strain–displacement choices listed in Fig. 10.3 are included, as switch options, in the single script *B_matrix_elastic.m* in the functions library. Details on each of the five matrix forms, and their function names, are given in the Summary.

10.8 Constitutive relations

Hooke published (1678) material experiments that found a linear relation between strain and stress, which were extended by Poisson in 1811. The corrected relationship is still called Hooke's law, and the material experiments are called 'constitutive relations'. For example, the relation between the normal strain in the x-direction and the stress components, for an isotropic material, with initial thermal strain is

$$\varepsilon_x = [\sigma_x - \upsilon(\sigma_y - \sigma_z)]/E + \alpha_x \Delta T,$$

where E is the material's elastic modulus (the slope of a strain–stress diagram), υ denotes the material's Poisson ratio (the ratio of $\varepsilon_y/\varepsilon_x$ and $\varepsilon_z/\varepsilon_x$), α_x is the coefficient of thermal expansion of the material, and ΔT is the temperature increase from the material's stress-free state.

Today, the usual practice is to invert the experimental definitions and have the stresses defined by the strains, and thus by the displacement gradients. For example, for an isotropic material (properties independent of direction), an initial thermal strain is

$$\sigma_x = \frac{E}{(1+\upsilon)(1-2\upsilon)}[(1-\upsilon)\varepsilon_x + \upsilon(\varepsilon_y + \varepsilon_z)] - \frac{E\alpha\Delta T}{(1-2\upsilon)}, \tau_{xy} = \tau_{yx} = G\gamma_{xy},$$

where $G = E/2(1+\upsilon)$ is the (theoretically dependent) shear modulus. In matrix form, the stresses are

$$\boldsymbol{\sigma} = \boldsymbol{E\varepsilon} - \boldsymbol{E\varepsilon}_0, \quad \boldsymbol{\varepsilon}_T = \boldsymbol{\alpha}\Delta T \subset \boldsymbol{\varepsilon}_0. \tag{10.7}$$

For an isotropic solid, this matrix form of Hooke's law is

$$\begin{Bmatrix} \sigma_x \\ \sigma_y \\ \sigma_z \\ \tau_{xy} \\ \tau_{xz} \\ \tau_{yz} \end{Bmatrix} = \frac{E}{(1+\upsilon)(1-2\upsilon)} \begin{bmatrix} a & \upsilon & \upsilon & 0 & 0 & 0 \\ \upsilon & a & \upsilon & 0 & 0 & 0 \\ \upsilon & \upsilon & a & 0 & 0 & 0 \\ 0 & 0 & 0 & b & 0 & 0 \\ 0 & 0 & 0 & 0 & b & 0 \\ 0 & 0 & 0 & 0 & 0 & b \end{bmatrix} \begin{Bmatrix} \varepsilon_x \\ \varepsilon_y \\ \varepsilon_z \\ \gamma_{xy} \\ \gamma_{xz} \\ \gamma_{yz} \end{Bmatrix} - \frac{E\alpha\Delta T}{(1-2\upsilon)} \begin{Bmatrix} 1 \\ 1 \\ 1 \\ 0 \\ 0 \\ 0 \end{Bmatrix}.$$

$$a = (1-\upsilon), \quad b = (1-2\upsilon)/2. \tag{10.8}$$

The specific constitutive matrices for common 1D and 2D approximations are given in the Summary.

10.9 Stiffness matrices

At this point, the strain energy, or potential energy, of any elastic solid can be written in terms of the unknown displacements. From the scalar definition in (10.1),

$$U = \frac{1}{2}\int_V \boldsymbol{\sigma}^T \varepsilon dV = \frac{1}{2}\int_V (\boldsymbol{E}\varepsilon)^T \varepsilon dV = \frac{1}{2}\int_V (\varepsilon)^T (\boldsymbol{E})^T \varepsilon dV$$
$$= \frac{1}{2}\int_V \varepsilon^T \boldsymbol{E}\varepsilon dV,$$

and substituting the strain–displacement matrix, $\{\boldsymbol{\varepsilon}^e\} = [\boldsymbol{B}^e(r,s,t)]$ $\{\boldsymbol{\delta}^e\}$, its contribution is

$$U^e = \frac{1}{2}\int_{V^e} ([\boldsymbol{B}^e(r,s,t)]\{\boldsymbol{\delta}^e\})^T [\boldsymbol{E}]^T ([\boldsymbol{B}^e(r,s,t)]\{\boldsymbol{\delta}^e\}) dV,$$
$$U^e = \frac{1}{2}\{\boldsymbol{\delta}^e\}^T \left[\int_{V^e} [\boldsymbol{B}^e(r,s,t)]^T [\boldsymbol{E}]^T [\boldsymbol{B}^e(r,s,t)] dV\right] \{\boldsymbol{\delta}^e\}.$$

This contribution has the same matrix form in the square brackets as the previous heat conduction matrix, but it is larger and the entries have different physical meanings. The symmetry square matrix is the iElement stiffness matrix:

$$[\boldsymbol{K}^e] = \int_{V^e} [\boldsymbol{B}^e(r,s,t)]^T [\boldsymbol{E}]^T [\boldsymbol{B}^e(r,s,t)] dV.$$

Thus,

$$U^e = \frac{1}{2}\{\boldsymbol{\delta}^e\}^T [\boldsymbol{K}^e]\{\boldsymbol{\delta}^e\}.$$

This potential energy measure is similar to the expression for a single elastic spring shown in Figure 10.1. Summing all of the scalar energies gives the system potential energy as

$$U = \frac{1}{2}\{\boldsymbol{\delta}\}^T [\boldsymbol{K}]\{\boldsymbol{\delta}\}. \tag{10.9}$$

10.10 Matrix equilibrium equations

Having converted all of the work and strain energy contributions in their interpolated matrix forms, the 'Total Potential Energy' of (10.1) is

$$\Pi = \frac{1}{2}\{\boldsymbol{\delta}\}^T[\boldsymbol{K}]\{\boldsymbol{\delta}\} - \{\boldsymbol{\delta}\}^T\{\mathbf{F}\}.$$

The requirement for equilibrium becomes

$$\frac{\partial \Pi}{\partial \boldsymbol{\delta}} = \mathbf{0}, \quad \text{or} \quad \frac{\partial \Pi}{\partial \delta_k} = 0_k, \quad \text{for } k = 1, 2, \ldots, n_d.$$

Thus, the matrix equilibrium equations are $\mathbf{0} = \frac{2}{2}[\boldsymbol{K}]\{\boldsymbol{\delta}\} - \{\mathbf{F}\}$ or

$$[\boldsymbol{K}]\{\boldsymbol{\delta}\} = \{\boldsymbol{F}\}. \tag{10.10}$$

But, the theorem states that the displacements must satisfy the EBC as well. Those conditions have not yet been enforced, so the assembled system stiffness matrix, $[\boldsymbol{K}]$, is singular (not invertible) and this system cannot be solved until reduced to a smaller size by enforcing the essential (Dirichlet) boundary conditions. This process of using partitioned matrices is exactly the same as the procedures used earlier for heat conduction.

Once calculated, remember that the analysis control point displacements are not physical displacements. The physical displacements on the solid are obtained by post-processing the basis functions to get

$$\{\boldsymbol{u}(x, y, z)\} = [\boldsymbol{N}(r, s, t)]\{\boldsymbol{\delta^e}\}$$

at physical locations determined from the spatial coordinates of the analysis control points:

$$(x, y, z) = [\boldsymbol{N}(r, s, t)][\boldsymbol{x^e}\ \boldsymbol{y^e}\ \boldsymbol{z^e}].$$

Coordinate evaluation can be useful for displaying color plots of components, displacements, strains, and stresses, and the scalar values of yield criteria. To find the displacement at a particular physical point, (x_p, y_p, z_p), in a knot span, the above-mentioned relation is back-solved using a Newton–Raphson iteration to quickly find the needed parametric point, (r_p, s_p, t_p), where

$$(x_p, y_p, z_p) \equiv [\boldsymbol{N}(r_p, s_p, t_p)][\boldsymbol{x^e}\ \boldsymbol{y^e}\ \boldsymbol{z^e}].$$

10.11 Procedure for performing an IGA stress analysis

The steps in an IGA displacement-based stress analysis for 2D are as follows:

(1) Allocate memory and clear the system stiffness matrix and force vectors for n_g displacement dof per analysis control points in the system.

(2) Loop over the iElements ===>>===>>====>> ====>>=====>>

(i) Allocate, and clear, the matrices needed for all iElement calculations, such as the iElement stiffness matrix, force vector, stress–strain matrix, strain–displacement matrix, basis functions and their derivatives, iElement AP connections, and iElement AP physical coordinates.

(ii) Determine B-spline or NURBS coordinates for each knot span, $[r_I, r_{I+1}) \times [s_J, s_J)$, and compute their ranges, range_r, and range_s.

(iii) Gather the 2D quadrature rule non-dimensional coordinates and their weights, say (u_q, v_q) and w_q, in 2D.

There are several options; remember that the sum of the quadrature weights will define the measure of their quadrature space, $\mathcal{M}_q = \sum_q w_q$, which along with the non-dimension measure of the knot span, $\mathcal{M}_{rs} = (r_{I+1} - r_I) \times (s_{J+1} - s_J)$, defines the first (and only constant) Jacobian scalar $\mathbb{J}_{rs:q} = \mathcal{M}_{rs}/\mathcal{M}_q$.

Herein, the default Gaussian quadrature data are evaluated in unit coordinates, so $\mathcal{M}_q \equiv 1$, but most IGA and FEA studies follow Gauss's original publication and use natural coordinates, $-1 \leq a \leq 1$, etc., in each direction and then $\mathcal{M}_q \equiv 2, 4$, or 8 for 1D, 2D, or 3D quadratures.

(iv) Gather the AP connectivity list for the current iElement, their physical coordinates, and the iElement material properties data.

(v) Loop over the quadrature points —> —> —> —> —> —> —>

(a) Map the quadrature point location into the B-spline, NURBS, or T-spline knot span (r_q, s_q), as well as to the physical location, (x_q, y_q). Evaluate the scalar Jacobian matrix between the non-dimensional quadrature space and the non-dimensional knot span, $\mathbb{J}_{rs:q} = \mathcal{M}_{rs}/\mathcal{M}_q$.

(b) Compute the basis functions, say the row matrix $[\boldsymbol{N}(r,s)]$, and the number of derivatives needed at the knot span location (r_q, s_q).

(c) Use the product of the first partial derivatives of the basis functions, $[\boldsymbol{d}N(r,s)]$, with respect to parametric coordinates, r and s, with the two physical spatial coordinates of the analysis control points in the rectangular matrix, say $[\boldsymbol{XY}^{iE}]$, to compute the 2×2 square Jacobian matrix from mapping from the non-dimensional quadrature space into the physical space $[\mathbb{J}_{\boldsymbol{xy:q}}]$. Compute its scalar determinant $|\mathbb{J}_{xy:q}|$, and its inverse $[\mathbb{J}_{\boldsymbol{xy:q}}]^{-1}$.

(d) Compute the two rows of physical gradients of the basis functions at the non-dimensional parametric point (r_q, s_q) by multiplying their local gradient by the inverse Jacobian matrix of the mapping from non-dimensional quadrature space into physical space, say $[\boldsymbol{DN}(r,s)] = [\mathbb{J}_{\boldsymbol{xy:q}}]^{-1}[\boldsymbol{dN}(r,s)]$. The resulting matrix, $[\boldsymbol{DN}(r,s)]$, has the x- and y-partial derivatives of the basis functions needed to populate the strain–displacement matrix.

(e) Insert selected components from the two rows of the physical derivatives, $[\boldsymbol{DN}(r,s)]$. Place them into the three rows (four if axisymmetric) of the rectangular strain–displacement matrix, $[\boldsymbol{B}(r,s)]$, for a planar analysis.

(f) Select and populate which of the three elasticity matrices, $[\boldsymbol{E}]$, is appropriate.

(g) Add this quadrature point's increment to the iElement stiffness formed by the triple matrix product, the two Jacobian determinates, and the weight of the quadrature point:

$$[\boldsymbol{K}^{iE}] = [\boldsymbol{K}^{iE}] + [\boldsymbol{B}(r_q, s_q)]^T[\boldsymbol{E}][\boldsymbol{B}(r_q, s_q)]\mathbb{J}_{rs:q}|\mathbb{J}_{xy:q}|w_q.$$

If there is a body force, like gravity, accelerations, or electromagnet forces per unit volume, $[\boldsymbol{f}_B] = [f_x, f_y]$, likewise integrate for their resultant force vector:

$$\{\boldsymbol{F}_B^{iE}\} = \{\boldsymbol{F}_B^{iE}\} + [\boldsymbol{N}(r_q, s_q)]^T[\boldsymbol{f}_B]\mathbb{J}_{rs:q}|\mathbb{J}_{xy:q}|w_q.$$

(h) Save arrays evaluated at the quadrature point to a file for later use in post-processing and plotting. At least save the physical coordinate on the part (x_q, y_q), the basis functions $[\boldsymbol{N}(r_q, s_q)]$, the strain–displacement matrix $[\boldsymbol{B}(r_q, s_q)]$, and, if variable, the stress–strain matrix $[\boldsymbol{E}(r_q, s_q)]$ at the point.

(i) If the mass matrix is expected to be needed for later vibrations or dynamics, use the mass density to form it:

$$[\boldsymbol{M}^{iE}] = [\boldsymbol{M}^{iE}] + [\boldsymbol{N}(r_q, s_q)]^T \rho^{iE} [\boldsymbol{N}(r_q, s_q)] J_{rs:q} |\mathbb{J}_{xy:q}| w_q.$$

End loop over the quadrature points <— <— <— <— <— <— <—

(vi) Gather the (two) degree-of-freedom numbers at each analysis control point supporting the basis functions and scatter (assembly) the iElement area integrals into the larger system matrices:

$$[\boldsymbol{K}] = [\boldsymbol{K}] + [\boldsymbol{K}^{iE}], [\boldsymbol{M}] = [\boldsymbol{M}] + [\boldsymbol{M}^{iE}],$$
$$\{\boldsymbol{F}_B\} = \{\boldsymbol{F}_B\} + \{\boldsymbol{F}_B^{iE}\}.$$

End loop over all the elements. <<==== <<==== <<==== <<==== <<====

(3) If there are non-zero Neumann or Robin conditions on portions of the edges bounding the area, then a similar 1D quadrature summation must be used for their force resultants. Then, the Jacobian matrix from the non-dimensional quadrature line to the physical planar curve, $[\mathbb{J}_{\boldsymbol{xy:q}}]$, is a scalar, 1×1, involving a square root of differential line lengths, as discussed elsewhere. Also remember that the planar concept includes the common axisymmetric structures which have a fourth 'hoop strain' and a fourth 'hoop stress', causing the strain–displacement matrix, $[\boldsymbol{B}(r, s)]$, and the square stress–strain matrix, $[\boldsymbol{E}]$, to have four rows instead of three for plane stress and plane strain applications.

10.12 Plane stress and plane strain

The IGA method easily models three-dimensional solids, including the most general anisotropic and non-homogeneous materials, with general loadings and boundary conditions. Here, the lack of 3D graphical post-processing tools limits applications to those that are two-dimensional.

The plane stress simplification is the most common of all planar simplifications and is taught to most aerospace, civil, and mechanical engineers in required courses in solid mechanics. Plane stress analysis approximates a very thin solid ($t \ll L$) where its bounding z-planes are free of stress ($\sigma_z = \tau_{xz} = \tau_{yz} = 0$), but the normal strain is not zero, $\varepsilon_z \neq 0$. The normal strain can be recovered, including the effects of an initial thermal strain, in post-processing calculations from the equation

$$\varepsilon_z = \alpha\Delta T - \upsilon(\sigma_x + \sigma_y)/E.$$

The active displacements at all points are the x- and y-components of the displacement vector, $\boldsymbol{u}^T = [u(x,y)\ v(x,y)]$. The retained stress and strain components needed for the plane stress analysis are $\boldsymbol{\sigma}^T = [\sigma_x\ \sigma_y\ \tau_{xz}]$ and $\varepsilon^T = [\varepsilon_x\ \varepsilon_y\ \gamma_{xy}]$. For plane stress, Hooke's law reduces to

$$\begin{Bmatrix} \sigma_x \\ \sigma_y \\ \tau_{xy} \end{Bmatrix} = \frac{E}{1-\upsilon^2}\begin{bmatrix} 1 & \upsilon & 0 \\ \upsilon & 1 & 0 \\ 0 & 0 & (1-\upsilon)/2 \end{bmatrix}\begin{Bmatrix} \varepsilon_x \\ \varepsilon_y \\ \gamma_{xy} \end{Bmatrix} - \frac{E\alpha\Delta T}{1-\upsilon}\begin{Bmatrix} 1 \\ 1 \\ 0 \end{Bmatrix}.$$

After any IGA stress study is completed, the stresses and/or strains are substituted into the appropriate material-dependent failure criteria to determine if any point in the material has yielded or 'failed' and thus violates the assumptions that justified a linear analysis. The reader is simply reminded that just computing the displacements, strains, and stresses *does not* complete any elasticity analysis.

10.13 Structural analysis data files

The extension of the scalar field analysis simulations highlights the increase in the complexity of applications having more than one degree of freedom per analysis control point by changing how the degrees of freedom are numbered. The prior examples of directly placing problem-dependent data into a script that carries out all of the matrix operations are not practical when there are more than two iElements in each parametric direction.

For field problems, having only a single generalized unknown, $n_g = 1$, at each AP made specifying a temperature boundary condition simply meant identifying the AP number where it was to be expressed. But, with structures, there are one, two, or three displacement components at each AP and a systematic way of identifying a vector component at an AP is needed. The same is true for an external force acting in one of the spatial directions and applied to an analysis control point. For planar stress analysis problems considered here, there are two displacement components per AP, $n_g = 1$. There is a clear need for identifying the system, or local, equation number associated with any dof at any analysis control point in the solution domain. This not only allows for a clear definition of specified displacements and external forces but is also critical to defining the equation numbers for gathering and scattering data between individual iElements and corresponding system data.

There are different choices for numbering the items in such a list. The most common approach, followed here, is to sequentially number all the same kind of dof at an AP before going to the next AP and repeating that process for all the APs in the domain. Consider the common case where the same number of quantities, say n_g (for number of generalized dof), occurs at every analysis control point (AP). Then, the local list of dof at each AP of any subregion or iElement can be defined as a sub-set of the same type of dof at all the APs. Let the total number of APs in a simulation be n_m; let the number of APs connected to each iElement be n_n. Then, the number of independent dofs active in an iElement is $n_i = n_n n_g$. Likewise, the total number of system dof is $n_d = n_m n_g$. At any AP, denote the number of the jth item of interest as $j = 1 : n_g$, and let the i-th local AP number of an element be $i = 1 : n_n$. The iElement AP 'connection list', say AP_list(i), identifies the system AP number in the full domain which corresponds to that local AP dof number. The local iElement dof number, $n(i, j)$, is

$$n(i, j) = n_g * (i - 1) + j, \quad 1 \le i \le n_n, \quad 1 \le j \le n_g, \qquad (10.11)$$

and, defining the corresponding system AP number in the mesh as I, the corresponding system dof number is

$$\mathrm{N}(I, j) = n_g * (I - 1) + j, \quad 1 \leq I \leq n_m, \quad 1 \leq j \leq n_g, \tag{10.12}$$

but since the corresponding system AP number in the mesh is $I = \text{AP\_list}(i)$, the corresponding system dof number is

$$N(i, j) = n_g * (\text{AP\_list}(i) - 1) + j, \quad 1 \leq i \leq n_n, \quad 1 \leq j \leq n_g. \tag{10.13}$$

Here, n_g is used to denote the number of generalized unknowns at each AP, also called the dof at each AP. These local and system equation numberings are employed in gathering from the system unknown dofs to local unknown dofs, and for scattering (adding) local known square and column arrays into locations within the corresponding system arrays. The same direct subscript conversions can be used to gather the physical coordinates of the AP, or application-dependent data, like AP temperatures for calculating thermal loads in an IGA stress study.

Equation (10.13) is the major key to automating the essential process of assembling (scattering) the sub-region or iElement arrays into the system arrays. The equations create a list of all of the system equation numbers associated with a specific sub-region or iElement. Here, that vector subscript (list) is often given as the name row, or dof, since it is initially used in examples to show how the rows of a local column matrix are scattered into the system column matrix. In other words, a vector subscript of that name is used to convert the row subscript of a local array to the corresponding row in a system array. The same process is used to assemble local square matrices which have their column subscripts converted using that same vector subscript. Since the vector subscript serves a dual purpose, it is sometimes also called the iElement index list.

Script 10.1 shows how (10.13) is implemented as a Matlab script. The input arguments to the function *get_element_index.m* are the number of APs connected to the iElement, n_n $\longleftrightarrow n_n$, the number of generalized unknowns per AP, n_g $\longleftrightarrow n_g$, and the list of system AP numbers connected to the iElement. The return from the function is the list of corresponding system equation numbers (which are used later as vector subscripts to achieve efficient programming).

Script 10.1. Gather the dof numbers for an IElement

```
function [rows] = get_element_index (n_g, n_n, AP_list)
% calculate DOF numbers of iElement, for gather, scatter
% AP_list = maximum connectivity list for an iElement
% n_g     = number of DOF per analysis control point
% n_n     = maximum number of Aps per iElement
% rows    = vector changing iEl to system eq numbers
rows = zeros (1, n_g * n_n)                    ; % allocate
for k = 1:n_n  ; % loop over iElement APs  ==>> ==>> ==>>
 global_AP = round (AP_list (k)) ; % corresponding sys AP
 for i = 1:n_g; % loop over DOF at AP ---> ---> ---> --->
  eq_global  = i + n_g*(global_AP - 1); % sys DOF, if any
  eq_element = i + n_g*(k         - 1) ; % iEl DOF number
  rows (1, eq_element) = eq_global      ; % valid DOF > 0
 end   ; % for DOF number i <--- <--- <--- <--- <--- <---
end ; % for each iElement AP <== <<== <<== <<== <<== <<==
% end get_element_index    % ============================
```

Establishing a unique number for every degree of freedom allows for structured input and output data files to be created for any IGA study. The input files can be manually created or generated by external software. Those files are text files, named msh_*.txt, with other names inserted between those two ends. If such a file is intended for an IGA study only, then the letters _iga_appear in the file name. If not, the name implies its use can be for either IGA or FEA studies.

Since IGA studies have to assign a weight to each AP if the data are to be used to construct NURBS or T-splines, one external data file used from here on is 'msh_AP_wxyz.txt', which sequentially contains one line per Analysis control Point, giving its weight, x, y, and z-coordinates. The z-coordinate defaults to zero if the file has only three columns. In the following plane stress example study, the file is shown as being read by script get_AP_wxyz. The AP connections list is read by script get_iga_AP_list which reads file msh_iga_AP_list.txt having one line per iElement, giving the list of APs which support it.

The example has EBCs on some of the displacements. They are read by script get_ebc_values from the text data file msh_ebc.txt containing one line for each essential boundary condition, which contains the AP_number, dof_number, and value_assigned for IGA. The first two integers identify the displacement dof number being specified as

$$\text{Eq}(\text{AP}, \text{dof}) = n_g * (\text{AP} - 1) + \text{dof}.$$

Likewise, the example plane stress study will include external point loads acting on some APs. They are identified in the same way to give the row number where they are to be added to any other internally created forces at that AP. Their data are taken from file msh_load_pt.txt which has one line per point force with the AP_number, dof_number, and the value assigned for IGA. The file is read by function get_point_sources.

Since a user's memory of studies fades with time, it is wise to keep a summary of a study stored in the same directory. Here, the file is named msh_remarks.txt and it is displayed with the call echo_user_remarks. For the plane stress cantilever example, it contains the remarks in Fig. 10.5. A similar scaled plot of the AP with support and load regions noted is provided in Fig. 10.6.

This problem used a relatively deep rectangular analysis domain with four bi-quadratic basis functions to predict the end displacement compared to Euler thin beam theory, which does not include the effects of Poisson's ratio. The Euler theory for deflections and stresses assumes that the depth/length ratio is generally $\leq 1/10$.

```
%  Plane-stress displacement based IGA study
% Here using bi-variant basis functions, p=q=2
%       13 o------o14--+--o15-----o16 Ly  o = AP
%          |iE3        |iE4       |       u_x, u_y
%        9 o       o10 |  o11     o12     16 CP
% Fixed    |-----------+----------|       Body_x,y
%        5 o       o6  |  o7      o8      F_x, F_y
%          |iE1        |iE2       |       R_x, R_y
%        1 o------o2---+--o3------o4  Lx  32 dof
%
% Problem data and geometry,  V=[0 0 0 0.5 1 1 1]
% Beam theory cantilever tip disp.   Lx=100 Ly=20
% YM = 1000.; PR = 0.00; Body = [0 0]; Properties
% Fixed = [1,2 9,10 17,18 25,26] from msh_ebc.txt
% Given = [0,0, 0,0   0,0   0,0] from msh_ebc.txt
% Loads = [16 24], Values = [1, 1] get_pt_sources
% Give exact displacements of u_y = 1 @ 4,8,12,16
```

Figure 10.5. Text with cantilever beam study summary

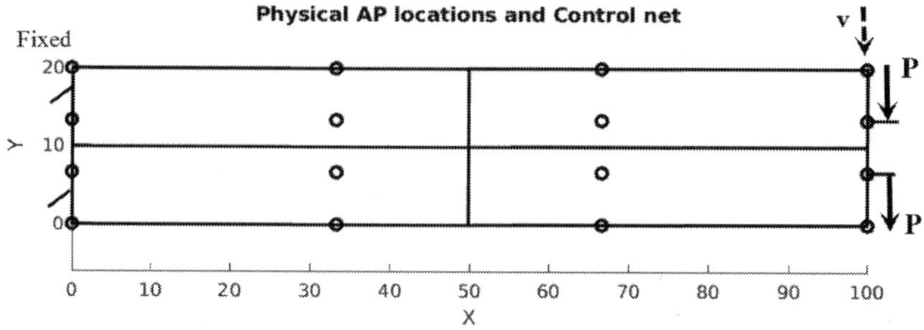

Figure 10.6. Physical shape of control region studied

Then, both with and without Poisson's ratio, the plane stress deflections are essentially the same as the value given by Euler beam theory. However, the stresses are more accurate in the plane stress since bending is still the main effect. When the beam depth exceeds a third of the beam length, the transverse shear effect governs and such a beam is called a 'deep beam' as a reminder that Euler beam theory is no longer accurate for predicting the stress distributions. For deep beam applications, plane stress models are much more accurate than Euler beam models.

The Euler beam theory for the deflection of a cantilever with a point load at the free end is

$$v = \frac{FL^3}{EI} \quad \text{with } I = \frac{tL_y^3}{12}.$$

For this plane stress example, the thickness, t, is unity. The theory of elasticity shows that to avoid infinite stresses at a point load, the end loading should be an equivalent line load which is zero at the top and bottom of the beam and varies parabolically to a maximum at the horizontal centerline of the beam. Such a line load distribution can, and should, be integrated with the basis functions to give the corresponding forces at the four analysis control points at the end. For the initial validation of this plane stress model, the end load is just split in half and placed at the two control points nearest the centerline. In other words, here, $F = 2P$, where P is the value of the forces in the plane stress model in Fig. 10.6.

Script 10.2 executes the numerically integrated plane stress model, and it uses the data files for the AP coordinates, the Dirichlet displacement conditions, and the two externally applied forces. Functions get_AP_wxyz, get_iga_AP_list, and ge_iga_ebc were used to create the vector subscripts of Fixed and Given, used before and new vectors Loads and Values since their creations are not obvious when there are several dofs per AP and when there is a large number of APs. The determination of the Free dof is standard once the Fixed dofs are specified and it is evaluated by the function *get_Free_from_Fixed.m*.

Script 10.2. Main analysis and checking steps for IGA plane stress studies

```
%                    Plane_stress_cantilever
%==============================================================
diary Plane_stress.txt                        % save the results
clear all         % using bivariant basis functions, p=q=2
echo_user_remarks ; % information about this application
n_g = 2; n_q = 9; n_r = 3; echo=1    ; % control integers

% Gather coords, connections, EBC, & external  pt forces
[n_m, n_s, w, AP_x, AP_y, z] = get_AP_wxyz (echo)
[n_e, n_n, AP_list]         = get_iga_AP_list(echo)
[Fixed, Given]             = get_iga_ebc (n_g, n_m, echo)
n_i = n_n * n_g; n_d = n_m * n_g  ; % iElement & sys dof
[Free] = get_free_from_fixed (n_d, Fixed)   % not manual
[Loads, Values]         = get_iga_pt_source (n_g,n_m,echo)

YM = 1000.; PR = 0.00; Body = [0 0]        ; % Properties
[E] = E_plane_stress (YM, PR)        ; % stress-strain law

AP_xy (:,1) = AP_x; AP_xy (:,2) = AP_y; % reshape coords

cS = zeros(n_d,1); SS = zeros(n_d,n_d)  ; % allocate sys
cF = zeros(n_d,1); cF(Loads) = Values             ; % clear

%        Knot vector data in parametric and index spaces
range_I=[0 0.5; 0.5 1]; range_J=[0 0.5; 0.5 1] ; % index
range_U=[0 0.5; 0.5 1]; range_V=[0 0.5; 0.5 1] ; % param
span_U = range_U(:,2)- range_U(:,1); % parametric length
span_V = range_V(:,2)- range_V(:,1); % parametric length
Index=[1 1; 2 1; 1 2; 2 2]          ; % index I,J for iEl
```

```
%   Open units needed for post-processing operations
el_unit = fopen ('el_store.bin', 'w')     ; % for writing
qp_unit = fopen ('qp_store.bin', 'w'); % for quadratures

%          Compute and assemble system of equations
Area  = 0        ; % Initialize domain area (for checking)
AreaP = 0        ; % Initialize domain area (for checking)
[r_q,s_q,w_q] = qp_rule_unit_quad (n_q);% in [0,1]x[0,1]
for e = 1:n_e ; % Loop over iEl (knot spans) ===>> ===>>
 fprintf('iEl %i is at corner %i, %i \n', e, Index(e,:))
 Area_e = 0          ; % initialize area for this iElement
 AP_e = AP_list (e, :) ; % Gather AP AP_e # for this iEl
 xy_e = AP_xy(AP_e, :)          ; % Gather their x- and y-
 I = Index (e ,1); J = Index (e ,2) ; % iEl column & row
 B =zeros(n_r,n_i);Se = zeros(n_i,n_i);cB =zeros(n_i,1);

 for q = 1:n_q ; % loop over Gauss points ---> ---> --->
  % Compute quadrature pt coordinates in parameter space
  R = range_I(I,1) + span_U(I)*r_q(q)    ; % in knot span
  S = range_J(J,1) + span_V(J)*s_q(q)    ; % in knot span
  J2 = span_U(I)*span_V(J) / 1       ; % unit Gauss [0,1]

  % Evaluate basis functions and derivative; for n_g = 1
  if ( I == 1 );% In practice use N, dN numeric function
    Nr=[(2*R-1)^2,-2*R*(3*R-2), 2*R^2]   ; % first I span
    dNr_dr = [(8*R-4), (4-12*R),4*R] ; % dNr/dr dNr/ds=0
  else                                  ; % second I span
   Nr=[(2*R-2)*(R-1),(-6*R^2+8*R-2),(2*R-1)^2];
   dNr_dr=[(4*R-4),(8-12*R),(8*R-4)];  % dNr/dr dNr/ds=0
  end                              ; % if which r-span to use
  if ( J == 1 )
   Ns=[(2*S-1)^2,-2*S*(3*S-2),2*S^2];
   dNs_ds=[(8*S-4),(4-12*S),4*S]     ; % dNr/ds, dNr/dr=0
  else                                  ; % second I span
   Ns=[(2*S-2)*(S-1),(-6*S^2+8*S-2),(2*S-1)^2];
   dNs_ds=[(4*S-4),(8-12*S),(8*S-4)];
  end                              ; % if which s-span to use
  %  Tensor products of basis functions & derivatives
  Nrs      = Nr'*Ns     ; Nrs      =reshape(Nrs,     1,n_n);
  dNrs_dr = dNr_dr'*Ns; dNrs_dr =reshape(dNrs_dr,1,n_n);
```

```
 dNrs_ds = Nr'*dNs_ds; dNrs_ds =reshape(dNrs_ds,1,n_n);
 Grad_rs = [dNrs_dr; dNrs_ds] ; % Local Gradient matrix

  %     Form physical Jacobian, determinate, inverse
  Jac = Grad_rs*xy_e; det_J = det(Jac); inv_J =inv(Jac);
  Area_e = Area_e + J2 * det_J * w_q(q) ; % iEl A update
  Grad_xy= inv_J*Grad_rs; % Nrs physical gradient matrix

  %    ** Convert to displacement vectors; n_g = 2 **
  % Set the non-zero terms in the vector interpolation
  for k = 1: n_g        ; % alternate with 0 for each dof
    H_q (k, k:n_g:n_i) = Nrs (1:n_n); % scalars into vec
  end       ; % for shifted rows of scalar basis functions

% Compute global position of quadrature point; For plots
  xy_q = Nrs * xy_e (1:n_n,:) ; % 1 x n_s  global (x, y)

% Insert gradient terms in Strain-Displacement Law, save
  B_q (1, 1:2:n_i) = Grad_xy (1, 1:n_n); % for esplion x
  B_q (2, 2:2:n_i) = Grad_xy (2, 1:n_n); % for esplion y
  B_q (3, 1:2:n_i) = Grad_xy (2, 1:n_n) ; % for gamma xy
  B_q (3, 2:2:n_i) = Grad_xy (1, 1:n_n) ; % for gamma xy

%          Update element square stiffness matrix
  Se = Se + (B_q' * E * B_q) * J2 * det_J * w_q (q)

%            Constant body force vector source
  cB = cB + H_q'*Body'*J2*det_J*w_q (q)    ; % body source

%      save data needed for post-processing stresses
  fwrite (qp_unit, xy_q, 'double')    ; % save coordinates
  fwrite (qp_unit, E_e,  'double')       ; % save material
  fwrite (qp_unit, B_q,  'double'); % strain-displacement
end  ; % for q loop over quadrature points <--- <--- <---
Area = Area + Area_e      ; % update total area as a check

 %         Scatter iElement cQ into c & Se into S
 dof    = get_element_index (n_g, n_n, AP_e)  ; % iEl eqs
 cS(dof)=cS(dof)+cB(:);SS(dof,dof) = SS(dof,dof)+Se(:,:);
end        ; % for e loop over iElements <<=== <<=== <<===
```

```
% Save iElement data needed for post-processing stresses
fwrite (el_unit, xy_e, 'double')      ; % save coordinates
fwrite (el_unit, E,    'double')       ; % save properties
fwrite (el_unit, Se,   'double')        ; % save stiffness
fwrite (el_unit, cB,   'double') ; % body force resultant

Lx = max (AP_x); Ly = max(AP_y)         ; % rename lengths
fprintf('Area of domain is %g \n', Area)     ; % checking
fprintf('Expected value is %g \n', Lx*Ly)    ; % checking
Fx_B =sum(cS(1:n_g:n_d)); Fy_B =sum(cS(2:n_g:n_d));% info
fprintf('Sum of body forces are    %g %g \n',Fx_B, Fy_B);
fprintf('Expected total values are %g %g \n',Lx*Ly*Body);

%    Solve for all displacements, and force reactions
cS       = cS + cF                    ; % Add external forces
[D, DR] = Solve_Free_DOF (SS, cS, Free, Fixed, Given);
list_save_results (n_g, n_m, D)         ; % save and print
Energy =  D(Free)'*SS(Free, Free)*D(Free) % strain energy
Work   = -D(Free)'*cS(Free)                          % work
disp('Leaving Plane_stress_canti_msh')
% end Plane_stress_cantilever ===========================
```

There are parts of this plane stress script that would have to be changed for use in a large practical IGA study. The analytic evaluation of the basis functions at the quadrature points should be replaced by a call to the function *num_basis_rect.m*. The recurrence relations therein automatically establish into which knot span (iElement) a quadrature point falls. Recall that with the default arguments, the function constructs the two knot vectors and then runs the recurrence relations to numerically evaluate the basis functions at the quadrature point. Thus, it is much more efficient in a large application to write, or recover, the two knot vectors which defined the rectangular group of iElements once and supply them as the optional arguments on every call instead of recomputing those thousands of times, or more.

This plane stress example can be used to find the displacements of any 1D, 2D, or 3D elasticity problem by using Matlab switch command to pick the appropriate stress–strain matric and the strain–displacement matrix. For example, it currently uses

$$[E] = E\_\text{plane\_stress}(\text{YM}, \text{PR}); \%\text{stress-strain law}$$

which was developed for FEA studies and is only one of the four stress–strain laws for elasticity models listed in detail in the Summary of this chapter. The stress–strain law is given as Script 10.3.

Script 10.3. The plane stress stress–strain version

```
function [E_e] = E_plane_stress (E, nu, G)
% * * * * * * * * * * * * * * * * * * * * * * * * * * *
%      PLANE STRESS CONSTITUTIVE MATRIX DEFINITION
%          STRESS COMPONENT ORDER: XX, YY, XY
% * * * * * * * * * * * * * * * * * * * * * * * * * * *
% E   = Young's modulus of elasticity
% G   = shear modulus (not independent)
% nu  = Poisson's ratio
% E_e = constitutive matrix
% n_r = number of rows in B and E matrices = 3

if ( nargin == 2 )                 ; % classic definition
  G  = 0.5 * E / (1 + nu)                    ; % in theory
end                               ; % if special material
c_1 = E / (1 - nu^2) ; c_2 = c_1 * nu        ; % constants

E_e = zeros (3, 3)                           ; % largely 0
E_e(1, 1) = c_1 ; E_e(2, 1) = c_2 ;            ; % normals
E_e(1, 2) = c_2 ; E_e(2, 2) = c_1 ;            ; % normals
E_e(3, 3) = G                                    ; % shear
% end E_plane_stress ====================================
```

Currently, the non-zero entries in the strain–displacement matrix are hard-coded as

```
% Insert gradient terms in Strain-Displacement Law, save
  B_q (1, 1:2:n_i) = Grad_xy (1, 1:n_n); % for esplion x
  B_q (2, 2:2:n_i) = Grad_xy (2, 1:n_n); % for esplion y
  B_q (3, 1:2:n_i) = Grad_xy (2, 1:n_n) ; % for gamma xy
  B_q (3, 2:2:n_i) = Grad_xy (1, 1:n_n) ; % for gamma xy
```

which is one of the four relations that occur in elasticity listed in detail in the Summary of this chapter. The above-mentioned matrix evaluation is the same for plane stress and plane strain and is given as the function *B_planar_elastic.m* listed in Script 10.4.

Script 10.4. The strain–displacement relation for planar elasticity

```
function [B] = B_planar_elastic (DGH, n_n)
% * * * * * * * * * * * * * * * * * * * * * * * * * * *
%       ELASTICITY STRAIN-DISPLACEMENT RELATIONS  (B)
%  For plane-stress or plane-strain: n_g = 2, n_r = 3
% * * * * * * * * * * * * * * * * * * * * * * * * * * *
% B   = strain-displacement matrix   (n_r, n_n * n_g)
% DGH = global derivatives of H      (n_s, n_n * n_g)
% n_g = number of displacements per node
% n_n = number of nodes per element
%       Stress and strain order is: xx, yy, xy

n_g = 2 ; n_r = 3    ; % for planar stress
n_i = n_n * n_g      ; % number of element dof
B = zeros (n_r, n_i) ; % initialize strain-displacement

for J = 1:n_n ; % loop over nodes ==>> ==>> ===>> ===>>
  K = n_g * (J - 1) + 1     ; % u-column contribution
  L = K + 1                 ; % v-column contribution
  B (1, K) = DGH (1, J)     ; % du/dx for normal xx
  B (3, K) = DGH (2, J)     ; % du/dy for shear  xy
  B (2, L) = DGH (2, J)     ; % dv/dy for normal yy
  B (3, L) = DGH (1, J)     ; % dv/dx for shear  xy
end ; % for J nodes of element  <<=== <<=== <<=== <<===
% end B_planar_elastic ================================
```

Using just some of the integer constants that define the size of the matrices in an IGA or FEA study, the appropriate [**B**] matrix for the stress study can be made with a single call to the FEA library function *B_matrix_elastic.m* shown in Script 10.5.

Script 10.5. Selecting and populating the elasticity stress–strain relation

```
function [B] = B_matrix_elastic (DGH, n_g, n_n, n_r, ...
                                 i_opt, H, R)
% * * * * * * * * * * * * * * * * * * * * * * * * * * * * *
%        ELASTICITY STRAIN-DISPLACEMENT RELATIONS  (B)
%                  Default to elastic solid
% * * * * * * * * * * * * * * * * * * * * * * * * * * * * *
```

```
% B     = STRAIN-DISPLACEMENT MATRIX       (n_r, n_n * n_g)
% DGH   = GLOBAL DERIVATIVES OF H          (n_s, n_n * n_g)
% H     = ELEMENT INTERPOLATION FUNCTIONS (n_n * n_g)
% i_opt = ELASTICITY CLASS
%         = 1, AXIAL BAR,     n_g = 1, n_r = 1, n_s = 1
%         = 2, PLANE STRESS, n_g = 2, n_r = 3, n_s = 2
%         = 3, PLANE STRAIN, n_g = 2, n_r = 3, n_s = 2
%         = 4, AXISYMMETRIC, n_g = 2, n_r = 4, n_s = 2
%         = 5, 3-D SOLID,     n_g = 3, n_r = 6, n_s = 3
% n_g   = number of PARAMETERS PER NODE
% n_i   = number of element dof = n_n * n_g
% n_n   = number of NODES PER ELEMENT
% n_r   = number of STRAINS (ROWS IN B)
%            XX, YY, XY, ZZ, XZ, YZ, 1 OR 3 OR 4 OR 6
% n_s   = DIMENSION OF SPACE
% R     = axisymmetric radius at the point

if ( nargin == 4 ) ; % default to solid
  i_opt = 5 ; end  ; % number of input arguments
if ( nargin == 5 ) ; % a non-solid element
  if ( i_opt < 1 | i_opt > 5 )
    error (' i_opt < 1 or > 5 in B_matrix_elastic')
  end ; % if invalid argument
end  ; % number of input arguments

n_i = n_n * n_g      ; % set constant
B = zeros (n_r, n_i) ; % initialize strain-displacement
% matrix

for J = 1:n_n ; % loop over nodes ===>> ===>> ===>> ===>>
  K = n_g * (J - 1) + 1           ; % "FIRST" COLUMN, U

%-->     ONE-DIMENSIONAL AXIAL BAR, i_opt = 1
  B (1, K) = DGH (1, J)           ; % DU/DX  FOR XX NORMAL
  if ( i_opt == 1 )
    continue ; % TO NEXT J VALUE
  end ; % if elastic bar
%-> PLANE STRESS, PLANE STRAIN, AXISYMMETRIC, 3D
%   STRESS & STRAIN COMPONENT ORDER: XX, YY, AND XY
```

```
  L = K + 1                           ; % "SECOND" COLUMN, V
  B (3, K) = DGH (2, J)               ; % DU/DY  FOR XY SHEAR
  B (2, L) = DGH (2, J)               ; % DV/DY  FOR YY NORMAL
  B (3, L) = DGH (1, J)               ; % DV/DX  FOR XY SHEAR
  if ( i_opt == 2 | i_opt == 3 )
    continue ; % TO NEXT J VALUE
  end  ; % if plane stress or plane strain

%-> AXISYMMETRIC ONLY
%   STRESS & STRAIN COMPONENT ORDER: XX, YY, XY, AND TT
  if ( i_opt == 4 ) ; % then axisymmetric
    if ( R <= 0 )
      error ('zero radius in axisymmetric elastic B_e')
    end ; % if division by zero
    B (4, K) = H (J) / R              ; % U/R HOOP, ZZ NORMAL
    continue ; % TO NEXT J VALUE
  end ; % if axisymmetric i_opt = 4

%-> 3D SOLID ONLY
%   STRESS & STRAIN COMPONENT ORDER: XX, YY, XY, ZZ, XZ, YZ
  M = L + 1                           ; % "THIRD" COLUMN, W
  B (5, K) = DGH (3, J)               ; % DU/DZ  FOR XZ SHEAR
  B (6, L) = DGH (3, J)               ; % DV/DZ  FOR YZ SHEAR
  B (4, M) = DGH (3, J)               ; % DW/DZ  FOR ZZ NORMAL
  B (5, M) = DGH (1, J)               ; % DW/DX  FOR XZ SHEAR
  B (6, M) = DGH (2, J)               ; % DW/DY  FOR YZ SHEAR
end ; % for all J nodes of the element
% end B_matrix_elastic ====================================
```

The data for the simple cantilever study were chosen such that an applied total force of 2 lb. would cause a horizontal centerline vertical deflection of 1 inch, which is predicted by Euler beam theory. Figures 10.7 and 10.8 verify that the plane stress program has done that the system reaction forces of Fig. 10.8 are shown graphically in Fig. 10.9 to verify that they increase with distance from the neutral axis so as to balance the moments of the vertical end forces. It has also provided the expected equal and opposite reaction forces. In IGA studies, the reactions are always equal and opposite to the applied forces. If they look wrong, it means the user has missed a decimal point and/or used the incorrect units when providing the loading data.

```
Computed Solution:
AP, 2 displacements per AP
1,      0.000e+00  0.000e+00
2,      7.096e-02  1.887e-02
3,      1.467e-01  6.005e-01
4,      1.522e-01  9.992e-01
5,      0.000e+00  0.000e+00
6,      3.436e-02  1.892e-02
7,      7.318e-02  6.004e-01
8,      7.474e-02  9.994e-01
9,      0.000e+00  0.000e+00
10,    -3.436e-02  1.892e-02
11,    -7.318e-02  6.004e-01
12,    -7.474e-02  9.994e-01
13,     0.000e+00  0.000e+00
14,    -7.096e-02  1.887e-02
15,    -1.467e-01  6.005e-01
16,    -1.522e-01  9.992e-01

APs with extreme values
DOF, AP  Maximum    AP   Minimum
1,   4   1.522e-01  16  -1.522e-01
2,   12  9.994e-01  1    0.000e+00

Energy = 2.0189  Work = -2.0189
```

Figure 10.7. Displacement values from the cantilever model

```
Reactions at essential BCs
AP, DOF, Reaction Value
1    1 -7.8941e+00
1    2 -2.8814e-01
5    1 -4.4118e+00
5    2 -7.2186e-01
9    1  4.4118e+00
9    2 -7.2186e-01
13   1  7.8941e+00
13   2 -2.8814e-01
Total reaction values = [0.00 -2.00]
```

Figure 10.8. Cantilever reaction force components

10.14 Post-processing for stresses

After computing the system displacements and recovering the reaction forces at the EBC, an IGA stress study continues to calculate, and save for plots, all of the strain components, stress components, and material yield (failure) criteria. For Lagrange polynomials, it

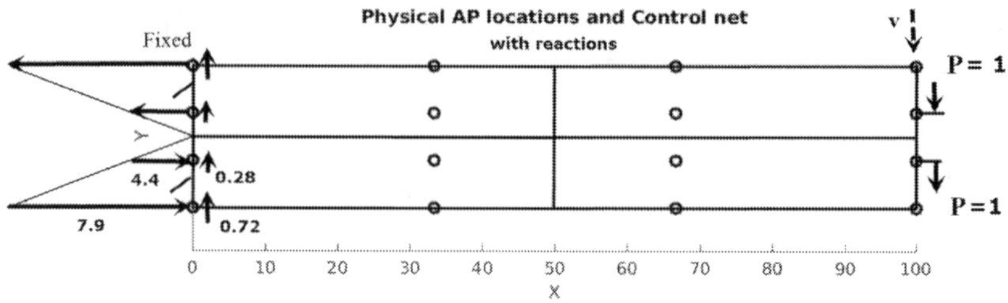

Figure 10.9. Sketch of cantilever AP reaction values

has been shown that the most accurate parametric locations for stress recovery are the Gauss quadrature points. Hopefully, this is true for B-spline basis functions too. Calculating the matrices necessary for such post-processing has just been done while calculating the iElement stiffness matrices, and force resultants. The choice is whether to save them now to sequential storage files or to recompute them all again in the post-processing loops. In the above-mentioned code segment, the decision was made to use sequential binary files for storing items at the quadrature points and some completed items after exiting the loop over iElements. Binary file read/write operations are about the second-fastest operations on most digital computers.

The binary data written above are read in the following segments given in Script 10.6 to calculate the stresses and strains and yield criteria at the quadrature points. The data have been saved to huge inefficient sequential text files for two reasons: (1) Text files are human readable and (2) the IGA techniques do not yet have good assessable graphical processes to view such results. This contrasts with FEA studies where sixty years of use have led to many efficient graphical post-processing tools.

Script 10.6. Typical post-processing steps for stress analysis

```
disp('AP Displacements and Reactions computed')
fprintf('================================ \n')
fprintf('Begin iElement Post-processing: \n')
fprintf('================================ \n')
fprintf('strain order is e_xx, e_yy, e_xy, & e_hoop \n')
fprintf('stress order is S_xx, S_yy, S_xy, & S_hoop \n')
```

```
fprintf('Yield criterion: von Mises & maximum shear \n')
show_qp = 0
if ( show_qp == 1 )              ; % will create printouts
disp('Above saved to iga_qp_xyz_stresses.txt & printed')
else                            ; % use binary storage only
  disp('Above only saved to iga_qp_xyz_stresses.txt')
end                               ; % data storage choice
strain = zeros (n_r, 1)          ; % pre-allocate strains
stress = zeros (n_r, 1)          ; % pre-allocate stresses
max_max_VM = 0             ; % max of all von Mises stress
max_max_MS = 0             ; % max of all max shear stress
max_VM_pt  = 0; max_VM_el = 0;% initialize peak location
max_MS_pt  = 0; max_MS_el = 0;% initialize peak location

%     Open the binary files from Plane-Stress
qp_unit = fopen ('qp_store.bin', 'r')  ; % open for read
el_unit = fopen ('el_store.bin', 'r')  ; % open for read

%      Open (inefficient) text files for plotting
q_id = fopen('iga_qp_xyz_stresses.txt','w'); % save file

for k = 1:n_e; % loop over iElements====>> ====>> ====>>

%                    Gather answers
  AP_e     = AP_list (k, 1:n_n)            ; % connectivity
  [rows]   = get_element_index (n_g, n_n, AP_e);% eq numb
  Ans_e (1:n_i) = D(rows)      ; % gather iElement results

%  Recover previously calculated iElement arrays
  [xy_e] = fread (el_unit,[n_n, n_s], 'double');% coords
  [E]    = fread (el_unit,[n_r, n_r], 'double') ; % prop
  [S_e]  = fread (el_unit,[n_i, n_i], 'double'); % stiff
  strain_0 = 0 ; % Are initial thermal strains present ?

%  Get the strains and stresses at each quadrature point
  for q = 1:n_q ; % begin quadrature loop ---> ---> --->

%  Recover previous calculations here (or re-compute)
    [xy_q]=fread(qp_unit,[1,  n_s],'double'); % coords
    [E_e] =fread(qp_unit,[n_r,n_r],'double');%material
    [B_q] =fread(qp_unit,[n_r,n_i],'double');%strain-d
    [H_q] =fread(qp_unit,[1,  n_i],'double')     ; % BF
```

```
%  Compute the generalized strains & stresses
    strain = B_q*Ans_e'            ; % in-plane strains
    stress = E_e*(strain-strain_0);% in-plane stresses

%    compute the material failure estimates
    vonMise = sqrt( (stress(1)-stress(2))^2  ...
                  + (stress(2))^2 + (stress(1))^2  ...
                  + 6*(stress(3))^2 )* 0.70710678 ;
    maxTau = vonMise / sqrt(3)          ; % upper bound

    if ( show_qp == 1 )     ; % list, strains, stresses
      fprintf('\n')
      fprintf('El, Pt, X, Y: \n')
      fprintf('    %i, %i, %g  %g  \n', ...
                  k, q, xy_q (1:2))
      fprintf('El, Pt, Strains:  \n')
      fprintf('  %i, %i, %g  %g  %g  \n', ...
               k, q, strain (1:n_r))          ; % list
      fprintf('El, Pt, Stresses: \n')
      fprintf('  %i, %i, %g  %g  %g  \n', ...
               k, q, stress (1:n_r))          ; % list
      fprintf('El, Pt, Von Mises: %i, %i, %g \n',  ...
               k, q, vonMise)          ; % list
      fprintf('El, Pt, max Shear: %i, %i, %g \n',  ...
               k, q, maxTau)           ; % list
    end                      ; % if list failure criteria

% Save coord, stresses, failure criteria for plots
   fprintf(q_id,'%g %g %g %g %g %g %g %g \n',  ...
         xy_q(1:n_s), stress(1:n_r), vonMise, maxTau);

    if ( vonMise > max_max_VM ) ; % update worst value
      max_max_VM = vonMise              ; % current value
      max_VM_el  = k                 ; % current iElement
      max_VM_pt  = q        ; % current quadrature point
    end                      ; % if current max VM value
    if ( maxTau > max_max_MS )  ; % update worst value
      max_max_MS = maxTau               ; % current value
      max_MS_el  = k                 ; % current iElement
      max_MS_pt  = q        ; % current quadrature point
    end                      ; % if current max MS value
```

```
end % for q quadrature points <-- <-- <--- <--- <---
end; % for each k iElement in mesh <<=== <<==== <<====

fprintf('\n')
fprintf('======================================== \n')
fprintf('Recovered Strains,Stresses,YieldCriteria \n')
fprintf('======================================== \n')
fprintf('\n')
fprintf('Maximum von Mises stress =%g \n', max_max_VM)
fprintf('at pt %i of iElement %i \n', max_VM_pt, ...
                                      max_VM_el)
fprintf('Maximum shear stress     = %g \n', max_max_MS)
fprintf('at pt %i of iElement %i \n', max_MS_pt, ...
                                      max_MS_el)
fprintf ('X, Y coordinates & stress components \n')
fprintf ('XX, YY, XY, ZZ, XZ, and vonMise, maxTau \n')
fprintf ('saved at all quadrature points in text \n')
fprintf('file iga_qp_xyz_stresses.txt for plots \n')
fclose(q_id)                      ; % stress items saved
disp('Leaving Plane_stress_canti_msh_post')
% end Plane_stress_canti_msh_post ===================
```

10.15 Work done by initial strains

Initial strains have been mentioned in the stress–strain relations. When an initial strain is present, its name implies the fact that it is active at the beginning of an imposed mechanical strain (and stress) and continues until the end of that strain. Therefore, the work that it does differs from that of the strain energy contribution by a factor of two and is

$$W_0 = \int_V \boldsymbol{\sigma}^T \boldsymbol{\varepsilon_0} dV,$$

$$W_0 = \int_V \boldsymbol{\sigma}^T \boldsymbol{\varepsilon_0} dV = \sum_{e=1}^{n_e} \int_{V^e} \boldsymbol{\sigma}^{e^T} \boldsymbol{\varepsilon}_{\mathbf{0}}^e dV = \sum_{e=1}^{n_e} \int_{V^e} (\boldsymbol{B}^e \boldsymbol{\delta}^e)^T \boldsymbol{E}^e \boldsymbol{\varepsilon}_{\mathbf{0}}^e dV.$$

This defines the element initial strain resultant force vector as

$$W_0 = \sum_{e=1}^{n_e} \boldsymbol{\delta}^{e^T} \boldsymbol{c}_0^e \quad \text{with } \boldsymbol{c}_{\mathbf{0}}^{\boldsymbol{e}} = \int_{V^e} \boldsymbol{B}^{e^T} \boldsymbol{E}^{\boldsymbol{e}} \boldsymbol{\varepsilon}_{\mathbf{0}}^e dV.$$

$$(n_i \times 1) = (n_i \times n_r)(n_r \times n_r)(n_r \times 1).$$

An important aspect of initial strains is that when they are included as resultant loads in the early stage of assembling, the system loads then must also appear in the post-processing for the mechanical stresses, $\boldsymbol{\sigma}$, and yield criterions by subtracting them from the mechanical strains, $\boldsymbol{\varepsilon}$, at a point. Usually, the recovery points are quadrature points:

$$\boldsymbol{\sigma}(r_q, s_q) = \boldsymbol{E}(r_q, s_q)(\boldsymbol{\varepsilon}(r_q, s_q)\boldsymbol{\varepsilon}_0(r_q, s_q)).$$

By far, the most common initial strain is due to a temperature change, $\boldsymbol{\nabla} T$, from the material's stress-free state:

$$\varepsilon_0(r_q, s_q) = \alpha \nabla T(r_q, s_q).$$

10.16 Kinetic energy

Later, in the examination of the dynamics of solids, it will be necessary to include the kinetic energy converted to the same matrix notation. The displacement components at any time, τ, at any point in an element are

$$\begin{Bmatrix} u(x, y, z, \tau) \\ v(x, y, z, \tau) \\ w(x, y, z, \tau) \end{Bmatrix} = \boldsymbol{u}(x, y, z, \tau) \equiv \boldsymbol{N}(x, y, z)\boldsymbol{\delta}^{\mathbf{e}}(\tau).$$

The velocity is

$$\dot{\boldsymbol{u}} = \frac{\partial \boldsymbol{u}(x, y, z, \tau)}{\partial \tau} \equiv \boldsymbol{N}(x, y, z)\frac{\partial \boldsymbol{\delta}^{\mathbf{e}}(\tau)}{\partial \tau} \equiv \boldsymbol{N}(x, y, z)\dot{\delta^e},$$

where a dot above a term denotes its partial derivative with respect to time. The kinetic energy of a point mass is one half the mass times the velocity squared. For a material of mass density of ρ, the kinetic energy of an iElement volume is

$$KE^e = \frac{1}{2}\int_{V^e} \dot{\boldsymbol{u}} \cdot \dot{\boldsymbol{u}} \rho dV = \frac{1}{2}\int_{V^e} \dot{\boldsymbol{u}}^T \rho \dot{\boldsymbol{u}} dV$$

$$= \frac{1}{2}\int_{V^e} (\boldsymbol{N}(x,y,z)\dot{\boldsymbol{\delta}}^e)^T \rho (\boldsymbol{N}(x,y,z)\dot{\boldsymbol{\delta}}^e) dV,$$

$$KE^e = \frac{1}{2}\dot{\boldsymbol{\delta}}^{e^T}\left[\int_{V^e} \boldsymbol{N}(x,y,z)^T \rho^e \boldsymbol{N}(x,y,z) dV\right]\dot{\boldsymbol{\delta}}^e \equiv \frac{1}{2}\dot{\boldsymbol{\delta}}^{e^T}\boldsymbol{m^e}\dot{\boldsymbol{\delta}}^e.$$

This defines the square symmetric element mass matrix, which is the same size as the stiffness matrix:

$$\boldsymbol{m^e} \equiv \int_{V^e} \boldsymbol{N}(x,y,z)^T \rho^e \boldsymbol{N}(x,y,z) dV.$$

10.17 Summary

$n\_b \longleftrightarrow n_b \equiv$ Number of boundary segments,
$n\_d \longleftrightarrow n_d \equiv$ Number of system unknowns $= n\_g \times n\_m$,
$n\_e \longleftrightarrow n_e \equiv$ Number of elements,
$n\_g \longleftrightarrow n_g \equiv$ Number of generalized dofs per AP,
$n\_i \longleftrightarrow n_i \equiv$ Number of unknowns per element $= n\_g \times n\_n$,
$n\_m \longleftrightarrow n_m \equiv$ Number of mesh APs,
$n\_n \longleftrightarrow n_n \equiv$ Number of APs per element or segment,
$n\_P \longleftrightarrow n_P \equiv$ Number of point loads,
$n\_p \longleftrightarrow n_p \equiv$ Dimension of parametric space,
$n\_q \longleftrightarrow n_q \equiv$ Number of total quadrature points,
$n\_r \longleftrightarrow n_r \equiv$ Number of rows in the $\boldsymbol{B^e}$ matrix (and material matrix $\boldsymbol{E}$),
$n\_s \longleftrightarrow n_s \equiv$ Dimension of physical space,

$b =$ boundary segment number, $e =$ element number,
$\subset =$ subset, $\cup =$ union of sets,
Boundary, element, and system unknowns: $\boldsymbol{\delta^B} \subset_b \boldsymbol{\delta^e} \subset_e \boldsymbol{\delta}$,

Geometry: $\Omega^e \equiv$ Element domain, $\Omega = \cup_e \Omega^e \equiv$ Solution domain,
$\Gamma^b \subset \Omega^e \equiv$ Boundary segment, $\Gamma = \cup_b \Gamma^b \equiv$ Domain boundary.

Displacement interpolation: $\boldsymbol{u}(\boldsymbol{x}) = \boldsymbol{N}(r)\boldsymbol{\delta}^{\boldsymbol{e}}, \boldsymbol{u}(\boldsymbol{x})^{\boldsymbol{T}} = \boldsymbol{\delta}^{\boldsymbol{e}^{\boldsymbol{T}}}\boldsymbol{N}(r)^{\boldsymbol{T}}$.
Displacement interpolation: $\boldsymbol{u}(\boldsymbol{x}) = \boldsymbol{N}(r)\boldsymbol{\delta}^{\boldsymbol{e}}, \boldsymbol{u}(\boldsymbol{x})^{\boldsymbol{T}} = \boldsymbol{\delta}^{\mathbf{e}^{\mathbf{T}}}\boldsymbol{N}(r)^{\boldsymbol{T}}$.
Strain–displacement matrix: $\boldsymbol{B}^e(\boldsymbol{r}), \boldsymbol{\varepsilon}(\boldsymbol{r}) \equiv \boldsymbol{B}^e(\boldsymbol{r})\boldsymbol{\delta}^{\boldsymbol{e}}$.
Mechanical work: $W = \int_V \boldsymbol{\delta}^T \boldsymbol{f} dV + \int_S \boldsymbol{\delta}^T \boldsymbol{T} dS + \sum_{k=1}^{k=n_P} \boldsymbol{\delta}_k^T \boldsymbol{F}_k$.
Engineering strain components: $\boldsymbol{\varepsilon}^T = [\varepsilon_x\ \varepsilon_y\ \varepsilon_z\ \gamma_{xy}\ \gamma_{xz}\ \gamma_{yz}]$.
Stress components: $\boldsymbol{\sigma}^T = [\sigma_x\ \sigma_y\ \sigma_z\ \tau_{xy}\ \tau_{xz}\ \tau_{yz}]$.
Principal axis thermal expansion coeff.:

$$\boldsymbol{\alpha}^T = [\alpha_x\ \alpha_y\ \alpha_z\ \alpha_{xy}\ \alpha_{xz}\ \alpha_{yz}]$$

Principal orthotropic CTE: $\boldsymbol{\alpha}^T = [\alpha_x\ \alpha_y\ \alpha_z\ 0\ 0\ 0]$.
Isotropic thermal expansion coeff. $\boldsymbol{\alpha}^T = \alpha[1\ 1\ 1\ 0\ 0\ 0]$.
Mechanical strain–stress relation:

$$\boldsymbol{\varepsilon} = \mathbb{E}\boldsymbol{\sigma} + \boldsymbol{\alpha}\Delta T, \quad \mathbb{E} = \mathbb{E}^T, \quad |\mathbb{E}| > 0.$$

Mechanical stress–strain relation: $\boldsymbol{\sigma} = \mathbb{E}\boldsymbol{\varepsilon} - \mathbb{E}\boldsymbol{\alpha}\Delta T,\ \boldsymbol{E} = \mathbb{E}^{-1}$.
Strain energy: $U = \frac{1}{2}\int_V \boldsymbol{\sigma}^T \boldsymbol{\varepsilon} dV = \frac{1}{2}\int_V \boldsymbol{\varepsilon}^T \boldsymbol{\sigma} dV = \frac{1}{2}\int_V \boldsymbol{\varepsilon}^T \boldsymbol{E}\boldsymbol{\varepsilon} dV$.
Work of initial strains, $\boldsymbol{\varepsilon}_0$:

$$W_0 = \int_V \boldsymbol{\sigma}^T \boldsymbol{\varepsilon}_0\, dV = \int_V \boldsymbol{\varepsilon}^T \boldsymbol{E}\boldsymbol{\varepsilon}_0 dV, \quad \boldsymbol{\varepsilon}_T = \alpha\Delta T \subset \boldsymbol{\varepsilon}_0.$$

iElement strain energy: $U^e = \frac{1}{2}\boldsymbol{\delta}^{\boldsymbol{e}^T}[\boldsymbol{K}^e]\boldsymbol{\delta}^{\boldsymbol{e}}$.
Stiffness matrix: $\boldsymbol{K}^{\boldsymbol{e}} = \int_{\Omega^e} \boldsymbol{B}^{e^T}\boldsymbol{E}^e\boldsymbol{B}^e d\Omega^e$.
Stress tensor:

$$\sigma^{\boldsymbol{T}} = [\sigma_x\ \sigma_y\ \sigma_z\ \tau_{xy}\ \tau_{xz}\ \tau_{yz}] \Leftrightarrow [\boldsymbol{\sigma}] = \begin{bmatrix} \sigma_{xx} & \tau_{xy} & \tau_{xz} \\ \tau_{xy} & \sigma_{yy} & \tau_{yz} \\ \tau_{xz} & \tau_{yz} & \sigma_{zz} \end{bmatrix}.$$

Principal stresses:

$$|[\boldsymbol{\sigma}] - \lambda_k[\boldsymbol{I}]| = 0,\ k = 1, \ldots, n_s \text{ and } \sigma_1 > \sigma_2 > \sigma_3$$

Hydrostatic pressure:

$$p = (\sigma_1 + \sigma_2 + \sigma_3)/3 = (\sigma_{xx} + \sigma_{yy} + \sigma_{zz})/3.$$

Experimental yield stresses: σ_t tensile, σ_c compressive,
$\kappa \equiv \sigma_c/\sigma_t$ ratio.

Von Mises criterion ($\kappa \equiv 1$ only):

$$\sigma_E = \frac{1}{\sqrt{2}}\sqrt{(\sigma_1 - \sigma_2)^2 + (\sigma_1 - \sigma_3)^2 + (\sigma_2 - \sigma_3)^2},$$

or

$$\sigma_E = \frac{1}{\sqrt{2}}\sqrt{\begin{array}{c}(\sigma_{yy} - \sigma_{xx})^2 + (\sigma_{zz} - \sigma_{xx})^2 + (\sigma_{zz} - \sigma_{yy})^2 \\ +6(\tau_{xy}^2 + \tau_{xz}^2 + \tau_{yz}^2).\end{array}}$$

Burzynski criterion (any κ):

$$\sigma_B = [3p(\kappa - 1) + \sqrt{9p^2(\kappa - 1)^2 + 4\kappa\sigma_E^2}]/2\kappa, \sigma_B(\kappa = 1) = \sigma_E.$$

Ductile material failure criterion: $\sigma_E \geq \sigma_t$ or $\sigma_B \geq \sigma_t$.
Absolute maximum shear stress criterion: $\tau_{\max} = (\sigma_1 - \sigma_3)/2$.
Strain–displacement relation, $\boldsymbol{\varepsilon}^{\boldsymbol{e}} = \boldsymbol{B}^{\boldsymbol{e}}(x, y, z)\boldsymbol{\delta}^{\boldsymbol{e}}$, for elastic solid:

$$\begin{Bmatrix} \varepsilon_x \\ \varepsilon_y \\ \varepsilon_z \\ \gamma_{xy} \\ \gamma_{xz} \\ \gamma_{yz} \end{Bmatrix}^e, \quad \boldsymbol{B}^{\boldsymbol{e}}(x, y, z) = [\boldsymbol{B}_{\boldsymbol{1}}^{\boldsymbol{e}} \vdots \boldsymbol{B}_{\boldsymbol{2}}^{\boldsymbol{e}} \vdots \cdots \boldsymbol{B}_{n_n}^{\boldsymbol{e}}],$$

$$\boldsymbol{B}_{\boldsymbol{k}}^{\boldsymbol{e}} = \begin{bmatrix} N_{k,x} & 0 & 0 \\ 0 & N_{k,y} & 0 \\ 0 & 0 & N_{k,z} \\ N_{k,y} & N_{k,x} & 0 \\ N_{k,z} & 0 & N_{k,x} \\ 0 & N_{k,z} & HN_{k,y} \end{bmatrix}.$$

Function name: *B_solid_elastic.m*
Strain–displacement relation, $\boldsymbol{\varepsilon}^{\boldsymbol{e}} = \boldsymbol{B}^{\boldsymbol{e}}(x, y)\boldsymbol{\delta}^{\boldsymbol{e}}$, for plane stress and plane strain:

$$\begin{Bmatrix} \varepsilon_x \\ \varepsilon_y \\ \gamma_{xy} \end{Bmatrix}^e, \quad \boldsymbol{B}_{\boldsymbol{k}}^{\boldsymbol{e}}(x, y) = \begin{bmatrix} N_{k,x} & 0 \\ 0 & N_{k,y} \\ N_{k,y} & N_{k,x} \end{bmatrix}.$$

Function name: *B_planar_elastic.m*
Strain–displacement relation, $\boldsymbol{\varepsilon}^{\boldsymbol{e}} = \boldsymbol{B}^{\boldsymbol{e}}(x,y)\boldsymbol{\delta}^{\boldsymbol{e}}$,
for axisymmetric stress:

$$\begin{Bmatrix} \varepsilon_R \\ \varepsilon_y \\ \gamma_{Ry} \\ \varepsilon_\theta \end{Bmatrix}^e, \quad \boldsymbol{B}_{\boldsymbol{k}}^{\boldsymbol{e}}(x,y) = \begin{bmatrix} N_{k,x} & 0 \\ 0 & N_{k,y} \\ N_{k,y} & N_{k,x} \\ N_k/R & 0 \end{bmatrix}, \quad R = \boldsymbol{N}(x,y)\boldsymbol{R}^{\boldsymbol{e}}.$$

Function name: *B_axisym_elastic.m*
Constitutive law, isotropic plane stress:

$$\begin{Bmatrix} \sigma_x \\ \sigma_y \\ \tau_{xy} \end{Bmatrix} = \frac{E}{1-\upsilon^2}\begin{bmatrix} 1 & \upsilon & 0 \\ \upsilon & 1 & 0 \\ 0 & 0 & (1-\upsilon)/2 \end{bmatrix}\begin{Bmatrix} \varepsilon_x \\ \varepsilon_y \\ \gamma_{xy} \end{Bmatrix} - \frac{E\alpha\Delta T}{1-\upsilon}\begin{Bmatrix} 1 \\ 1 \\ 0 \end{Bmatrix}.$$

Function name: *E_plane_stress.m*
Constitutive law, anisotropic plane stress:

$$\begin{Bmatrix} \varepsilon_{nn} \\ \varepsilon_{ss} \\ \varepsilon_{tt} \\ \gamma_{ns} \end{Bmatrix} = \begin{bmatrix} 1/E_n & -\upsilon_{sn}/E_s & -\upsilon_{tn}/E_t & 0 \\ -\upsilon_{ns}/E_n & 1/E_s & -\upsilon_{ts}/E_t & 0 \\ -\upsilon_{nt}/E_n & -\upsilon_{st}/E_s & 1/E_t & 0 \\ 0 & 0 & 0 & 1/G_{ns} \end{bmatrix}$$

$$\begin{Bmatrix} \sigma_{nn} \\ \sigma_{ss} \\ \sigma_{tt} \\ \tau_{ns} \end{Bmatrix} + \Delta T\begin{Bmatrix} \alpha_n \\ \alpha_s \\ \alpha_t \\ 0 \end{Bmatrix}.$$

Constitutive law, isotropic plane strain:

$$\begin{Bmatrix} \sigma_x \\ \sigma_y \\ \tau_{xy} \end{Bmatrix} = \frac{E}{(1+\upsilon)(1-2\upsilon)}\begin{bmatrix} a & \upsilon & 0 \\ \upsilon & a & 0 \\ 0 & 0 & b \end{bmatrix}\begin{Bmatrix} \varepsilon_x \\ \varepsilon_y \\ \gamma_{xy} \end{Bmatrix} - \frac{E\alpha\Delta T}{1-\upsilon}\begin{Bmatrix} 1 \\ 1 \\ 0 \end{Bmatrix},$$

$$a = (1-\upsilon), \quad b = (1-2\upsilon)/2.$$

Function name: *E_plane_strain.m*
Constitutive law, isotropic axisymmetric stress:

$$\begin{Bmatrix} \sigma_r \\ \sigma_z \\ \tau_{rz} \\ \sigma_\theta \end{Bmatrix} = \frac{E}{(1+\upsilon)(1-2\upsilon)} \begin{bmatrix} a & \nu & 0 & \nu \\ \nu & a & 0 & \nu \\ 0 & 0 & b & 0 \\ \nu & \nu & 0 & a \end{bmatrix} \begin{Bmatrix} \varepsilon_r \\ \varepsilon_z \\ \gamma_{rz} \\ \varepsilon_\theta \end{Bmatrix}$$

$$-\frac{E\alpha\Delta T}{(1-2\upsilon)} \begin{Bmatrix} 1 \\ 1 \\ 0 \\ 1 \end{Bmatrix},$$

$$a = (1-\upsilon), \quad b = (1-2\upsilon)/2, \quad \upsilon \neq 0.5.$$

Function name: *E_axisym_stress.m*
Constitutive law, isotropic solid stress:

$$\begin{Bmatrix} \sigma_x \\ \sigma_y \\ \sigma_z \\ \tau_{xy} \\ \tau_{xz} \\ \tau_{yz} \end{Bmatrix} = \frac{E}{(1+\upsilon)(1-2\upsilon)} \begin{bmatrix} a & \upsilon & \upsilon & 0 & 0 & 0 \\ \upsilon & a & \upsilon & 0 & 0 & 0 \\ \upsilon & \upsilon & a & 0 & 0 & 0 \\ 0 & 0 & 0 & b & 0 & 0 \\ 0 & 0 & 0 & 0 & b & 0 \\ 0 & 0 & 0 & 0 & 0 & b \end{bmatrix}$$

$$\begin{Bmatrix} \varepsilon_x \\ \varepsilon_y \\ \varepsilon_z \\ \gamma_{xy} \\ \gamma_{xz} \\ \gamma_{yz} \end{Bmatrix} - \frac{E\alpha\Delta T}{(1-2\upsilon)} \begin{Bmatrix} 1 \\ 1 \\ 1 \\ 0 \\ 0 \\ 0 \end{Bmatrix},$$

$$a = (1-\upsilon), \; b = (1-2\upsilon)/2, \; \upsilon \neq 0.5.$$

Function name: *E_elastic_solid.m*

Chapter 11

Eigenanalysis

11.1 Introduction

Another common type of analysis problem is where a coefficient in the governing differential equation, λ, is an unknown global constant to be determined, called an eigenvalue. This class of analysis is called an eigen-problem from the German word *eigen*, meaning belonging distinctly to a group. Eigen-problem solutions require much more computation effort than solving a linear system of equations. There is a group of solutions to such a differential equation because for each computed eigenvalue which satisfied the PDE, there is an associated vector, called the eigenvector, whose length is the same as the number of degrees of freedom in the simulation.

One example of such a PDE is the scalar Helmholtz differential equation:

$$k\nabla^2 u(x, y, t) + \lambda u(x, y, t) = 0,$$

which is solved as a group for multiple eigenvalues, λ_n, and a corresponding set of solution eigenvectors to define the group of solutions. Another common application leading to an eigen-problem analysis is the vibration of elastic solids where the eigenvalues and their corresponding deflected mode shapes are to be determined. In theory, a continuous system has an infinite number of eigenvalues, but an IGA approximation has as many eigenvalues as there are degrees of freedom (after enforcing the EBC). Most engineering applications only require just the first few or the last few eigenvalues and eigenvectors. There are efficient algorithms for both approaches.

For waveguides and other applications, the eigenvalue can be a complex number, but for elastic vibrations, it is a non-negative number corresponding to the square of the natural frequency of vibration, $\lambda = \omega^2$, where any zero value corresponds to a rigid body (non-vibrating) motion.

There are handbook analytic solutions for the natural frequencies for the most common circuits, bars, beams, shafts, membranes, plates, and shells for numerous boundary conditions, including point spring supports and local point masses. The IGA method allows for the fast modeling of the vibration of any elastic solid.

For IGA studies, the typical PDE gets reduced to a matrix system of linear homogeneous equations. For those equations to have a non-zero solution, the determinant of the square matrix in the brackets must vanish, i.e.,

$$\det[\boldsymbol{K} - \lambda \boldsymbol{M}] = |\boldsymbol{K} - \lambda \boldsymbol{M}| = 0.$$

This condition leads to a group of solutions (eigen-problems) equal in number to the number of [Free] unknowns in $\boldsymbol{\delta}$ after the EBC have been enforced:

$$[\boldsymbol{K} - \lambda_j \boldsymbol{M}]\boldsymbol{\delta_j} = 0, \quad j = 1, 2, \ldots,$$

where λ_j are the eigenvalues corresponding to the eigenvector $\boldsymbol{\delta_j}$. The usual convention is to normalize the eigenvector so that the absolute value of its largest term is unity. Here, the examples are presented utilizing MATLAB, and thus, some specific features of that environment need to be noted. In general, for numerical simulations, the accuracy of the eigenvalues improves much faster than the shape of the corresponding eigenvectors.

11.2 MATLAB eigen-solvers

MATLAB provides the two functions *eig* and *eigs* for eigen-problem solutions. The IGA method provides the two matrix arguments, say the stiffness matrix, $\boldsymbol{K}$, and the mass matrix, and $\boldsymbol{M}$, that (after the EBCs are imposed) are real, symmetric, and non-negative matrices that define the problem as solving

$$[\boldsymbol{K} - \lambda_j \boldsymbol{M}]\boldsymbol{\delta_j} = 0, \quad j = 1, 2, \ldots$$

For a 'small' number of degrees of freedom, the MATLAB function call $[\boldsymbol{V}, \boldsymbol{\Lambda}] = \text{eig}(\boldsymbol{K}, \boldsymbol{M})$ returns two square matrices where the jth column of the first square matrix, $\boldsymbol{V}$, contains the jth eigenvector, $\boldsymbol{\delta_j}$, and its corresponding eigenvalue, λ_j, is found at the jth diagonal term of the diagonal matrix $\boldsymbol{\Lambda}$. In other words, the returned row number, j, of any eigenvector, λ_j, is also the column number of its corresponding eigenvector, $\boldsymbol{\delta_j}$. The eigenvalues can be conveniently placed in a vector, with the same row numbers, by using the MATLAB function *diag* as $\boldsymbol{\lambda} = diag(\boldsymbol{\Lambda})$. Numerical round-off errors can make the theoretical positive number have a tiny complex part. Thus, the safe thing to do is to use the MATLAB function *real* to ensure that $\boldsymbol{\omega}^2 = \boldsymbol{\lambda} = real(\boldsymbol{\lambda})$ in vibration studies.

However, the MATLAB eigenvalues, $\boldsymbol{\lambda}$, are not always in a sequential order either increasing or decreasing. For a buckling study, only the single smallest eigenvalue is needed. This can be extracted by using the MATLAB function *min*. But, it is also necessary to extract its eigenvector that shows the buckled shape of the structure. This can be done using the extended *min* function that returns two arguments:

$$[\lambda_1, row_1] = min(\boldsymbol{\lambda}),$$

where the smallest eigenvalue λ_1 was found in row_1 of the vector $\boldsymbol{\lambda}$ and the corresponding eigenvector column can be extracted by next using that row number as a column number and setting $\boldsymbol{\delta_1} = \boldsymbol{V}(:, row_1)$.

For a medium-sized problem, where only the first few smallest values in the vector $\boldsymbol{\lambda}$ are needed, they are found using the extended MATLAB function *sort* as

$$[\boldsymbol{\lambda_{\text{new}}}, \boldsymbol{Order}] = sort(\boldsymbol{\lambda}),$$

where the vector $\boldsymbol{\lambda_{\text{new}}}$ contains all of the eigenvalues in ascending order such that the smallest eigenvalue is $\lambda_{\text{new}}(1) = \lambda_{\text{small}}$ and the vector subscript $\boldsymbol{Order}$ gives the row number where the new sorted value was found in the original random list of eigenvalues. The vector subscript $\boldsymbol{Order}$ is very important since it is the key to finding the eigenvector that corresponds to a particular eigenvalue. For example, the smallest eigenvalue and its eigenvector are $\lambda_{\text{new}}(1)$, and $\boldsymbol{\delta_1} = \mathbf{V}(:, \boldsymbol{Order}(1))$ and the fifth pair is $\lambda_{\text{new}}(5)$ and $\boldsymbol{\delta_5} = \mathbf{V}(:, \boldsymbol{Order}(5))$.

For very large problems where the analysts need only a small number of the smallest eigenvalues, as in mechanical vibrations, or a small number of the largest eigenvalue, then the call to the MATLAB function *eigs* provides the much more efficient eigenvalue control with

$$[\boldsymbol{V}, \boldsymbol{\Lambda}] = eigs(\boldsymbol{K}, \boldsymbol{M}, n', sm'),$$

where the number n is the number of the eigenvalues required and the string '*sm*' requests the smallest eigenvalues. However, the eigenvalues on the diagonal of $\boldsymbol{\Lambda}$ are still usually in a random order and it is still necessary to sort them and to use the ***Order*** subscripts to extract the proper eigenvectors. The *eigs* function defaults to giving the largest eigenvalues if the optional string is omitted.

11.3 IGA eigen-problems

When a differential equation having an unknown global constant is solved by an IGA method, the unknown global constant factors out of all of the element matrices and appears unchanged in the assembled governing matrix system. The typical form of the matrix system becomes the 'general eigen problem':

$$[\boldsymbol{K} - \lambda \boldsymbol{M}]\boldsymbol{\delta} = \boldsymbol{0},$$

where $\boldsymbol{K}$ is typically a stiffness matrix or conduction matrix, $\boldsymbol{M}$ is typically a generalized mass matrix or a literal mass matrix for vibration problems, $\boldsymbol{\delta}$ corresponds to the analysis control point values of the primary unknowns (displacements, acoustical pressure, water elevation, etc.), and λ is the global unknown constant to be determined.

The consistent IGA theories create a full symmetric element mass matrix. But, based on prior finite difference methods which always create a diagonal mass matrix, some users prefer to 'diagonalize' the consistent mass matrix by using a diagonal matrix constructed from scaling up the original diagonal so that the sum of the new diagonal equals the total mass. Some numerical experiments have shown improved numerical accuracy for both eigenvalue and time history solutions when the average of the consistent mass matrix and its diagonalized form are employed (the averaged mass matrix).

For vibration studies, the eigenvalue is the square of the natural frequency: $\lambda = \omega^2$. When calculated by an IGA formulation, the frequency (in radians per second) ω_n is more accurate than the next higher frequency, ω_{n+1}. As more degrees of freedom are added, each natural frequency estimate becomes more accurate. Solutions should continue to be repeated with an increased number of degrees until the highest desired eigenvalue is unchanged to a desired number of significant figures after additional degrees of freedom are included in the model.

11.4 Stress stiffening

The vibration eigen-problem is derived from the assumption of 'free' simple harmonic motion. Here, the word 'free' means that the system is free of any external forces. For solid vibrations having a non-zero forcing term, the overall stiffness matrix of the system must be revised to include the stresses due to the external forces prior to beginning an eigen-solution. The external forces must first be imposed on a static solution, with the same EBC, to determine the stresses they impose. This stress state is used in turn to calculate the additional 'geometric stiffness matrix' of each iElement which is assembled into the new system geometric stiffness matrix, say $\boldsymbol{K_G}$ (also known as the initial stress matrix), which is added to the original structural stiffness matrix to yield the net system stiffness matrix, $\boldsymbol{K}$, so the updated eigen-problem is

$$[(\boldsymbol{K} + \boldsymbol{K_G}) - \lambda \boldsymbol{M}]\boldsymbol{\delta} = \boldsymbol{0}.$$

This process is known as including stress stiffening effects on the vibration problem. A classic example where such an update is needed is in the calculation of the vibration frequencies of turbine blades in aircraft engines that are subjected to very high forces due to angular and radial acceleration as well as braking deaccelerations.

The inclusion of the geometric stiffness matrix, also called the stress stiffening matrix, is mandatory for structural buckling problems covered later.

11.5 Vibrating string

The differential equation of the transverse motion of a tensioned elastic string, without bending resistance, damping, or external transverse loads is

$$T\frac{\partial v^2}{\partial x^2} - \rho\frac{\partial v^2}{\partial t^2} = 0, \tag{11.1}$$

where $v(x,t)$ is the transverse displacement of the string with a tension of T and a mass density per unit length of ρ and t denotes time. Later, when time histories are studied in detail, this will be referred to as a wave equation. This PDE is of the hyperbolic class. From physics, it is known that strings vibrate with simple harmonic motion (SHM). Applying this assumption here, define a separation of variables

$$v(x,t) = v_o(x)\sin(\omega t), \tag{11.2}$$

where $v_o(x)$ is a 'mode shape' defining the shape of the string along the x-direction as it changes between positive and negative values that change with a frequency of ω. The second time derivative of $v(x,t)$ is

$$\frac{\partial v^2}{dt^2} = -\omega^2 v_o(x)\sin(\omega t) = -\omega^2 v(x,t).$$

The assumption of SHM changes the hyperbolic PDE over time and space into an elliptic ODE in space:

$$T\frac{dv^2}{dx^2} + \rho\omega^2 v = 0.$$

This form of the ODE has an unknown global constant (ω^2) multiplying the solution value, and it is called a scalar Helmholtz equation. This has the usual EBC and NBC options to define a unique solution. For a guitar string, the usual EBCs are that both ends have zero transverse displacements. The use of the Galerkin method and integration by parts (the introduction of the NBCs) gives the governing integral form

$$I = \left[v\left(T\frac{dv}{dx}\right)\right]_0^L - \int_0^L \frac{dv}{dx}\left(T\frac{dv}{dx}\right)dx + \int_0^L v(x)\rho\omega^2 v(x)dx = 0.$$

Next, an IGA study of a user-supplied knot vector introduces a NURBS or B-spline parametric domain and the basis functions in each knot span to describe the transverse displacements as

$$v(x) = [\boldsymbol{N}(r)]\{\boldsymbol{v^r}\} = ([\boldsymbol{N}(r)]\{\boldsymbol{v^r}\})^{\boldsymbol{T}} = \{\boldsymbol{v^r}\}^{\boldsymbol{T}}[\boldsymbol{N}(r)]^{\boldsymbol{T}}.$$

Noting that the frequency, ω, is an unknown global constant, it is pulled outside any spatial integral where it appears and the governing matrix system: $[\boldsymbol{K} - \omega^2\boldsymbol{M}]\{\boldsymbol{v}\} = 0$. When the EBCs of zero end displacements, $v(0) = 0 = v(L)$, are enforced, the remaining external forces are zero and the reduced system matrix problem is

$$[\boldsymbol{K} - \omega^2\boldsymbol{M}]\boldsymbol{\delta} = \boldsymbol{0}.$$

These system matrices are assembled from the typical iElement stiffness, and consistent mass matrices are

$$\boldsymbol{K^r} = \int_{L^r} \left[\frac{d\boldsymbol{H}(r)}{dx}\right]^T T^r \left[\frac{d\boldsymbol{H}(r)}{dx}\right] dx, \quad \boldsymbol{M^r} = \int_{L^r} [\boldsymbol{N}(r)]^{\boldsymbol{T}} \rho^r [\boldsymbol{N}(r)] dx.$$

For constant properties and user AP locations of $AP\_x = [0, 1/3, 2/3, 1]L$ to cause a constant knot span Jacobian, the closed-form cubic stiffness and mass matrices for the knot vector of $\mathbb{A}u = \{0, 0, 0, 0, 1, 1, 1, 1\}$ on the single knot span of $[0, 1)$ are

$$[\boldsymbol{K}] = \frac{T^r}{10L^r}\begin{bmatrix} 18 & -9 & -6 & -3 \\ -9 & 12 & 3 & -6 \\ -6 & 3 & 12 & -9 \\ -3 & -6 & -9 & 18 \end{bmatrix} \text{ and}$$

$$[\boldsymbol{M}] = \frac{\rho^r L^r}{140}\begin{bmatrix} 20 & 10 & 4 & 1 \\ 10 & 12 & 9 & 4 \\ 4 & 9 & 12 & 10 \\ 1 & 4 & 10 & 20 \end{bmatrix},$$

respectively. The resulting vibration shapes (mode shapes) come in sets that are either symmetric or anti-symmetric with respect to the center of the string.

Note that the assumptions imply that the string has slope continuity along its entire length (with unknown values at the support points). This is the default condition of a B-spline or NURBS approximation.

Example 11.1.

Given: A guitar string is needed for a musical instrument. The string has both ends fixed, a mass per unit length of ρ, and a constant tension for of T. Determine how the tension of the string affects the first natural frequency of the string.

Solution: The first symmetric mode is the fundamental one with respect to the center of the string. Approximate it using a single cubic B-spline iElement having four AP, with both end APs fixed, i.e., the fixed dof are at [Fixed] = [1, 4] with displacements of [Given] = [0, 0] and the unknown AP dof numbers are [Free] = [2, 3]. Also, for a single iElement $L^r = L$ and the system matrix eigen-system, before enforcing the two zero end displacements, is

$$\left(\frac{T}{10L}\begin{bmatrix}18 & -9 & -6 & -3\\ -9 & 12 & 3 & -6\\ -6 & 3 & 12 & -9\\ -3 & -6 & -9 & 18\end{bmatrix} - \omega^2\frac{\rho L}{140}\begin{bmatrix}20 & 10 & 4 & 1\\ 10 & 12 & 9 & 4\\ 4 & 9 & 12 & 10\\ 1 & 4 & 10 & 20\end{bmatrix}\right)\begin{Bmatrix}v_1\\ v_2\\ v_3\\ v_4\end{Bmatrix}$$
$$= \begin{Bmatrix}-c(0)_{\text{NBC}}\\ 0\\ 0\\ c(L)_{\text{NBC}}\end{Bmatrix}.$$

Enforcing the two zero displacement conditions leaves only the two center rows and columns available:

$$\left(\frac{T}{10L}\begin{bmatrix}12 & 3\\ 3 & 12\end{bmatrix} - \omega^2\frac{\rho L}{140}\begin{bmatrix}12 & 9\\ 9 & 12\end{bmatrix}\right)\begin{Bmatrix}v_2\\ v_3\end{Bmatrix} = \begin{Bmatrix}0\\ 0\end{Bmatrix}.$$

For a non-trivial solution, $v_2 \neq 0 \neq v_3$, the determinant of the square matrix in the brackets must vanish. This happens only for specific values of ω_k, $1 \leq k \leq n_d$ that are equal in number to the number of free dof in the model, n_d. As discussed above, for a small problem like this the MATLAB function call $[\boldsymbol{V}, \boldsymbol{\Lambda}] = eig(\boldsymbol{K}, \boldsymbol{M})$ returns two square matrices where each eigenvector, $\boldsymbol{v_j}$, as the jth column of the first square matrix, $\boldsymbol{V}$, and the corresponding eigenvalue, λ_j, is placed in the jth-row diagonal term of the matrix.

In small problems, a faster result is obtained by defining $[\boldsymbol{A}] = [\boldsymbol{M}]^{-1}[\boldsymbol{K}]$ and using $Vals = eig(\boldsymbol{A})$. This gives

$$\{Vals\} = \begin{Bmatrix} \omega_1^2 \\ \omega_2^2 \end{Bmatrix} = \frac{T}{\rho L^2} \begin{Bmatrix} 10 \\ 42 \end{Bmatrix},$$

where the first 'natural' frequency is $\omega_1^2 = 10T/\rho L^2$ or $\omega_1 = 3.1623\sqrt{T/\rho \mathrm{L}^2}$. The exact solution for the kth mode is $\omega_k = k\pi\sqrt{T/\rho L^2}$ rad/sec, where odd values of k are symmetric modes and even values are non-symmetric. The answer to the given question is that the frequency of the string vibration increases with the square root of its tension.

Here, the single cubic iElement has only 0.66% error in both the first and second natural frequencies. The exact first mode shape is a half-sine curve, the amplitude of which is normalized to unity. This mode shape is approximated spatially here by the cubic B-spline in a single iElement model.

As a more practical example, consider the tensioned spring simulated by two or more cubic B-splines using the special AP coordinate locations and illustrating the MATLAB *eig* function. Script 11.1 shows such a MATLAB vibration simulation for two such elements and the natural frequencies of the string motion compared to the exact frequencies. Note that the results show that the frequency error rapidly increases from near zero for the first mode to almost 33% for the fifth mode. The visual accuracy in the vibration mode shapes usually degrades even faster. To obtain more accuracy, the number of iElements, n_e, along the length of the string has to be increased along with the number of AP points and connections. For example, doubling the number of iElements to four reduces the error in the fifth mode from about 33% to about 1%, as shown in Fig. 11.1.

This study actually uses two C^0 patches joined at AP number 4. An FEA study using the same number of AP and cubic interpolation would have had that too, but an FEA quadratic would have had two such slope discontinuities. Again, an alternative to using two patches is to insert more internal knots and use a single patch over the length of the string. Then no slope discontinuities occur, which yields better accuracy, especially for the mode shapes (eigenvectors) which are theoretically smooth.

Script 11.1. Tensioned string eigenvalue–eigenvector calculations

```
% Vibration of a tensioned string              iga_string_vib.m
% PDE: T*V_xx = rho*V_tt,          SHM T*V_xx = -omega^2*rho*V
%     via p=3 B-splines with AP_x = [0, 1/3, 2/3, 1]Le
%       with closed form stiffness and mass matrices
%
%       | (iE1), L1     | (iE2), L2     |
% T <== o----o----o----o----o----o----o ==> T --> x
% AP    1    2    3    4    5    6    7
%       |<-              L              ->|
%   increase number of iElements, n_e, for more accuracy
n_e = 2; n_g = 1; n_n = 4; n_m = 7       ; % system constants
n_i = n_n*n_g; n_d = n_m*n_g            ; % iEl and system sizes
AP_list= [1 2 3 4 ; 4 5 6 7]         ; % AP connections per iEl
Fixed = [1, 7]; Given = [0., 0.]; Free =[2:1:6]      ; % data
L  = 2. ; T= 6e5 ; rho = 0.0234                ; % m, N, kg/m
Le = L / n_e             ; % length of iElements, increase n_e

K   = zeros(n_d, n_d); M = zeros(n_d, n_d)       ; % allocate
K_e = [18  -9   -6   -3;              % stiffness, p3 J constant
       -9   12   3   -6;
       -6    3  12   -9;
       -3   -6  -9   18]*T/(10*Le)        ; % stiffness matrix
M_e = [20   10   4    1;                   % mass. p3 J constant
       10   12   9    4;
        4    9  12   10;
        1    4  10   20]*rho*Le/140             ; % mass matrix

%           Assemble (scatter) iElement matrices
for k = 1:n_e ; % loop on iElements ===> ===> ===> ===> ===>
  dof = AP_list (k, :)                  ; % ONLY for n_g = 1 here
  K(dof, dof) = K(dof, dof) + K_e              ; % sys stiffness
  M(dof, dof) = M(dof, dof) + M_e                   ; % sys mass
 end          ; % for iElements <=== <=== <=== <=== <=== <===

 %   Solve for general form eigenvalues and eigenvectors
 [Vec, Diag] = eig (K(Free, Free), M(Free, Free)) ; %
 %   Vec has eigenvectors, Diag has eigenvalue squared
 fprintf ('Mode, String natural frequencies \n')   % heading
 exact = pi() * sqrt (T/(rho*L^2))      ; % times mode number
```

```
 for m = 1:length(Free)     ; % loop on modes ---> ---> --->
   omega = sqrt( real (Diag(m,m)))              ; % iEl freqency
   true  = m * exact                  ; % exact natural frequency
   fprintf ('%i, IGA = %7.2e, Exact = %7.2e \n', ...
              m, omega, true)        ; % pretty print results
 end      ; % for active modes <--- <--- <--- <--- <--- <---
% end of iga_spring_vib                             Running gives
% Mode,  String natural frequencies                 Error
% 1, IGA = 7.95e+03, Exact = 7.95e+03               0.0%
% 2, IGA = 1.60e+04, Exact = 1.59e+04               0.6%
% 3, IGA = 2.45e+04, Exact = 2.39e+04               2.5%
% 4, IGA = 3.28e+04, Exact = 3.18e+04               3.1%
% 5, IGA = 5.29e+04, Exact = 3.98e+04               33.%
```

```
AP_list =         1        2        3        4
                  4        5        6        7
                  7        8        9       10
                 10       11       12       13
Mode, String natural frequencies
1,   IGA = 7.95e+03, Exact = 7.95e+03
2,   IGA = 1.59e+04, Exact = 1.59e+04
3,   IGA = 2.39e+04, Exact = 2.39e+04
4,   IGA = 3.20e+04, Exact = 3.18e+04
5,   IGA = 4.02e+04, Exact = 3.98e+04   #5 Error 1.0%
6,   IGA = 4.90e+04, Exact = 4.77e+04   down from 33%
7,   IGA = 5.85e+04, Exact = 5.57e+04
8,   IGA = 6.56e+04, Exact = 6.36e+04
9,   IGA = 8.84e+04, Exact = 7.16e+04
10,  IGA = 1.06e+05, Exact = 7.95e+04
11,  IGA = 1.24e+05, Exact = 8.75e+04
```

Figure 11.1. Larger numbers of iElements reduces lower mode errors

11.6 Torsional vibrations

The equation of motion of a vibrating torsional shaft or drill string is of the same form as the prior transverse string vibration:

$$GJ\frac{\partial^2\theta}{\partial x^2} - \rho J\frac{\partial^2\theta}{\partial t^2} = 0,$$

where G is the material shear modulus, J is the cross-section polar moment of inertia, ρ is the mass density per unit length, and θ is the (small) angle of twist of the shaft. The rotational inertia of the shaft cross-section is defined as $I = \rho J$. By analogy to all of the stiffness matrices in Section 11.4, the torsional stiffness matrix always includes the term $GJ/L \equiv k_t$ (divided by a number). This term is known as the torsional stiffness of the shaft and many problems supply that number to describe a shaft segment of length L.

It is common for equipment to be attached to a vibrating system. If the mass, or rotational inertia, of the equipment is large, then the usual practice is to treat the equipment as a point mass and attach it to a node at an element interface. That is, the point mass (or inertia) is added to the diagonal of the system mass matrix at the node where the equipment is located. The implementation of this SHM is listed in Script 11.2. Other than a change in physical interpretations, it just includes the addition of the point inertia at the line closing the loop over all of the iElements. However, in an IGA study, this is only physically correct since the inertia attachment point was at the interpolatory analysis control point. Again, the lowest frequencies are quite accurate.

Script 11.2. Adding point masses or inertias to a vibration simulation

```
% Ref:  Thomson, Theory of Vibrations Ex 56.2 drill with BHA
L     = 5000.    ; % domain lengths
G     = 1.728e9  ; % shear modulus
rho   = 15.22    ; % wt. density
J_e   = 9.4e-4   ; % string inertia
J_BHA = 29.3     ; % bit inertia
Le    = L/n_e    ; % uniform element length

K   = zeros(n_d, n_d); M = zeros(n_d, n_d)      ; % allocate
K_e = [18  -9   -6   -3;             % stiffness, p3 J constant
       -9  12    3   -6;
       -6   3   12   -9;
       -3  -6   -9   18]*G*J_e/(10*Le)  ; % stiffness matrix
M_e = [20  10    4    1;                  % mass. p3 J constant
       10  12    9    4;
        4   9   12   10;
        1   4   10   20]*J_e*rho*Le/140      ; % mass matrix

%          Assemble (scatter) iElement matrices
for j = 1:n_e ; % loop on iElements ===> ===> ===> ===> ===>
  dof = AP_list (j, :)                   ; % ONLY for n_g = 1 here
```

```
  K (dof, dof) = K (dof, dof) + K_e     ; % add to stiffness
  M (dof, dof) = M (dof, dof) + M_e          ; % add to mass
end           ; % for iElements <=== <=== <=== <=== <=== <===
M(n_d, n_d) = M(n_d, n_d) + J_BHA       ; % add point inertia

%   Solve for eigenvalues and eigenvectors, general form
[Vec, Diag] = eig(K(Free,Free),M(Free,Free))
%   Vec has eigenvectors, Diag has eigenvalue squared

 fprintf ('Mode, String natural frequencies \n')   % heading
 true = [2.42, 7.99]                        ; % rad/sec from Ref
 for m = 1:2                ; % loop on 2 modes ---> ---> --->
   omega = sqrt( real (Diag(m,m)))              ; % iEl freqency
   fprintf ('%i, IGA = %7.2e, Exact = %7.2e \n', ...
              m, omega, true(m))     ; % pretty print results
 end    ; % for 2 active modes <--- <--- <--- <--- <--- <---
% Mode, String natural frequencies
% 1, IGA = 2.42e+00, Exact = 2.42e+00
% 2, IGA = 7.93e+00, Exact = 7.99e+00       Error: 0.8%
```

11.7 Beam vibrations

The transverse vibration, v, of a beam yields the same matrix eigenproblem even though it begins a differential equation of motion containing fourth-order spatial derivatives:

$$EI\frac{\partial^4 v}{\partial x^4} - \rho A\frac{\partial^2 v}{\partial t^2} = 0,$$

where A and I are the area and moment of inertia of the cross-section, E and ρ are the elastic modulus and mass density of the material, so $\rho A = m$ is the mass per unit length. The bending stiffness matrix and consistent mass matrix are

$$[\boldsymbol{K}] = \int_L \frac{d^2[\boldsymbol{N}(u)]}{dx^2}^T EI\frac{d^2[\boldsymbol{N}(u)]}{dx^2}dx,$$

$$[\boldsymbol{M}] = \int_L [\boldsymbol{N}(u)]^T \rho(x)[\boldsymbol{N}(u)]dx.$$

In practice, these evaluations require numerical integration (quadratures), but as with the prior beam examples assuming a linear geometry mapping and using a quintic ($p = 5$) degree basis functions, it is possible to form both as analytic matrices:

$$[\boldsymbol{K}] = \frac{\mathrm{EI}}{7L^3}\begin{bmatrix} 400 & -600 & 80 & 60 & 40 & 20 \\ -600 & 1040 & -260 & -160 & -60 & 40 \\ 80 & -260 & 240 & 40 & -160 & 60 \\ 60 & -160 & 40 & 240 & -260 & 80 \\ 40 & -60 & -160 & -260 & 1040 & -600 \\ 20 & 40 & 60 & 80 & -600 & 400 \end{bmatrix}$$

$$[\boldsymbol{M}] = \frac{\rho L}{2772}\begin{bmatrix} 252 & 126 & 56 & 21 & 6 & 1 \\ 126 & 140 & 105 & 60 & 25 & 6 \\ 56 & 105 & 120 & 100 & 60 & 21 \\ 21 & 60 & 100 & 120 & 105 & 56 \\ 6 & 25 & 60 & 105 & 140 & 126 \\ 1 & 6 & 21 & 56 & 126 & 252 \end{bmatrix}.$$

Figure 11.2 shows a cantilever beam with an elastic spring support at the end point. If the point spring stiffness is zero, then the system becomes a standard cantilever and if the spring stiffness is infinite, the system becomes a propped cantilever. For all intermediate values of the stiffness, this is called an elastic support. Such a system could also have an end point rotational spring to control the end rotation of the beam. The ratio $k_b \equiv EI/L^3$ is known as the bending stiffness of a beam of span L. The analytic solution for the system in Fig. 11.2 depends on the ratio of the spring stiffness to the bending stiffness, say $R \equiv k/k_b$. A MATLAB script to find the frequencies of the beam in Fig. 11.2 is given in Script 11.3, and the results of the vibration script are in Fig. 11.4. In this example, adding

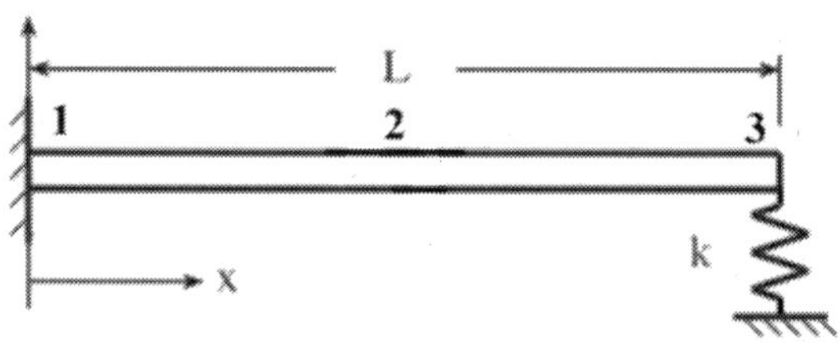

Figure 11.2. **Beam with a vertical spring support**

the spring increased the fundamental (first) frequency by more than a factor of 3.

Script 11.3. Cantilever frequencies with or without point stiffness

```
function [] = iga_cantilever_spring_freq (stiff)
%        Natural frequency of a cantilever via
%      quintic basis functions supported by 6 AP
%
%             Fixed (v, v'=0)           | vert spring
%   APs       1----2----3----4----5---6 if stiff = 1
%
% **** In IGA v'1 = 0 is enforced by v1 = 0 = v2 ****
n_e = 1; n_g = 1; n_m = 6; n_d = n_m * n_g ; AP_list = [1:6];
Fixed = [1:2]; Given = [0., 0.];
Free    = [3:6]                     ; % free AP after  EBC
n_modes = max(size (Free))             ; % number of modes
EI = 1. ; rho = 1. ; L = 1. ; m = rho*L ;
scale = sqrt (EI / m) / L^2          ; % exact in rad/sec
exact = [3.516  22.03  61.70  120.9]*scale  ; % exact

K = [400, -600,   80,   60,   40,   20;   % stiffness
    -600, 1040, -260, -160,  -60,   40;
      80, -260,  240,   40, -160,   60;
      60, -160,   40,  240, -260,   80;
      40,  -60, -160, -260, 1040, -600;
      20,   40,   60,   80, -600,  400]*EI/(7*L^3);

M = [252,  126,   56,   21,    6,    1;         % mass
     126,  140,  105,   60,   25,    6;
      56,  105,  120,  100,   60,   21;
      21,   60,  100,  120,  105,   56;
       6,   25,   60,  105,  140,  126;
       1,    6,   21,   56,  126,  252]*rho*L/2772;

if ( stiff == 1 ); k = 500*EI/L^3; else; k=0; end
K (n_m, n_m) = K (n_m, n_m) + k   ; % add end spring

% Get eigenvalues & compare to four exact ones
[Vec_eig, D_vals] = eig (K(Free,Free), M(Free,Free));
fprintf('Four exact frequencies (rad/sec) \n')%header
```

```
display (exact)       ; % exact cantilever frequencies
fprintf('IGA Mode 1-4 frequencies (rad/sec) \n');
RadPS = sqrt(real(diag(D_vals)))  ; % freq in rad/sec
disp(RadPS')                      ; % show first four freq

%  Scale to +-1 and list AP model shape values
Eigen_vecs = zeros (n_d, n_modes)      ; % for plots
Eigen_vecs (Free, :) = Vec_eig(:, :)   ; % for the EBC
big_abs = max(abs(Eigen_vecs(1:2:5, :))); % scale +-1
for n = 1:n_modes; % loop over free modes ==> ==> ==>
  temp = big_abs (n); % largest abs disp in this mode
  Eigen_vecs(:,n) = Eigen_vecs(:,n)/temp;% normalized
end ; %  for loop to normalize modes   <== <== <== <==
fprintf ('Normalized eigenvectors for modes: \n')
disp([1:n_modes])                    ; % free mode numbers
disp(Eigen_vecs)   ; % eigenvectors including the EBC
```

```
% Four exact frequencies (rad/sec) for no end spring
%      3.5160   22.0300   61.7000  120.9000
% IGA Mode 1-4 frequencies (rad/sec)
%      3.5160   22.1578   63.3466  281.5963
% Normalized eigenvectors for modes:
%          1,         2,         3,         4,
%          0          0          0          0 % EBC
%          0          0          0          0 % EBC
%    -0.1753   -1.0000    1.0000   -0.4759
%    -0.4490   -0.8626   -0.9231    1.0000
%    -0.7247    0.0351   -0.1553   -0.7877
%    -1.0000    0.7952    0.2733    0.2235
%
% IGA Mode 1-4 frequencies (rad/sec), with end spring
%    14.9787    45.0988   88.9748  299.3957
% Normalized eigenvectors for modes:
%          1,         2,         3,         4,
%          0          0          0          0
%          0          0          0          0
%     0.6773    1.0000    0.5425   -0.4199
%     1.0000   -0.6222   -1.0000    1.0000
%     0.4775   -0.2904    0.5784   -0.9878
%     0.0007   -0.0015    0.0081    0.4613
```

Figure 11.3. Comparing cantilever modes with and without an end spring

Figure 11.4 shows the first two normalized mode shapes from a classic cantilever beam on the left and with a supporting spring on the right. The solid dots denote the normalized IGA analysis control

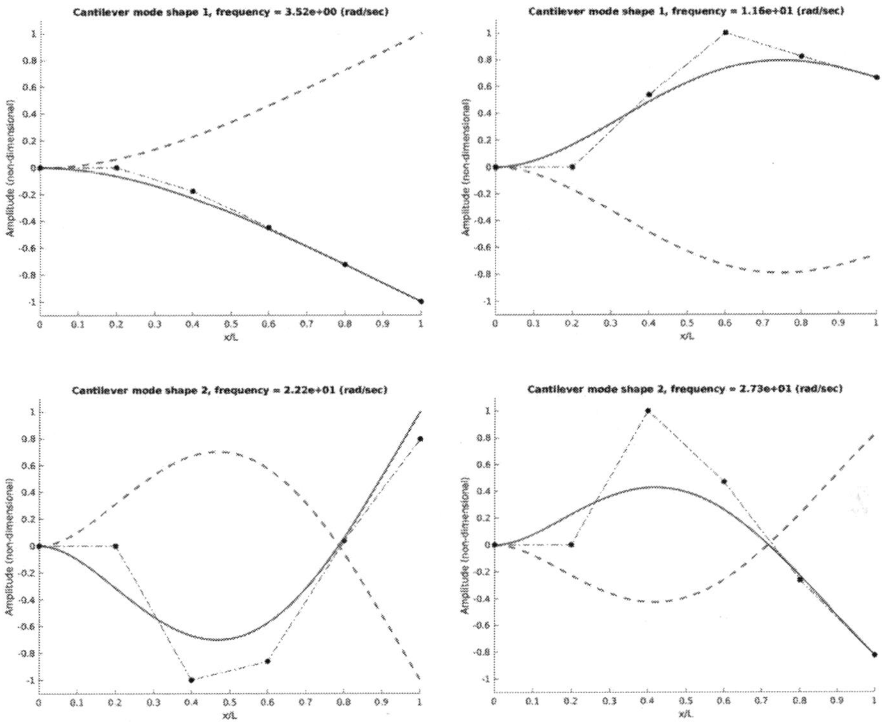

Figure 11.4. First two modes of a cantilever without and with an end spring (right)

point mode displacements. Figure 11.4 shows that a spring support at the cantilever point usually having the largest SHM displacement reduces that relative displacement and all others along the beam length. As a side note, in the second mode plot, there is a new point near 3/4 of the length that never moves during the simple harmonic motion. Each additional mode introduces another such point. In the vibration literature, such points are called 'nodes' and that terminology pre-dates IGA.

11.8 Membrane vibration

The extension of 1D IGA vibration problems to two-dimensional problems simply requires using the tensor product of the basis functions in two directions, as illustrated before in thermal analysis.

Basically, the 'generalized stiffness matrix' and the generalized mass matrix' must be created for each iElement and assembled in the system as done before. Instead of passing the system-level square matrices to a linear equation solver, they are sent to an eigen-solver.

Consider that a vibrating thin membrane with zero displacement on its boundary, of any shape, corresponds to the shape of a soap bubble. The MATLAB logo is one example that is very similar to the amplitude of the vibration of an L-shaped membrane (with the displacements enlarged). A membrane is a thin tensioned material undergoing transverse small displacements without bending resistance. Such a membrane can be regarded as generalizing the tensioned string to two dimensions. Many membranes require a vibration analysis. For a global uniform membrane tension per unit length on the boundary, T, the two-dimensional equation of motion (wave equation) is

$$T\nabla^2 v(x,y) + \rho h \frac{\partial^2 v(x,y)}{\partial t^2} = 0,$$

where $v(x, y)$ is the transverse displacement, ρ is the material mass density, h is the membrane thickness, so ρh is the mass per unit area. The first term is just an extension of the prior tensioned string by adding the y-derivative of the basis functions as the second row of the prior strain-displacement matrix, $\boldsymbol{B}^{\boldsymbol{e}}$, and gives a membrane stiffness matrix of

$$\boldsymbol{K}^{\boldsymbol{e}} = \int_{A^e} \boldsymbol{B}^{e^T} T \boldsymbol{B}^e dA = \int_{A^e} \begin{bmatrix} \partial \boldsymbol{N}/\partial x \\ \partial \boldsymbol{N}/\partial y \end{bmatrix}^{\boldsymbol{e}T} T \begin{bmatrix} \partial \boldsymbol{N}/\partial x \\ \partial \boldsymbol{N}/\partial y \end{bmatrix}^{\boldsymbol{e}} dA.$$

The membrane of thickness, h, also has a mass per unit area, $m = \rho h$, and the corresponding consistent membrane mass matrix has the same symbolic form as the string, but it is a larger size matrix:

$$\boldsymbol{M}^{\boldsymbol{e}} = \int_{A^e} \boldsymbol{N}^{e^T} \rho^e h^e \boldsymbol{N}^e dA.$$

In this case, the 'generalized stiffness' is just a planar conduction matrix with the tension, T, replacing the thermal conductivity, k. The 'generalized mass matrix' is just the planar face convection matrix with the convection coefficient, h, replaced by the mass per unit area. When both are assembled into two symmetric square matrices, they are viewed as having the form

$$[\boldsymbol{K} - \omega_j^2 \boldsymbol{M}]\boldsymbol{\delta_j} = 0.$$

The 'free' dof portions of the assembled matrices are sent to an eigen-solver, which returns the eigenvalue, ω_j^2, and the corresponding normalized displacement mode shapes, $\boldsymbol{\delta_j}$, for $j = 1, 2, \ldots$ length (Free).

It is not uncommon to have to extend the above vibration modes to determine the natural frequencies of a structure that is also subjected to an axial load. Then the eigen-problem is

$$|(\boldsymbol{K_E} + \boldsymbol{K_N}) - \omega^2 \boldsymbol{M}| = 0,$$

where $\boldsymbol{K_E}$ is the usual elastic stiffness matrix, $\boldsymbol{K_N}$ is the geometric stiffness matrix associated with the axial load, $\boldsymbol{F}$, and $\boldsymbol{M}$ is the usual mass matrix. When the axial load is a constant, say F_B, then it factors out of each element geometric stiffness matrix and becomes a global constant. Then the formulation changes to

$$|(\boldsymbol{K_E} + F_B \boldsymbol{K_n}) - \omega^2 \boldsymbol{M}| = 0,$$

where the 'unit initial stress matrix', $\boldsymbol{K_n}$, is formed with a unit positive load. This equation shows that any axial load, F_B, has an influence on the natural frequency, ω. In general, a tensile force (+) increases the natural frequency while a compression force (−) lowers the natural frequency.

One of the most common applications of this state is finding the natural frequencies of turbine blades in jet engines. Their high angular velocity imposes a tension centripetal force on the blades that increases with radial position. A static solution at a fixed rotational speed defines the blade stresses. Those stresses in turn define the blade geometric (initial stress) matrix. Then the eigen-problem is solved to compute the blade's natural frequencies at each rotational speed of the engine.

11.9 Structural buckling

There are two major categories leading to the sudden catastrophic failure of a mechanical or structural component: structural instability, which is often called buckling, and brittle material failure. Buckling failure is primarily characterized by a sudden loss of structural stiffness that usually renders the structure unusable. The buckling estimate is computed from an IGA eigen-problem solution. Slender or thin-walled components under compressive stress are susceptible to buckling. Buckling studies are much more sensitive to the component restraints than a normal stress analysis. A buckling mode describes the deformed shape the structure assumes (but not the direction) when it buckles, but (like a vibration mode) says nothing about the numerical values of the displacements.

Undergraduate linear buckling studies usually focus on the buckling of columns using the Euler small deflection theory. Those ideal estimates depend heavily on the EBCs (end constraints) and handbooks give the solution for ideal beams with various end, and mid-span, constraints. Many of those critical column buckling loads can be written as $P_{\text{cr}} = k\pi^2 \text{EI}/L^2$, where k represents the effects of the Dirichlet BCs. The five most common EBC constants for an ideal beam are shown in Fig. 11.5.

This section addresses the IGA implementation of the common linear theory of buckling, which tends to greatly overestimate the buckling forces for 'short' columns. Any accurate buckling estimate depends heavily on the geometry of the model. For example, a beam is assumed to be a perfect straight line with the axial load applied exactly along that line. It is also assumed that the axial load acts exactly through the centroid of the cross-section of the

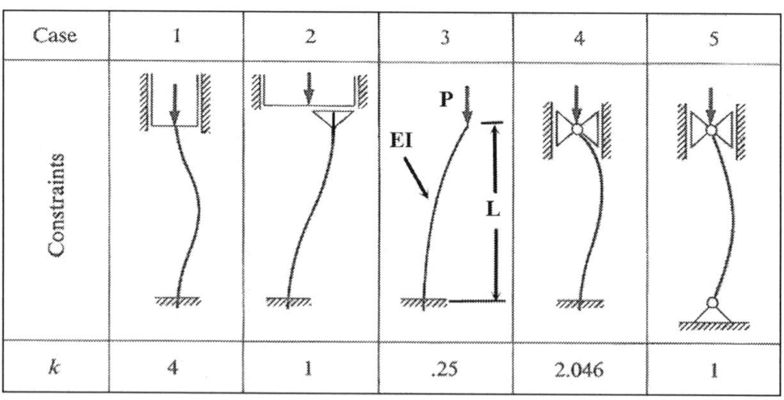

Figure 11.5. **Restraints influence the critical buckling load, $P_{\text{cr}} = k\pi^2 EI/L^2$**

beam. Otherwise, the loading becomes eccentric, causing a transverse moment to be applied as well as the axial load. If the part has small irregularities that have been omitted from the geometric data, they can drastically reduce the actual buckling load.

When a constant axial load acts along the entire beam, say F_B, it factors out of the element initial stress (geometric stiffness) matrices and the assembly process to appear as a constant in the system matrices. The goal is to find the scaling up or down of the applied load that causes linear buckling. This scale factor is called the 'buckling load factor (BLF)' and replacing F_B with BLF $\times$ F_B leads to a buckling eigen-problem to determine the value of BLF (see Table 11.1).

For a general structure with a set of loads (called a load case), the system 'geometric stiffness matrix' depends on the stress in each element. Thus, it depends on all of the assembled loads, say F_{ref}. In general, a linear static analysis is completed and the deflections are determined from the system equilibrium using the elastic stiffness matrix: $[\boldsymbol{K_E}]\{\boldsymbol{v}\} = \{\boldsymbol{F}_{\text{ref}}\}$. Those displacements are post-processed to determine the stresses in each iElement. Then the geometric stiffness matrix is calculated from those stresses. The stress state, and thus its geometric stiffness matrix, is directly proportional to the resultant system load, so as the loads are scaled by the buckling factor, the system geometric stiffness matrix increases by the same amount

$$\boldsymbol{F} \rightarrow \text{BLF } \boldsymbol{F}_{\text{ref}}, \quad \boldsymbol{K_N} \rightarrow \text{BLF } \boldsymbol{K}_{\text{ref}}.$$

The scaling value, BLF, that renders the combined system stiffness to have a zero determinant (that is to become unstable) is the

Table 11.1. Interpretation of the buckling load factor

Value	Status	Note
$\lvert\text{BLF}\rvert > 1$	No Buckling	The applied loads are less than the theoretical critical loads
$\lvert\text{BLF}\rvert \leq 1$	Buckling predicted	The applied loads exceed the theoretical critical loads
$\text{BLF} < 0$	Buckling possible	Buckling occurs if the directions of the applied loads are reversed

increase (or decrease) of all of the applied loads that will cause the structure to buckle. This critical value is calculated from a buckling eigen-problem using the previous matrices plus the geometric stiffness matrix based on the current load case:

$$|\boldsymbol{K_E} - \text{BLF}\ \boldsymbol{K_{\text{ref}}}| = 0.$$

This equation is solved for the value of the BLF. Then the system load case that would theoretically cause buckling is BLF $\boldsymbol{F_{\text{ref}}}$. The solution of the eigen-problem also yields the relative buckled mode shape (eigenvector), $\boldsymbol{v_{\text{BF}}}$.

In theory, there are as many buckling modes as there are free dof in a structure. Due to the limitations of linear buckling theory, only the first buckling mode is of practical importance. Structural buckling is usually instantaneous and catastrophic. A factor of safety of at least 4 is often applied to linear buckling estimates. There are software systems that can solve the more accurate nonlinear post-buckling behavior of structures.

The estimated Euler buckling force can be unrealistically large as the length of the member decreases. A linear buckling calculation implies that the compression load–deflection relation and thus the material compression stress–strain relation are linear. This is true only up to where the stress reaches the material compressive yield stress, say S_{yc}. In other words, the critical compressive stress, $\sigma_{\text{cr}} = P_{\text{cr}}/A$, caused by the critical force, P_{cr}, must also be considered. To do this, the critical stress must be rewritten in terms of the cross-sectional area. This is done by using the geometric 'radius of gyration', r_g, defined as $I \equiv Ar_g^2$. Then for typical boundary conditions

$$P_{\text{cr}} = k\pi^2 EI/L^2 = k\pi^2 E(Ar_g^2)/L^2 = k\pi^2 EA/(L/r_g)^2,$$

where the ratio L/r_g is known as the 'slenderness ratio' of the cross-section. It is evaluated at the location of the smallest I found on any plane perpendicular to the axis of the load. The critical compressive stress is $\sigma_{\text{cr}} = P_{\text{cr}}/A = k\pi^2 E/(L/r_g)^2 \leq S_{\text{yc}}$. Once the critical stress exceeds half of the yield stress, other empirical theories, such as the Johnson theory in Fig. 11.6, are used.

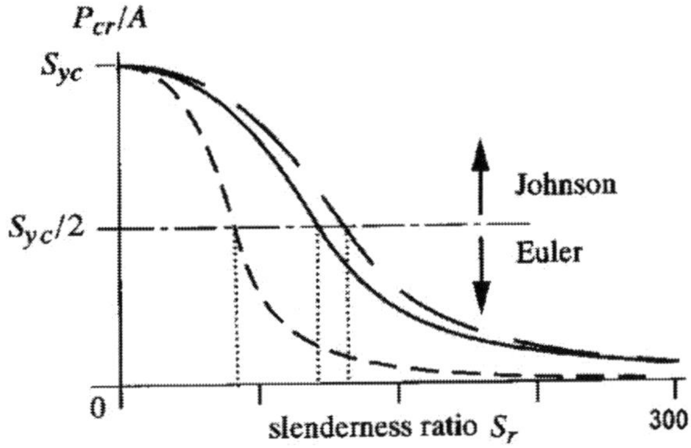

Figure 11.6. **Euler and Johnson buckling curves for three Dirichlet BC**

Example 11.2.

Given: Case 1 of Fig. 11.5 represents a fixed–fixed beam column of length $2L$ which has an unknown lateral displacement at its center (L) along with a zero slope. Use a single-span cubic basis function for a beam, of length L, to estimate its axial buckling force.

Solution: The governing ODE and matrix forms are

$$\frac{d^2}{dx^2}\left[EI\frac{d^2v}{dx^2}\right] - P\frac{d^2v}{dx^2} = q(x) \rightarrow [\boldsymbol{K_{EI}}]\{\boldsymbol{\delta}\} - [\boldsymbol{K_P}]\{\boldsymbol{\delta}\} = \{\boldsymbol{c_q}\}$$

$$([\boldsymbol{K_{EI}}] - P[\boldsymbol{K_P^*}])\{\boldsymbol{\delta}\} = \{\boldsymbol{c_q}\}.$$

Factoring out the unknown global constant, P, in the final matrix form more clearly shows that an axial compression force reduces the net stiffness, and strength, of the beam column. For no transverse load per unit length, $q = 0$, and constant coefficients $(EI,\ P)$ the problem converts this into an eigen-problem:

$$|[\boldsymbol{K_{EI}}] - P[\boldsymbol{K_P^*}]| = 0,$$

where the *single* value of P which causes the loss of effective stiffness is sought. For IGA formulations, it has been shown that the axial force coupling matrix, $-P[\boldsymbol{K_P^*}]$, after integration by parts is exactly the same as heat conduction matrices, but with the thermal conductivity, k, replaced by the compressive load, $-P$. For the cubic basis functions defined by the analysis knot vector

$\mathbb{A}u = \{0, 0, 0, 0, 1, 1, 1, 1\}$ Appendix 1 gives them as

$$[\boldsymbol{N}(u)] = [-(u-1)^3, 3u(u-1)^2, -3u^2(u-1), u^3],$$

which yields the single knot span matrix buckling system matrices:

$$\left| \frac{EI}{3L^3} \begin{bmatrix} 12 & -18 & 0 & 6 \\ -18 & 36 & -18 & 0 \\ 0 & -18 & 36 & -18 \\ 6 & 0 & -18 & 12 \end{bmatrix} - \frac{P}{10L} \begin{bmatrix} 18 & -9 & -6 & -3 \\ -9 & 12 & 3 & -6 \\ -6 & 3 & 12 & -9 \\ -3 & -6 & -9 & 18 \end{bmatrix} \right| = 0,$$

and applying the three Dirichlet conditions that $v_1 = 0$ and the symmetry condition, for an open knot vector, that $v_4 = v_3 = 0$ leaves only the second transverse displacement, i.e., only the term in the second row and column:

$$\left| \frac{EI^e}{3L^3}[36] - \frac{P}{10L}[12] \right| = \{0\}.$$

Thus, the estimated buckling force is $P = 10\,EI/L^2$. The exact symmetric buckling force is $P_{\text{exact}} = \pi^2 EI/L^2$ giving a 1.3% error in the IGA buckling force.

11.10 Modes and frequencies of 2D continua

The natural frequency calculations for plane stress, plane strain, and axisymmetric solids are almost identical. Thus, the plane-stress case will be discussed here. By also evaluating the kinetic energy, the definitions of elastic stiffness matrix and the consistent mass matrices were given in Chapter 10 as

$$\boldsymbol{S}^e = \int_{V^e} \boldsymbol{B}^{e^T} \boldsymbol{E}^e \boldsymbol{B}^e dV \quad \boldsymbol{M}^e \equiv \int_{V^e} \boldsymbol{N}^T \rho^e \boldsymbol{N} dV.$$

Plane-stress and plane-strain applications have the same strain–displacement matrices, $\boldsymbol{B}^e$, but their constitutive matrices, $\boldsymbol{E}^e$, have different entries. The axisymmetric solids have one more row in their $\boldsymbol{B}^e$ and $\boldsymbol{E}^e$ matrices. For planar solids, the differential volume is $dV = t^e\, dA$ for a plane of thickness of t^e, and for the axisymmetric solid, it is $dV = 2\pi\, R\, dA$.

Empirical studies have shown that alternate forms of the mass matrix are available. Special exact integration rules that use the APs as quadrature points produce a diagonal mass matrix. The same diagonal matrix, $\boldsymbol{M}^e_{\text{diag}} = \lceil \boldsymbol{M}^e \rfloor$, can be obtained by scaling up the diagonal of the consistent mass matrix:

$$\boldsymbol{M}^{\boldsymbol{e}}_{\textbf{diag}} = \text{diag}(\boldsymbol{M}^{\boldsymbol{e}}) \times \text{sum}(\boldsymbol{M}^e)/\text{sum}(\text{diag}(\boldsymbol{M}^{\boldsymbol{e}}))$$

so as to preserve the total of mass present.

11.11 Summary

PDE eigenvalue form:	$\nabla^2 u(x, y, t) + \lambda u(x, y, t) = 0.$
IGA basis functions:	$u(x) = \boldsymbol{N}(r)\boldsymbol{u}^{\boldsymbol{e}}.$
Matrix form:	$\det[\boldsymbol{K} + \lambda \boldsymbol{M}] = \lvert\boldsymbol{K} + \lambda \boldsymbol{M}\rvert = 0.$
Eigengroup:	$[\mathrm{K} + \lambda_{\mathrm{j}}\mathrm{M}]\delta_{\mathrm{j}} = 0,\ j = 1, 2, \ldots$ length (free).
Iron's bound theorem:	$\lambda^e_{\text{min}} \leq \lambda_{\text{min}} \leq \lambda_{\text{max}} \leq \lambda^e_{\text{max}}.$
Vibration:	$\lambda = \omega^2$, ω real.
Rigid body motion:	$\omega = 0.$
Spring-mass:	$\omega = \sqrt{k/m}.$
Stiffnesses:	$\boldsymbol{K_E}$ = elastic, $\boldsymbol{K_k}$ = foundation, $\boldsymbol{K_G}$ = geometric.
Buckling:	$[\boldsymbol{K_E} + \boldsymbol{K_k} + BLF\ \boldsymbol{K_G}]\boldsymbol{\delta} = \boldsymbol{0}.$
Beam-Column:	$\lvert(\boldsymbol{K_E} + F_B\boldsymbol{K_n}) - \omega^2 \boldsymbol{M}\rvert = 0$, $\boldsymbol{K_G}{=}F\boldsymbol{K_n}.$

Chapter 12

Transient and Dynamic Solutions

12.1 Introduction to transient systems

The general linear model PDE is extended into the transient domain by including the first time derivative of the solution, T:

$$-D(x,t)\frac{\partial^2 T(x,t)}{\partial x^2} + A(x,t)\frac{\partial T(x,t)}{\partial x} + C(x,t)T(x,t) + F(x,t)$$
$$= G(x,t)\frac{\partial T(x,t)}{\partial t}.$$

Including time can be accomplished by using higher-dimensional space-time basis functions with the previous methods to use a separation of the variables approach. That is, let the interpolated AP degrees of freedom be time dependent, with the space interpolations unchanged, and use a separation of variables to find an IGA approached in each iElement using

$$T(x,y,z,t) \equiv \boldsymbol{N}(x,y,z)\boldsymbol{T^e}(t).$$

Then, the time derivative is

$$\frac{\partial T(x,y,z,t)}{\partial t} \equiv \boldsymbol{N}(x,y,z)\frac{\partial \boldsymbol{T^e}(t)}{\partial t} \equiv \boldsymbol{N}(x,y,z)\dot{T}^e(t).$$

To obtain the integral form, use the Galerkin approach and multiply the PDE by $T(x, y.z, t)$ and integrate by parts. Then, a new integral appears giving a generalized mass matrix for each iElement:

$$\int_{\Omega} T(x,y,z,t)\; G(x,y,z,t)\; \partial T/\partial t\, d\Omega$$
$$\Rightarrow \sum_{e=1}^{n_e} \boldsymbol{T}^{e^T} \left[\int_{\Omega^e} \boldsymbol{N}^{e^T} G^e\; \boldsymbol{N}^e\; d\Omega \right] \dot{T}^e(t)$$
$$= \sum_{e=1}^{n_e} \boldsymbol{T}^{e^T} [\boldsymbol{M}^e] \dot{T}^e(t)$$

with

$$\boldsymbol{M}^e(t) = \int_{\Omega^e} \boldsymbol{N}(x,y,z)^T\; G(x,y,z,t)\; \boldsymbol{N}(x,y,z) d\Omega.$$

In transient heat transfer, the unknown, T, is temperature and the time coefficient, G, is the product of the material's mass density and its specific heat at constant pressure, $G \equiv \rho c_p$. Then, the new square matrix coupled with the solution velocity, $\dot{T}^e$, is called the heat capacity matrix.

This PDE and other first-order time-dependent scalar field problems solved by using IGA basis functions in space lead to a matrix ordinary differential equation in time of the form

$$[\boldsymbol{K}]\{\boldsymbol{T}(t)\} + [\boldsymbol{M}]\{\dot{\boldsymbol{T}}(t)\} = \{\boldsymbol{c}(t)\}, \dot{\{\}} = \frac{\partial}{\partial t}\{\},$$

where the first three terms in the PDE contribute a square conduction matrix to $[\boldsymbol{K}]$, the time derivative term in the PDF contributes the square capacity matrix, $[\boldsymbol{M}]$, and the source term F contributes to the source vector $\{\boldsymbol{c}(t)\}$.

This matrix ODE is subject to the initial condition (IC), $\{\boldsymbol{T}(0)\} = \boldsymbol{T}_{IC}$, and the time-dependent Dirichlet essential boundary conditions (EBCs), $\boldsymbol{T}_{\mathbf{EBC}}(t)$. After the matrix system has been modified to include the EBCs, by adding $\boldsymbol{c}_{\mathbf{EBC}}(t)$ to $\boldsymbol{c}(t)$, only the time history values of the remaining free degrees of freedom need to be computed. Here, the approach is simplified by assuming that the Dirichlet conditions are time independent.

Langer *et al.* (2016) applied IGA space-time to parabolic problems like those in this section.

Single-step transient histories: Many engineering applications involve sharp transients or nonlinear responses, $[\boldsymbol{K}] = [\boldsymbol{K}(\{\boldsymbol{T}(t)\})]$. Such applications require a direct time integration to obtain the solution time history. Some direct methods require an estimate of the initial time rate of change of the solution. It is often approximated as zero, but it can be found from the IC as

$$\{\dot{\boldsymbol{T}}(0)\} = [\boldsymbol{M}]^{-1}(\{\boldsymbol{c}(0)\} - [\boldsymbol{K}]\{\boldsymbol{T}(0)\}),$$

where the capacitance matrix, $[\boldsymbol{M}]$, can be approximated by its scaled diagonal to obtain a quick approximation of the initial velocity of the system.

After any EBCs are applied, the original system will be reduced to the size of the 'Free' degrees of freedom:

$$[\boldsymbol{K}]\{\boldsymbol{T}(t)\} + [\boldsymbol{M}]\{\dot{\boldsymbol{T}}(t)\} = \{\boldsymbol{c}(t)\} - \{\boldsymbol{c}_{\mathbf{EBC}}(t)\} \equiv \{\boldsymbol{p}(t)\},$$

where $\{\boldsymbol{c}_{\mathbf{EBC}}(t)\}$ comes from any non-zero essential boundary condition (EBC) multiplying the columns of the larger original matrices. There are hundreds of algorithms for integrating this system in time. These include using finite differences in time, finite elements in time, or even space-time finite elements. Since many engineering applications involve discontinuous changes at certain points in time, single-step methods are more likely to capture and respond accurately to the discontinuity. If it is known that the driving source terms, $\{\boldsymbol{c}(t)\}$, and the EBC, $\{\boldsymbol{c}_{\mathbf{EBC}}(t)\}$, are continuous in time, then multi-step methods, like the Runge–Kutta four-step method, become more efficient. Many young engineers have used the Matlab ODE solvers, like ODE45, to solve for scalar time histories. Since all arguments in Matlab are matrices, those prior uses of ODE45 were solving the above-mentioned matrix system, but just with 1×1 matrices.

Most temporal integration schemes use finite differences in time or finite elements in time to derive the algorithm using constant increments of time, say Δt, to define time as $t_n = t_0 + n\,\Delta t$ at time step number n. Most of the algorithms form a static linear system to be solved at each time step as

$$[\boldsymbol{S}(\Delta t, \beta, \boldsymbol{K}, \boldsymbol{M})]\{\boldsymbol{T}\}_n = \{\boldsymbol{F}(\beta, t_n, t_{n-1}, \boldsymbol{K}, \boldsymbol{M}, \boldsymbol{c}, \{\boldsymbol{T}\}_{n-1})\},$$

where β is an algorithm constant. In this form, the left square matrix is constant if it is not nonlinear and if the time step, Δt, is kept constant. Assembling the linear system matrix $[\boldsymbol{S}]$ is computationally expensive and obtaining its inverse or factorization is very expensive. Thus, it is desirable to do those operations only once or twice. The right-hand column changes with every time step and the result at time step n is

$$\{\boldsymbol{T}\}_n = [\boldsymbol{S}(\Delta t, \beta, \boldsymbol{K}, \boldsymbol{M})]^{-1}\{\boldsymbol{F}\}_n.$$

12.2 Generalized Trapezoidal Algorithms

The popular Hilbert–Hughes–Taylor α-Method (Hilbert *et al.*, 1977) that generalized the trapezoidal one-step method in time leads to an updated square matrix defined as

$$[\boldsymbol{S}] = ([\boldsymbol{M}]/\Delta t + \beta[\boldsymbol{K}]),$$

and the source vector updated at every time step is

$$\{\boldsymbol{f}\}_n = ([\boldsymbol{M}]/\Delta t + (\beta - 1)[\boldsymbol{K}])\{\boldsymbol{T}\}_{n-1} + (1 - \beta)\{\boldsymbol{p}\}_{n-1} + \beta\{\boldsymbol{p}\}_n.$$

These combine to give a linear matrix system to be solved at every time step:

$$[\boldsymbol{S}]\{\boldsymbol{T}\}_n = \{\boldsymbol{f}\}_n.$$

Note that so long as the time step Δt is kept constant, the computationally expensive assembly and factorization (or inversion) of the linear system square matrix $[\boldsymbol{S}]$ only need to be done once. This contrasts to the source vector $\{\boldsymbol{f}\}_n$, which must be updated at every time step. The forward and backward substitution of $\{\boldsymbol{f}\}_n$ into a factorization of $[\boldsymbol{S}]$ is relatively cheap and fast. The time step is usually smallest at the start of the transient solution. As a steady state solution is approached, a large time step could be used, but often is not used. If the time step is changed, then the last solution from the prior small time step is used as the IC for the larger time step.

The common choices for the algorithm constant β are as follows:

$\beta = 0$ the conditionally stable forward difference (Euler) method,
$\beta = \frac{1}{2}$ the unconditionally stable trapezoidal (Crank–Nicolson) method,

$\beta = \frac{2}{3}$ the unconditionally stable Galerkin-in-time method,
$\beta = 1$ the unconditionally stable backward difference method.

The above-mentioned four algorithms are all single-step methods since they only require storing the results from the last time step. They minimize the memory requirements and were developed when computers had extremely small amounts of memory, but their main asset is that they deal well with sudden changes in the source term or the EBCs. There are many other algorithms that lead to a linear system to be solved at each time step.

When the solution in time is known to be smooth, then multiple-step methods can lead to more efficient transient solutions. Of course, they require more memory in order to save the solution from each of the time steps needed to formulate the multi-step algorithm.

A typical Matlab script for running the generalized trapezoidal transient integration is *gen_trap_history.m*, and portions of it are given in Script 12.1. The first part of the illustrated calculation allocates the needed memory and extracts the independent dofs. Those steps could actually be done in the calling program. Here, they serve as reminder that the presence of EBCs also influences the transient solution as well. The second part of the calculations is quite compact when done in matrix notation. If the EBC were time dependent, then a minor extension would allow for that requirement. The presence of any EBC over time means that the reactions at those dofs will change with time. The optional calculation is included in the illustrated script.

The output from a transient solution can be listed, but that is not very practical; so the time history is usually saved to a file (binary is best) to be post-processed by graphing the time history of selected degrees of freedom, contouring the mesh results, or making mesh carpet plots at selected times. A typical listing of transient temperatures and reaction heat flows is given in Fig. 12.1, while time history graphs of the temperature at three dofs are seen in Fig. 12.2. From those time history graphs, it is seen that for most of the time steps there are only small changes at a point as the steady state solution is approached.

```
Fixed = [ 4     5     6]; Free  = [ 1     2     3]
Current  time step = 2.50e-02

Time history solution results at above 'Free' DOF
Step, Time,          Solution at DOF  1  2  3
   1, 0.0000e+00,  5.0000e+00  5.0000e+00  5.0000e+00
   2, 2.5000e-02,  5.1150e+00  5.2009e+00  5.1950e+00
   3, 5.0000e-02,  5.2737e+00  5.3791e+00  5.3568e+00
   4, 7.5000e-02,  5.4529e+00  5.5408e+00  5.4947e+00
   5, 1.0000e-01,  5.6391e+00  5.6892e+00  5.6152e+00
   6, 1.2500e-01,  5.8246e+00  5.8263e+00  5.7226e+00
   7, 1.5000e-01,  6.0050e+00  5.9536e+00  5.8196e+00
   8, 1.7500e-01,  6.1779e+00  6.0721e+00  5.9083e+00
   9, 2.0000e-01,  6.3420e+00  6.1824e+00  5.9900e+00
  10, 2.2500e-01,  6.4971e+00  6.2853e+00  6.0655e+00
  . . .
  76, 1.8750e+00,  8.7241e+00  7.7332e+00  7.1129e+00
  77, 1.9000e+00,  8.7258e+00  7.7343e+00  7.1137e+00
  78, 1.9250e+00,  8.7274e+00  7.7353e+00  7.1144e+00
  79, 1.9500e+00,  8.7289e+00  7.7363e+00  7.1151e+00
  80, 1.9750e+00,  8.7303e+00  7.7372e+00  7.1158e+00
NOTE: steady state 8.7500      7.7500      7.1250

Time history reaction results at 'Fixed' DOF
Step, Time,          Reactions at DOF 4   5   6
   1, 0.0000e+00,  0.0000e+00  0.0000e+00  0.0000e+00
   2, 2.5000e-02, -4.8037e+00 -1.3560e+01 -4.0000e+00
   3, 5.0000e-02, -5.5165e+00 -1.4854e+01 -4.0000e+00
   4, 7.5000e-02, -6.1631e+00 -1.5958e+01 -4.0000e+00
   5, 1.0000e-01, -6.7567e+00 -1.6922e+01 -4.0000e+00
   6, 1.2500e-01, -7.3053e+00 -1.7781e+01 -4.0000e+00
   7, 1.5000e-01, -7.8145e+00 -1.8557e+01 -4.0000e+00
   8, 1.7500e-01, -8.2882e+00 -1.9267e+01 -4.0000e+00
   9, 2.0000e-01, -8.7296e+00 -1.9920e+01 -4.0000e+00
  10, 2.2500e-01, -9.1414e+00 -2.0524e+01 -4.0000e+00
  . . .
  76, 1.8750e+00, -1.4933e+01 -2.8903e+01 -4.0000e+00
  77, 1.9000e+00, -1.4937e+01 -2.8909e+01 -4.0000e+00
  78, 1.9250e+00, -1.4941e+01 -2.8915e+01 -4.0000e+00
  79, 1.9500e+00, -1.4945e+01 -2.8921e+01 -4.0000e+00
  80, 1.9750e+00, -1.4949e+01 -2.8926e+01 -4.0000e+00
Steady state        -15.000     -29.000     -4.0000
```

Figure 12.1. Transient temperatures and reaction heat flows at three AP.

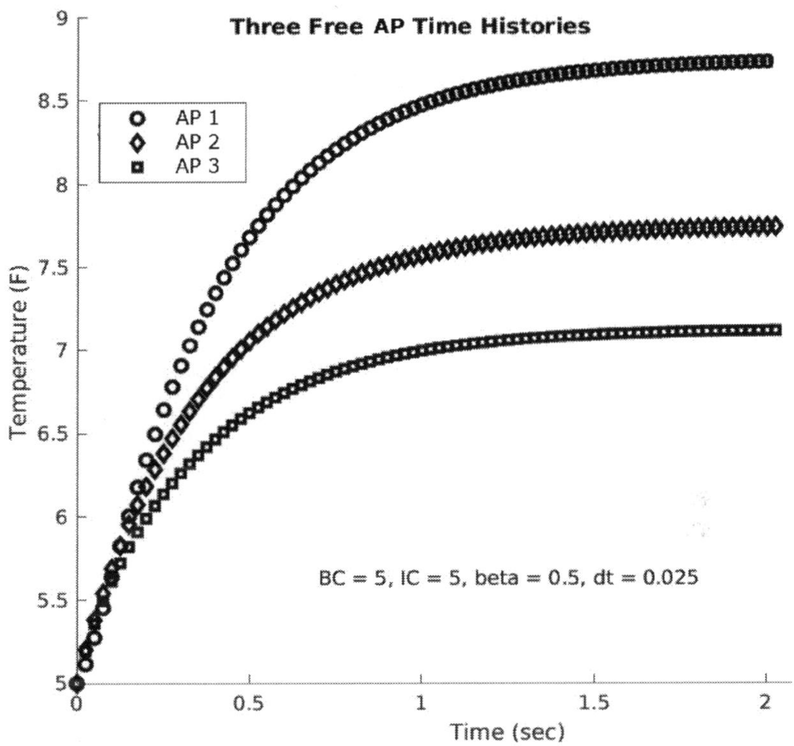

Figure 12.2. **Graph of transient temperatures at three AP.**

Script 12.1. Generalized trapezoidal time integration

```
function [R_u, React, time] = gen_trap_history (K, M, ...
               c_Q, beta, n_steps, delta_t, ...
               ic_value, Fixed, bc_value, Free) % ========
%  Define generalized trapezoidal time integration data
% for solution results R(t) from Initial Condition state
%       [beta * K + M / delta_t] * R_new = c_rhs
% c_rhs = [(beta - 1)K + M / delta_t] * R_old
%     + (1 - beta)c_old + beta c_new, here c_old=c_new
% beta = 1/2 Crank-Nicolson method,   2/3 Galerkin in time
% beta = 0   Foward difference,       1 Backward difference
% n_steps = number of steps,   delta_t = size of time step
% time = actual time values,        R_u = solution history
```

```
n_bc   = size (Fixed, 2)        ; % number of essentrial BC
n_d    = size (c_Q, 1)             ; % number of system DOF
n_unkn = n_d - n_bc                  ; % number of unknowns
time = [0:1:n_steps] * delta_t       ; % actual time values

% Allocate time history storage, inset initial condition
K_uu = zeros (n_unkn, n_unkn)      ; % allocate conduction
M_uu = zeros (n_unkn, n_unkn)        ; % allocate capacity
c_Qu = zeros (n_unkn, 1)               ; % allocate source
c_bcu= zeros (n_unkn, 1)           ; % allocate EBC source
c_old= zeros (n_unkn, 1)          ; % allocate time source
R_u  = zeros (n_unkn, n_steps+1) ; % allocate time history
R_bc = ones  (n_bc, 1) * bc_value; % known BC temperatures
React= zeros (n_bc, n_steps) ; % optional reaction history
R_u  = zeros (n_unkn, n_steps+1) ; % allocate time history
R_u (:, 1) = ic_value            % insert initial condition

% Extract the unknown equations active in the time history
%      M_uu * R_u_dot + K_uu * R_u = c_Qu(t) + c_bcu(t)
M_uu = M (Free, Free)                     ; % free capacity
K_uu = K (Free, Free)                   ; % free conduction
K_ub = K (Free, Fixed)                   ; % coupled to EBC
K_bb = K (Fixed, Fixed)                        ; % optional
c_Qu = c_Q (Free)                           ; % free source

%  Matrices for linear system K_hat * R_u(time) = F(time)
%     that will be solved at each time step for R_u

%       [beta * K + M / delta_t] * R_new = c_rhs
% c_rhs =[(beta - 1)K + M / delta_t] * R_old + c_Qu + c_bc
K_hat = beta * K_uu + M_uu / delta_t  ; % constant in time
K_inv = inv (K_hat)                         ; % invert once
% Square terms for RHS:    F = c_Qu + c_bc + K_RHS*R_u_old
K_RHS = K_hat - K_uu                  ; % constant in time
```

```
% Time marching solution, for constant data and delta_t
R_u (:, 1) = ic_value         ; % insert initial condition
for k = 2:1:n_steps+1;% march to next time ---> ---> ---->
  %               source from prior time step
  c_old = K_RHS * R_u (:, k-1)              ; % prior time
  %           update the effective source terms
  c_bc = -K_ub * R_bc              ; % if EBC time dependent
  F = c_Qu + c_bc + c_old                  ; % at this time
  % Backsubstitute current source for current solution
  R_u (:, k) = K_inv * F                         ; % answer
  % Recover optional reaction heat flow at EBC equations
  React(:,k) =K_ub' * R_u(:,k) + K_bb * R_bc - c_Q(Fixed);
end  ;  % for all time steps <--- <--- <--- <--- <--- <---
% end gen_trap_history =====================================
```

12.3 Accuracy and Control

The much older finite difference method for spatial approximations leads to the capacitance, $[\boldsymbol{M}]$, being diagonal, while the consistent IGA basis function form of $[\boldsymbol{M}]$ is a full matrix. The IGA system matrix is still sparse and diagonally dominant, but not diagonal. More recent studies recommend using the average of $[\boldsymbol{M}]$ and its scaled diagonal.

For a given time step, the temperature response depends on the system non-dimensional Fourier number, $Fo \equiv k\Delta t/\rho c L^2$, which represents the ratio of heat transfer by conduction through a distance of L, in a volume of L^3, to the rate of energy storage in the same volume. The larger the Fourier number, the larger the distance of penetration of a temperature change into a solid in a given time due to a surface thermal shock (sudden change in surface temperature). A large value of the Fourier number indicates a faster propagation of heat through a body.

Transient IGA solutions are often validated by comparing the time history to analytical solutions where a surface temperature changes instantly from its IC value to a different value set by an EBC, or where a sudden heat flow into the surface occurs. There are physical problems of that sort, like a surface hit with a powerful laser beam, but if the solution at $t = 0^+$ is very important, the problem should be formulated with non-Fourier heat transfer, which includes the second time derivative of the temperature.

Improper meshing and/or improper time step selection can lead to impossible physical temperatures in the first few time steps at locations just inside a surface subject to a thermal shock. Some authors say such a response is due to the consistent formulation of the capacitance matrix and the diagonal (finite difference) form should be used instead. If those early near-surface temperatures are important, then build the proper model and solve it in the consistent way. The diagonal matrix results always look physically possible and most analysts use it.

Consider a thermal shock where a temperature is suddenly raised on the surface of a body. The temperature along a line normal to that surface is sketched in Fig. 12.3. There are analytic solutions for the transient temperature along a line normal to the surface. They start with a very sharp temperature gradient that decreases as time proceeds. In any element, the element temperature value is like a least squares fit over the element domain to the exact value. In the direction normal to the surface, there is a very narrow region of changed temperature. If the iElements lengths at and near the boundary are too large in comparison to the temperature change region, then the calculation likely will give a physically impossible temperature decrease when heat is added to the surface. This response is caused by the time step being too small in relation to the mesh size and the material properties. The non-dimensional Fourier number for the system is too small and the remedy is to decrease the size of the elements in the direction normal to the shocked surface or to increase the time step size. In other words, the user has not built

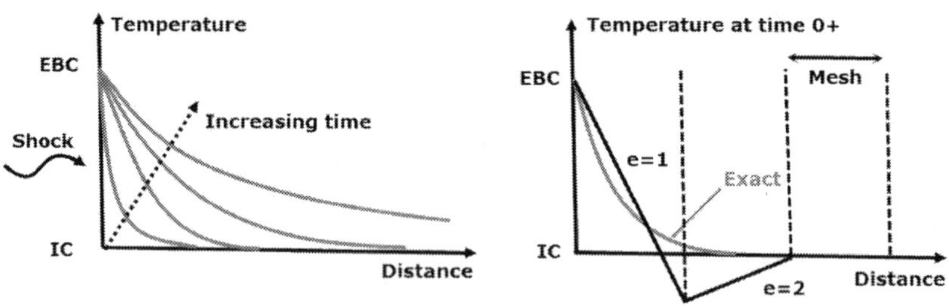

Figure 12.3. Thermal shock requires both mesh and time step control

a consistent mesh and/or time step for the physical problem being modeled. Diagonal capacitance matrices are in part popular because they can hide poor and/or inexperienced modeling skills and ensure what always looks like a physically correct solution.

Choosing the time step size affects the stability of the time history and the presence of oscillations in the solution. The Crank–Nicolson method is unconditionally stable (the solution will not blow up), but it still can oscillate wildly about the true solution. There are mathematical limits on the time step size needed to be stable and to avoid oscillations. When the parameter $\beta < 0.5$, the time step must be less than the critical value of $\Delta t_{\text{critical}} = 2/[(1 - 2\beta)\lambda_{\max}]$ in order to obtain a stable time history. There, the eigenvalue is the maximum found from $|[\boldsymbol{S}] - \lambda[\boldsymbol{M}]| = 0$, which can be bounded by the maximum of the element eigenvalues $|[\boldsymbol{S}]^e - \lambda^e[\boldsymbol{M}]^e| = 0$.

12.4 Introduction to Dynamic Solutions

In the sections on elasticity and stress analysis in Chapter 10, it was shown that the element stiffness and consistent mass matrices of

$$\boldsymbol{K}^e = \int_{\Omega^e} \boldsymbol{B}^{e^T} \boldsymbol{E}^e \boldsymbol{B}^e d\Omega, \quad \boldsymbol{M}^e = \int_{\Omega^e} \boldsymbol{N}^{e^T} \rho^e \boldsymbol{N}^e d\Omega$$

are assembled in exactly the same way to define the system stiffness and mass matrices, $[\boldsymbol{K}]$ and $[\boldsymbol{M}]$. Dynamic structures and wave propagation problems solved by IGA usually lead to a matrix ordinary hyperbolic differential equation in time of the form

$$[\boldsymbol{K}]\{\boldsymbol{\delta}(t)\} + [\boldsymbol{D}]\{\dot{\boldsymbol{\delta}}(t)\} + [\boldsymbol{M}]\{\ddot{\boldsymbol{\delta}}(t)\} = \{\boldsymbol{f}(t)\}, \dot{\{\}} = \partial\{\}/\partial t,$$

where $[\boldsymbol{D}]$ is the system damping matrix. The new term is the displacement acceleration

$$\frac{\partial^2\{\boldsymbol{\delta}(t)\}}{\partial t^2} \equiv \boldsymbol{N}(x, y, z)\frac{\partial^2\{\boldsymbol{\delta}(t)^e\}}{\partial t^2} \equiv \boldsymbol{N}(x, y, z)\{\ddot{\boldsymbol{\delta}}(t)^e\}.$$

These equations of motion for the system are subject to EBCs and the IC, on the displacements $\boldsymbol{\delta}(0) = \boldsymbol{\delta}_{\text{IC}}$, and the velocities $\dot{\boldsymbol{\delta}}(0) = \dot{\boldsymbol{\delta}}_{IC}$. Many engineering applications involve sharp transients or nonlinear stiffness responses, $[\boldsymbol{K}] = [\boldsymbol{K}(\{\boldsymbol{\delta}(t)\})]$. Such applications require a direct time integration to obtain the solution time history. Some direct methods require an estimate of the initial acceleration of the solution. It is often approximated as zero, but it can be found from the IC as

$$\{\ddot{\boldsymbol{\delta}}(0)\} = [\boldsymbol{M}]^{-1}(\{\boldsymbol{f}(0)\} - [\boldsymbol{K}]\{\boldsymbol{\delta}_{\text{IC}}\} - [\boldsymbol{D}]\{\dot{\boldsymbol{\delta}}_{\text{IC}}\}),$$

where the mass matrix, $[\boldsymbol{M}]$, can be approximated by its scaled diagonal

$$[\boldsymbol{M}] \approx [\boldsymbol{M}]_{\text{Diagonal}}.$$

The damping matrix, $[\boldsymbol{D}]$, definition varies with the application, but by far the most common form is a combination of structural damping and mass damping:

$$[\boldsymbol{D}] = \alpha[\boldsymbol{K}] + \beta[\boldsymbol{M}],$$

where the two proportions are usually related to the critical dampening, ξ, and a frequency range, $\omega_1 < \omega_2$ of interest:

$$\begin{aligned}\alpha &= 2(\xi_2\omega_2 - \xi_1\omega_1)/(\omega_2^2 - \omega_1^2),\\ \beta &= 2\omega_1\omega_2(\xi_1\omega_2 - \xi_2\omega_1)/(\omega_2^2 - \omega_1^2).\end{aligned}$$

After any EBCs are applied, the original system will be reduced in size to, say

$$[\boldsymbol{K}]\{\boldsymbol{\delta}(t)\} + [\boldsymbol{D}]\{\dot{\boldsymbol{\delta}}(t)\} + [\boldsymbol{M}]\{\ddot{\boldsymbol{\delta}}(t)\} = \{\boldsymbol{p}(t)\},$$

where $\{\boldsymbol{p}(t)\}$ includes the original time-dependent loads, $\boldsymbol{f}(t)$, plus the effects of the (constant or) time-dependent EBCs.

Tezduyar's research group applied IGA in space-time to several very challenging problems in computational mechanics including studies by Liu *et al.* (2022) and Takizawa *et al.* (2012).

Single-step dynamic histories: At this point, there are many algorithms for integrating this system in time. Most temporal integration schemes use finite differences in time or finite elements in time to derive the algorithm using constant increments of time, say Δt, to define time as $t_n = (t_0 + n\ \Delta t)$ at time step number n. Most of the algorithms form a linear system to be solved at each time step as

$$[\boldsymbol{S}(\Delta t, a_j, \boldsymbol{K}, \boldsymbol{D}, \boldsymbol{M})]\{\boldsymbol{\delta}\}_n = \{\boldsymbol{f}_n(a_j, t_n, t_{n-1}, \boldsymbol{K}, \boldsymbol{D}, \boldsymbol{M}, \boldsymbol{p}, \{\boldsymbol{\delta}\}_{n-1})\},$$

where the a_j are a set of algorithm constants (usually $1 \leq j \leq 8$). In this form, the left square matrix is constant if the time step, Δt, is constant. Assembling $[\boldsymbol{K}]$ is computationally expensive and obtaining its inverse or factorization is very expensive. Thus, it is desirable to do these operations only once or twice. The right-hand column changes with every time step and the result at time step n is

$$\{\boldsymbol{\delta}\}_n = [\boldsymbol{S}(\Delta t, a_j, \boldsymbol{K}, D, M)]^{-1}\{\boldsymbol{f}\}_n.$$

For example, the popular Newmark–Beta method (Newmark, 1952) has two constants, β and γ, that combine with the time step to define six other constants:

$$a_1 = \frac{1}{\beta \Delta t^2}, \quad a_2 = \frac{1}{\beta \Delta t}, \quad a_3 = \frac{1}{2\beta} - 1,$$

$$a_4 = \frac{\gamma}{\beta \Delta t}, \quad a_5 = \frac{\gamma}{\beta} - 1, \quad a_6 = \left(\frac{\gamma}{2\beta} - 1\right)\Delta t,$$

$$[\boldsymbol{S}] = ([\boldsymbol{M}] + a_4[\boldsymbol{D}] + a_1[\boldsymbol{K}]),$$

$$\begin{aligned}\{\boldsymbol{f}\}_n &= \{\boldsymbol{f}\}_{n-1} + [\boldsymbol{K}](a_1\{\boldsymbol{\delta}\}_{n-1} + a_2\{\dot{\boldsymbol{\delta}}\}_{n-1} + a_3\{\ddot{\boldsymbol{\delta}}\}_{n-1}) \\ &\quad + [\boldsymbol{D}](a_4\{\boldsymbol{\delta}\}_{n-1} + a_5\{\dot{\boldsymbol{\delta}}\}_{n-1} + a_6\{\ddot{\boldsymbol{\delta}}\}_{n-1}),\end{aligned}$$

then the velocity and acceleration are updated as

$$\begin{aligned}\{\dot{\boldsymbol{\delta}}\}_{n+1} &= a_4(\{\boldsymbol{\delta}\}_{n+1} - \{\boldsymbol{\delta}\}_n) - a_5\{\dot{\boldsymbol{\delta}}\}_n - a_6\{\ddot{\boldsymbol{\delta}}\}_n, \\ \{\ddot{\boldsymbol{\delta}}\}_{n+1} &= a_1(\{\boldsymbol{\delta}\}_{n+1} - \{\boldsymbol{\delta}\}_n - \Delta t\{\dot{\boldsymbol{\delta}}\}_n) - a_3\{\ddot{\boldsymbol{\delta}}\}_n.\end{aligned}$$

The four most common choices for β and γ are as follows:

(1) The Newmark average acceleration method: $\beta = 1/4$, $\gamma = 1/2$; it is unconditionally stable. The accuracy is $O(\Delta t^2)$.
(2) The linear acceleration method: $\beta = 1/6$, $\gamma = 1/2$; it is conditionally stable. The accuracy is $O(\Delta t^2)$.
(3) The Fox–Goodwin method: $\beta = 1/12$, $\gamma = 1/2$; it is conditionally stable. The accuracy is $O(\Delta t^4)$.
(4) The Hilber–Hughes–Taylor α-method: $\beta = (1-\alpha)^2/4$, $\gamma = (1-2\alpha)/2$ with the main constant range $-1/3 \leq \alpha \leq 0$; it is unconditionally stable. The accuracy is $O(\Delta t^2)$.

12.5 Wilson Single-Step Method

Another direct time history integration algorithm (used here) is the Wilson method. The Wilson (1972) method is an extension of the linear acceleration method with $\gamma = 1/2$, as illustrated in Fig. 12.4, in a manner that makes it stable. It assumes that the acceleration is linear from time t to $(t+\theta\Delta t)$ with $\theta \geq 1$, and calculates the displacement, velocity, and acceleration there. Then, the method interpolates back to find those values at $(t + \Delta t)$. Theoretical studies show it to be unconditionally stable for $\theta \geq 1.37$, and that the optimal value is $\theta = 1.4208$, so $\theta = 1.42$ is the commonly used value.

Integrating the linear acceleration assumption gives the predicted velocity and displacement:

$$\begin{aligned}
\dot{\boldsymbol{\delta}}(t+\theta\Delta t) &= \dot{\boldsymbol{\delta}}(t) + [\ddot{\boldsymbol{\delta}}(t+\theta\Delta t) + \ddot{\boldsymbol{\delta}}(t)]/2,\\
\boldsymbol{\delta}(t+\theta\Delta t) &= \boldsymbol{\delta}(t) + \theta\Delta t\dot{\boldsymbol{\delta}}(t) + (\theta\Delta t)^2[\ddot{\boldsymbol{\delta}}(t+\theta\Delta t) + 2\ddot{\boldsymbol{\delta}}(t)]/6,
\end{aligned}$$

and these are rewritten in terms of the displacement at the extended time:

$$\begin{aligned}
\ddot{\boldsymbol{\delta}}(t+\theta\Delta t) &= 6[\boldsymbol{\delta}(t+\theta\Delta t) - \boldsymbol{\delta}(t)]/(\theta\Delta t)^2 - \dot{\boldsymbol{\delta}}(t)/\theta\Delta t - 2\ddot{\boldsymbol{\delta}}(t),\\
\dot{\boldsymbol{\delta}}(t+\theta\Delta t) &= 3[\boldsymbol{\delta}(t+\theta\Delta t) - \boldsymbol{\delta}(t)]/\theta\Delta t - \theta\Delta t\,\ddot{\boldsymbol{\delta}}(t)/2.
\end{aligned}$$

These estimates are substituted back into the equations of motion to define the matrix equations for solving one time step. In other words, during one step, the Wilson method calculates the solution at $(t+\theta\Delta t)$ using the linear system

$$\boldsymbol{S}(t+\theta\Delta t)\ \boldsymbol{\delta}(t+\theta\Delta t) = \boldsymbol{c}(t+\theta\Delta t),$$

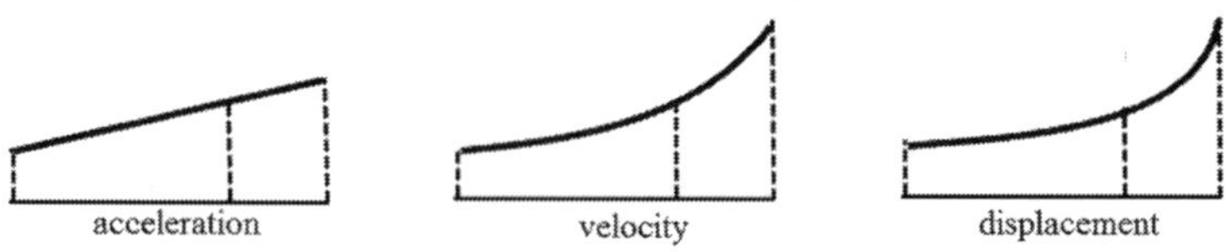

Figure 12.4. The Wilson method extends the linear acceleration step

where the effective matrices are

$$\boldsymbol{S}(t+\theta\Delta t) = 3\boldsymbol{G}/\theta\Delta t + \boldsymbol{K}, \quad \text{with } \boldsymbol{G} \equiv 2\boldsymbol{M}/\theta\Delta t + \boldsymbol{D},$$
$$\boldsymbol{c}(t+\theta\Delta t) = \boldsymbol{p}(t+\theta\Delta t) + \boldsymbol{G}\{3\ \boldsymbol{\delta}(t)/\theta\Delta t + 3\ \dot{\boldsymbol{\delta}}(t) + \theta\Delta t\ \ddot{\boldsymbol{\delta}}(t)\}$$
$$- \boldsymbol{D}\{\dot{\boldsymbol{\delta}}(t) + \theta\Delta t\ \ddot{\boldsymbol{\delta}}(t)/2\}.$$

Equation $\boldsymbol{S}\ \boldsymbol{\delta} = \boldsymbol{c}$ is solved once per time step for the new displacement, $\boldsymbol{\delta}(t+\theta\Delta t)$, and then the method interpolates backward to find the values at $(t+\Delta t)$ as

$$\ddot{\boldsymbol{\delta}}(t+\Delta t) = \ddot{\boldsymbol{\delta}}(t) + \{\ddot{\boldsymbol{\delta}}(t+\theta\Delta t) - \ddot{\boldsymbol{\delta}}(t)\}/\theta,$$
$$\dot{\boldsymbol{\delta}}(t+\Delta t) = \dot{\boldsymbol{\delta}}(t) + \theta\Delta t\ \ddot{\boldsymbol{\delta}}(t) + \theta\Delta t\{\ddot{\boldsymbol{\delta}}\ (t+\theta\Delta t) - \ddot{\boldsymbol{\delta}}(t)\}/2\theta,$$
$$\boldsymbol{\delta}(t+\Delta t) = \boldsymbol{\delta}(t) + \theta\Delta t\ \dot{\boldsymbol{\delta}}(t),$$

which are then used to start the next integration step. The Wilson algorithm is implemented in Script 12.2.

The above-mentioned algorithms are all single-step methods since they only require storing the results from the last time step. They minimize the memory requirements and were developed when computers had an extremely small memory, but their main asset is that they deal well with sudden changes in the solution such as earthquakes imposing rapid changes in the displacements or external impact forces. There are many other algorithms that lead to a linear system being solved at each time step.

Script 12.2. The Wilson θ-method for time validation

```
function [] = Wilson_direct (n_d, a, b, c, drp, r, dr, ...
             d2r, p, omega, dt, n_steps, i_print, n_c, ibc)
% * * * * * * * * * * * * * * * * * * * * * * * * * * * * * * *
% step by step integration of matrix equations:
%      a*d2r(t)/dt2 + b*dr(t)/dt + c*r(t) = p(t)
% from initial conditions: r(0), dr(0)/dt, to linear system
% s*r(t) = f(t) at each time. (by the Wilson method)
% * * * * * * * * * * * * * * * * * * * * * * * * * * * * * * *
% allocate a (n_d, n_d), b (n_d, n_d), ibc (n_c), ...
%          r (n_d), dr (n_d), d2r (n_d), p (n_d), ...
%          drp (n_d), c (n_d, n_d)
zero = 0. ;
```

```
% initial values of r and dr are passed thru arguments
% a         = mass (or capacitor) matrix
% b         = damping (or resistor) matrix
% c         = stiffness (or inductor) matrix
% drp       = value of dr at time = t + delt = tau
% dt        = time step size
% f         = linear system net forcing, stored over p
% ibc       = array containing the n_c dof's with zero bc
% i_print   = no. of integration steps between printing
% n_c       = no. dof with specified values of zero (if any)
% n_d       = total number of dof in system
% n_steps   = no. of integration steps, t_final = dt*n_steps
% omega     = 1.25 is suggested Wilson extrapolation amount
% p         = forcing function
% r,dr,d2r  = 0,1,2 order deriv. of r wrt t at time=t
% s         = linear system square matrix, stored over b
% tau       = extrapolated time step, omega*dt

% ** initial  calculations **
disp ('direct step by step integration')
if  i_print < 1
  i_print = 1 ;          % data check
end
n_steps = (n_steps/i_print)*i_print  ;
if  n_steps <= 0
  n_steps = i_print  ;   % data check
end % if validation
delt  = (omega - 1)*dt   ;   % extrapolation increment
tau   = omega*dt         ;   % extrapolation amount
if  n_c < 1
  disp ('warning: no constraints in direct \n')
end % if warning

% Set one node to display time history
node_h = 3              ; % node for valudation
h_id = fopen ('node_history.txt', 'w') ;

% initialize arrays for loads< velocity< acceleration
p = zeros(n_d,1); drp = zeros(n_d,1); d2r = zeros(n_d,1);

% overwrite "b" with "s" matrix to save storage
b = b + 2*a/tau + tau*c/3 ; % name change
```

```
% ** apply boundary conditions ( to s ) **
if  n_c > 0     % then apply essential boundary conditions
  for i = 1:n_c % BC value is zero here
    [b, p] = modify_full (n_d, ibc(i), zero, b, p) ;
  end % for on essential boundary conditions
end % if boundary conditions needed

%  ** invert b (once) **
b = inv (b);   % actually, factor if s is large

%    (approximate the initial value of d2r)
%      ** calculate solution at time t **
i_count = i_print - 2 ;
for i_step = 0: n_steps % step zero is initial acceleration ===>> ===>>
  i_count = i_count + 1 ;
  t       = dt*(i_step-1) ;
  t_plus  = t + delt ;                % extrapolate for the load
  if  i_step <= 1
    t_plus = zero ;                   % unless initial condition
  end % if
  if  i_count == i_print
    fprintf(1,'\n step number = %5i  time = %e \n', i_step, t);
    fprintf(1,'\n   i      r(i)         dr/dt          d2r/dt2 \n')
  end % if
% forcer is to define the forcing function p(t)
  p = forcer (t_plus, n_d) ; % new force value
% form modified forcing function, f, at t + delt
  for i = 1:n_d
    p(i) = p(i) + sum ( a(i,:)*( 2*dr(:)/tau + d2r(:)) ) ...
    - sum ( c(i,:)*(r(:) + 2*tau*dr(:)/3 + tau*tau*d2r(:)/6) );
  end % for net forcing vector

% ** apply boundary conditions ( to f ) **
  if  n_c > 0
    p( ibc(1:n_c) ) = 0 ; % for zero value EBC
  end % if bc needed

% solve for drp at time t+delt, s_inverse*f
  drp = b*p' ;  % forward and back substitute if s is large

% using data at t-dt and t+delt calculate values at t
  for i = 1:n_d
    drdt   = (1-1/omega)*dr(i) + drp(i)/omega  ;
    d2rdt2 = d2r(i)*(1-2/omega)+2*(drdt-dr(i))/omega/dt   ;
```

```
%   approximate the initial value of d2r, if i_step <= 1
    if  i_step > 1
      r(i)  = r(i) + dt*(2*dr(i)+drdt)/3 + dt*dt*d2r(i)/6 ;
      dr(i) = drdt ;
    end % if it is a starter step
    d2r(i) = d2rdt2 ;
  end % for each degree of freedom

% output results for time t
  if ( i_step > 1 )
   fprintf(h_id,'%e  %e  %e  %e \n', t, r(node_h), dr(node_h), d2r(node_h));
  end % if skip initial acceleration at -del_t
  if  i_count == i_print % then print this step
    for k = 1:n_d
      fprintf (1,'%5i %e %e %e \n', k, r(k), dr(k), d2r(k) )
    end % for dof result listing at this time
    i_count = 0 ;
  end % if print increment
end % for i_step over all times <<=== <<=== <<=== <<=== <<=== <<===
fclose (h_id) ;
fprintf ('Saved Wilson results in file node_history.txt \n')
% end function Wilson_direct ==========================
```

It is difficult to debug and gain experience with these, and other, direct time history integration methods. Exact solutions are rare, except for single dof models. Biggs, in *Introduction to Structural Dynamics*, gave the exact analytic time history for a three-dof spring-mass system with time-varying loads as shown in Fig. 12.5. The exact time history solution given by Biggs consists of two segments. The first segment covers the time when each load is decreasing to zero. The values of the displacement and velocities at the end of that time are used as initial conditions to continue the solution as a free vibration response. The example solution for the displacements, velocities, and accelerations was also computed by the Wilson method included in the provided script. Using a smaller time step leads to a more accurate solution, but requires a proportionally longer execution time. The IGA version of this model would use three linear basis function bars with lumped (point) masses. This includes neglecting the mass of the bars.

For example, using the Wilson method to solve the Biggs model with a time step of 0.00125 sec. gives the displacement and velocity curves compared to the exact results (dashed) in Fig. 12.6. Those

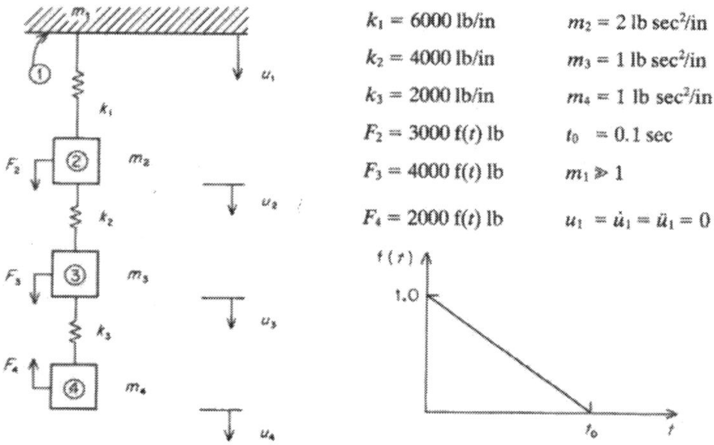

Figure 12.5. The Biggs analytic time history spring-mass system

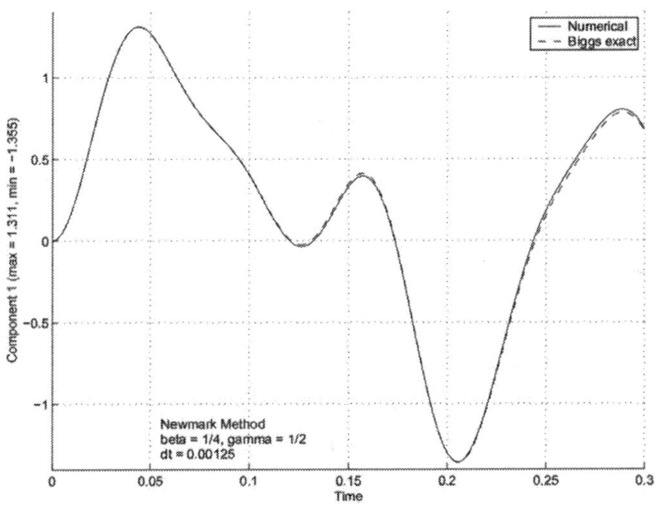

Figure 12.6. Biggs' model displacement for $\Delta t = 0.00125$ sec

results are fairly accurate, but doubling the time step to cut the run time in half significantly increases the errors in the results.

Applications like the Biggs example require the user to write a function describing the time history of the externally applied forces (or input points on a force–time curve). The Biggs example force–time relations are given in Script 12.3, and the control integers and starting the time history are provided in Script 12.4.

Script 12.3. Problem-specific time-varying force multipler

```
function [p] = forcer (t, n_d)
% p     = active force vector at time t
% t     = current time
% n_d   = number of terms in p
% * * * * * * * * * * * * * * * * * * * * * * * * * * *
% define forcing function for step by step integration
% NOTE: this requires application dependent functions
% * * * * * * * * * * * * * * * * * * * * * * * * * * *
% transient test problem Bigg's p. 123,
% linear decreasing time ramp for forces
f_of_t = 0.0                  ; % default forcing value at t
t_one  = 0.1                  ; % length of time ramp
if ( t < t_one )              ; % force changing with time
  f_of_t = 1-t/t_one          ; % current time multipler
end                             % if within time ramp

f2=3000; f3=4000; f4=-2000      ; % initial force values
p(1) = 0                        ; % new force at time t
p(2) = f2*f_of_t                ; % new force at time t
p(3) = f3*f_of_t                ; % new force at time t
p(4) = f4*f_of_t                ; % new force at time t
end % function forcer
```

Script 12.4. Defining problem-dependent data and starting the time stiffness

```
function [] = direct_forcer_validate ===================
% test of direct integration programs, full matrices
% am * d2u(t)/dt2 + ac * du(t)/dt + ak * u(t) = f(t)

% By WIlson algorithm (unconditionally stable method)
%     transient test problem Bigg's p. 123
%  This is a short driver to validate the direct time
```

```
%  integration functions. "direct" and "forcer" by
%  comparing the results to an analytic solution by Biggs
%  for a multi-degree of freedom system.
%      M*d2u(t)/dt2 + C*du(t)/dt + K*u(t) = F(t)
%  -----------------------------------------------------
%  The example involves 3 masses moving wrt a fixed mass
%  at joint 1.  The mass there is not important, but is
%  used to illustrate how a term is sometimes included
%  but then removed by a boundary condition.
%
%  u1=0     -->u2     -->u3     -->u4
%  M1---K1---M2---K2---M3---K3---M4
%  Fixed    ==>F2     ==>F3     <==F4
%
%  Briggs, J. M., Introduction to Structural Dynamics",
%  McGraw Hill, 1964.
% -------------------------------------------------------

n_d = 4                       ; % Number of degrees of freedom
n_c = 1                              ; % Number of constraints
k_print = 1      ; % Number of time steps between printouts
n_steps = 241                   ; % Total number of time steps
dt = 0.00125                                ; % Time step size
omega = 1.25                     ; % Wilson algorithm constant

ak  = zeros (n_d, n_d)    ; % zero stiffess matrix
am  = zeros (n_d, n_d)    ; % zero mass matrix
ac  = zeros (n_d, n_d)    ; % zero damping matrix
u   = zeros (n_d, 1)      ; % zero displacements
f   = zeros (n_d, 1)      ; % zero applied forces
vel = zeros (n_d, 1)      ; % zero velocity
acc = zeros (n_d, 1)      ; % zero acceleration
up  = zeros (n_d, 1)      ; % zero a working vector
ibc = zeros (n_d, 1)      ; % zero boundary condition flag

% fix the top mass  (binary flag)
ibc (1) = 1 ;

%       set non-zero stiffness terms
ak = [6000   -6000       0       0 ; ...
     -6000   10000   -4000       0 ; ...
         0   -4000    6000   -2000 ; ...
```

```
%       set non-zero mass terms
am = [4   0   0   0 ; ...
      0   2   0   0 ; ...
      0   0   1   0 ; ...
      0   0   0   1] ;
%
%     direct time integration, Wilson method
Wilson_direct (n_d, am, ac, ak, up, u, vel, acc,   ...
        f, omega, dt, n_steps, k_print, n_c, ibc)
end % direct_forcer_validate


function [s, c] = modify_full (n_d, n, value, s, c)
% * * * * * * * * * * * * * * * * * * * * * * * * * *
% apply an essential b.c. to full symmetric eqs
%             s*d = c,  d(n) = value
% * * * * * * * * * * * * * * * * * * * * * * * * * *
% c       = full column matrix
% n       = dof number of constrained parameter
% n_d     = total number of equations
% s       = full square matrix
% value   = given value of dof number n
% note:     reaction data are lost
%
  s_nn = s(n, n)              ; % get current diagonal term
  c = c - value*s (:, n)      ; % move given column to right
                                % side
  s (:, n) = zeros (n_d, 1)   ; % zero square matrix rows
  s (n, :) = zeros (1, n_d)   ; % zero square matrix columns
  s (n, n) = s_nn             ; % set identity in row n
  c (n)    = value*s_nn       ; % set identity in row n
end % function modify_full
```

12.6 Summary

12.6.1 *Transient summary*

Scalar field model ordinary differential equation:

$$\frac{\partial}{\partial x}\left(k_x\frac{\partial u}{\partial x}\right)+\frac{\partial}{\partial y}(k_y\frac{\partial u}{\partial y})+Tu+Q(x,y)=\rho_p\frac{\partial u}{\partial \tau}.$$

Equivalent integral form:

$$\int_{\Gamma^b} u\left(k_{nn}\frac{\partial u}{\partial n}\right)d\Gamma - \int_{\Omega}\left(\frac{\partial u}{\partial x}\left(k_x\frac{\partial u}{\partial x}\right) + \frac{\partial u}{\partial y}\left(k_y\frac{\partial u}{\partial y}\right)\right)d\Omega$$

$$-\int_{\Omega} uTud\Omega - \int_{\Omega} uQd\Omega - \int_{\Omega} u\rho c_p ud\Omega = 0.$$

Conduction, convection, advection, and source matrices:

$$\boldsymbol{S}^e_\kappa = \int_{\Omega^e} \vec{\nabla}\boldsymbol{N}^{\boldsymbol{T}}\boldsymbol{\kappa}\vec{\nabla}\boldsymbol{N}d\Omega, \boldsymbol{A}^e_v = \int_{\Omega^e} \boldsymbol{N}^{\boldsymbol{e}^T} m^e v^e \boldsymbol{B}^{\boldsymbol{e}}d\Omega,$$

$$\boldsymbol{M}^e_h = \int_{\Omega^e} \boldsymbol{N}^{\boldsymbol{e}^T} h^e \boldsymbol{N}^{\boldsymbol{e}}d\Omega, \boldsymbol{M}^e_\rho = \int_{\Omega^e} \boldsymbol{N}^{\boldsymbol{e}^T} \rho^e c^e_p \boldsymbol{N}^{\boldsymbol{e}}d\Omega,$$

$$\boldsymbol{c}^e_Q = \int_{\Omega^e} \boldsymbol{N}^{\boldsymbol{e}^T} Q^e d\Omega, \boldsymbol{c}^b_{NBC} = \int_{\Gamma^b} \boldsymbol{N}^{\boldsymbol{b}^T}\left(k_{nn}\frac{\partial u}{\partial n}\right)^b d\Gamma,$$

$$\boldsymbol{c}^b_q = \int_{\Gamma^b} \boldsymbol{N}^{\boldsymbol{b}^T} q_n \, d\Gamma.$$

Transient matrix system: $[\boldsymbol{S_k} + \boldsymbol{A_v} + \boldsymbol{M_h}]\boldsymbol{u}(\tau) + [\boldsymbol{M_\rho}]\dot{u}(\tau) = \boldsymbol{c}(\tau)$.
Steady state matrix system, $\dot{u}(\tau) = \boldsymbol{0}$: $[\boldsymbol{S_k} + \boldsymbol{A_v} + \boldsymbol{M_h}]\boldsymbol{u} = \boldsymbol{c}$.

12.7 Dynamic summary

System equation of motion:

$$[\boldsymbol{K}]\{\boldsymbol{\delta}(t)\} + [\boldsymbol{D}]\{\dot{\boldsymbol{\delta}}(t)\} + [\boldsymbol{M}]\{\ddot{\boldsymbol{\delta}}(t)\} = \{\boldsymbol{f}(t)\}, \dot{\{\}} = \partial\{\}/\partial t.$$

$[\boldsymbol{K}]$ = stiffness matrix, $[\boldsymbol{M}]$ = mass matrix (consistent, diagonal, or averaged),

$[\boldsymbol{D}] = \alpha[\boldsymbol{K}] + \beta[\boldsymbol{M}]$ = damping matrix for two critical dampening, ξ,

At two frequency ranges, $\omega_1 < \omega_2$ with

$\alpha = 2(\xi_2\omega_2 - \xi_1\omega_1)/(\omega_2^2 - \omega_1^2)$,

$\beta = 2\omega_1\omega_2(\xi_1\omega_2 - \xi_2\omega_1)/(\omega_2^2 - \omega_1^2)$.

Element stiffness matrix: $\boldsymbol{K}^e = \int_{\Omega^e} \boldsymbol{B}^{e^T}\boldsymbol{E}^e\boldsymbol{B}^e \, d\Omega$.

Element consistent mass matrix: $\boldsymbol{M}^e = \int_{\Omega^e} \boldsymbol{N^e}^T \rho^e \boldsymbol{N^e}\, d\Omega$.
Displacement interpolation: $\boldsymbol{u}(\boldsymbol{x})= N(r,s)\ \boldsymbol{\delta^e}$.
Linear system solved at each time step: with algorithm constants a_j

$$[\boldsymbol{S}(\Delta t, a_j, \boldsymbol{K}, \boldsymbol{D}, \boldsymbol{M})]\{\boldsymbol{\delta}\}_n = \{\boldsymbol{f}_n(a_j, t_n, t_{n-1,}\boldsymbol{K}, \boldsymbol{D}, \boldsymbol{M}, \boldsymbol{p}, \{\boldsymbol{\delta}\}_{n-1})\},$$
$$\{\boldsymbol{\delta}\}_n = [\boldsymbol{S}(\Delta t, a_j, \boldsymbol{K}, D, M)]^{-1}\{\boldsymbol{f}\}_n.$$

Commonly,

$$\begin{aligned}
[\boldsymbol{S}] &= ([\boldsymbol{M}] + a_4[\boldsymbol{D}] + a_1[\boldsymbol{K}]),\\
\{\boldsymbol{f}\}_n &= \{\boldsymbol{f}\}_{n-1} + [\boldsymbol{K}](a_1\{\boldsymbol{\delta}\}_{n-1} + a_2\{\dot{\boldsymbol{\delta}}\}_{n-1} + a_3\{\ddot{\boldsymbol{\delta}}\}_{n-1}),\\
&\quad +[\boldsymbol{D}](a_4\{\boldsymbol{\delta}\}_{n-1} + a_5\{\dot{\boldsymbol{\delta}}\}_{n-1} + a_6\{\ddot{\boldsymbol{\delta}}\}_{n-1}),\\
\{\dot{\boldsymbol{\delta}}\}_{n+1} &= a_4(\{\boldsymbol{\delta}\}_{n+1} - \{\boldsymbol{\delta}\}_n) - a_5\{\dot{\boldsymbol{\delta}}\}_n - a_6\{\ddot{\boldsymbol{\delta}}\}_n,\\
\{\ddot{\boldsymbol{\delta}}\}_{n+1} &= a_1(\{\boldsymbol{\delta}\}_{n+1} - \{\boldsymbol{\delta}\}_n - \Delta t\{\dot{\boldsymbol{\delta}}\}_n) - a_3\{\ddot{\boldsymbol{\delta}}\}_n.
\end{aligned}$$

Chapter 13

T-Spline Basis Functions

13.1 T-Mesh Origin

A possible point of confusion is that the T-mesh required to compute the basis functions can originate in three distinctly different ways. The vast majority of the literature assumes that the user has interactive access to CAGD NURBS software, or a data exchange file, with a rectangular NURBS mesh index space that is to be interactively converted to a T-mesh, T-junctions, and T-spline basis functions. Such a conversion process deletes most of the NURBS-knots and introduces new T-knots for refinement purposes. A very important step in IGA and CAGD was the development of an efficient algorithm for that process leading to a non-integer T-mesh index space. To master IGA, a reader should understand the well-documented process (Sederberg *et al.*, 2003) for creating a refined T-mesh from and original NURBS-mesh.

Another very common way to create a T-mesh, T-junctions, or T-spline basis functions from a NURBS-mesh is to start with an NURBS integer index space from an data exchange file (like STEPS and ISO 10303-21) and to carry out the same process beginning with those data. With either approach, the creation of a T-mesh and T-splines from an initial NURBS-mesh is discussed from the point of view where a user starts from an integer-based NURBS index space and interactively converts it to a T-mesh, for refinement purposes, by inserting new T-knots between the initial NURBS integer spaces, thereby forming a fractional index space for the final T-mesh.

A very much less common alternate viewpoint is offered where a beginner visualizes the final T-mesh and T-knots, identifies the NURBS knots deleted from its tensor product creation to make the final T-mesh, and uses a free-standing T-mesh generator to create the final T-mesh integer Index Space. This is analogous to common FEA mesh generators with additional data omitted from its final outputs. Since the usual creation of the final T-mesh involves repeatedly inserting T-knots at the mid-edges of line segments in the original integer NURBS index space, the teaching aid output can be thought of as the usual T-mesh fractional indices multiplied by the integer constant 2^n where n is the maximum number of knot insertions into any one unit spacing in the NURBS original indices.

Unfortunately, access to T-spline data containing exact geometry from CAGD software has greatly reduced. The T-spline technology is now covered by a US Patent. The author's understanding of US law means the patent owner's have a legal monopoly for the use of the T-spline technology for 20 years, unless they choose to license the technology to others. Upon the issuance of said patient, T-spline software and user interfaces, in existence before the patent was issued, had to be removed from several popular CAGD products, including Rhino-3D and SolidWorks.

13.2 T-Meshes

Herein, the majority of the T-meshes discussed will be illustrated in integer spaces with rectangular or square cells. Often, dashed lines in the index space will denote where NURBS control points are to be omitted in creating the T-mesh control points and T-knots. Occasionally, the corresponding physical T-mesh will also be shown, like in Fig. 13.1.

13.3 T-Spline Anchors

The process for creating the basis functions for a T-junction relies on an 'anchor' point in the T-mesh Index Space, which depends on whether the polynomial degree(s) is even, odd, or mixed. The

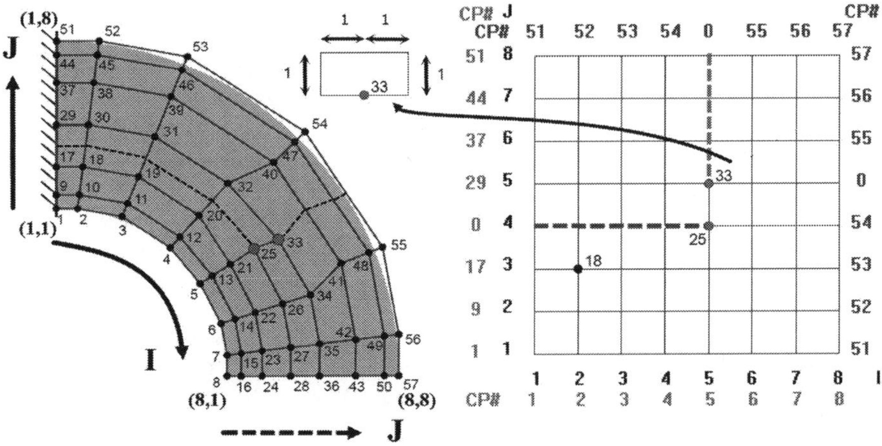

Figure 13.1. Shaded physical shape, its control point numbers, and with T-junctions at knots 25 and 33 mapped from an integer Index Space

directions of the 'local knot vectors' for a T-spline surface or plane in Index Space can be thought of as a stencil centered on the 'anchor' point and extending away from the anchor, in both directions, along constant index values for an acceptable edge or knot. A T-spline stencil looks similar to a finite difference stencil like the one sketched in Fig. 13.2. The support domain defining the stencil is obtained by marching from the T-junction in each direction until $(p+1)/2$-integer orthogonal edges, or the limit of the Index Space, are encountered. The simplest case of template identification is where the polynomial is odd in all directions, resulting in a 'local knot vector' of length $(p+2)$ in each direction. The dashed lines in Fig. 13.2 bound the domain of support around anchor point A created by the tensor product of the local knot vectors, which include knot A at their center. In other words, the T-spline basis functions for anchor A are zero on and outside of the dashed lines as are all of their spatial derivatives.

Examples of the anchor points for mixtures of polynomial orders in 2D are illustrated in Fig. 13.3 with the local knot vectors listed below the integer spaces. A cubic polynomial ($p = 3$) creates a symmetric stencil with two knots on each of the four lines extending from the anchor, like the figure. If an extension line reaches an edge line of the index space, then the last knot value is repeated.

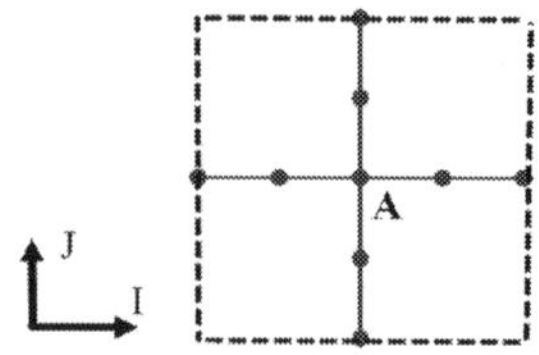

Figure 13.2. T-spline knot stencils about an odd anchor

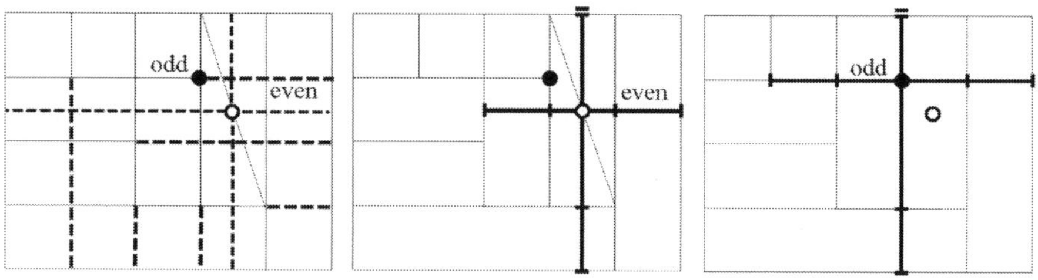

Figure 13.3. Basis function 'anchors' for even or odd polynomials in both directions at a T-junction

In the IGA literature, interactively deriving a T-spline basis function from a NURBS mesh is based on an associated Index Space 'anchor'. For odd-degree polynomials, the 'anchor' is simply at the T-junction vertex. For even-degree basis functions, the basis function anchor (or center) is at the center of the index space rectangle adjacent to the T-junction.

13.4 T-Spline Basis Construction

The notion of T-splines extends to any degree. Here, most of the discussions will be restricted to bi-cubic T-splines. Cubic T-splines are C^2 in the absence of multiple knots. Thus, they are useful analysis basis forms.

Consider the example of a bi-cubic (odd) T-mesh in the (rectangular) Index Space and a valid 'local knot vector' configuration in each index direction. The process of creating the T-splines basis functions will be illustrated with Fig. 13.4 as follows:

(1) Starting from a typical knot vertex **V** in the T-mesh, the horizontal, r, local knot vector is obtained by marching toward each

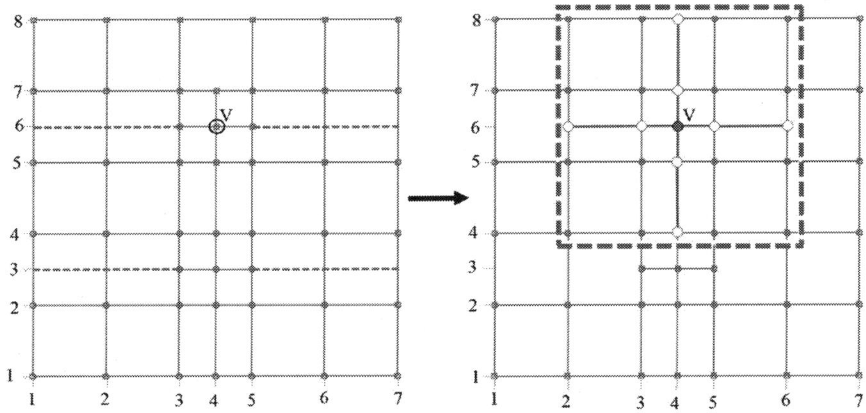

Figure 13.4. T-knot support region in the index space and parametric space

of the two horizontal directions until two sequential edges or vertices are encountered. The open circles on the horizontal extension lines denote those locations. The corresponding univariate basis is the cubic B-spline, $N_{0,3}(r)$, is defined by the 'local knot vector' $\{r_1, r_2, r_3, r_4, r_5\}$:

$$N_{0,3}(r) = N[r_2, r_3, r_4, r_5, r_6](r).$$

(2) Similarly, the open circles on the vertical lines denote those extension locations. In the vertical, s, direction in the Index Space, the univariate basis function in the cubic B-spline, $N_{0,3}(s)$, is defined by the 'local knot vector' $\{s_4, s_5, s_6, s_7, s_8\}$:

$$N_{0,3}(s) = N[s_4, s_5, s_6, s_7, s_8](s).$$

(3) The product, $T_{0,3}(r, s)$, of the two univariate basis functions gives the required T-spline bi-cubic basis function corresponding to the T-junction at knot **V**. This means that all of the T-knots within the dashed bounding square will contribute support to the basis function at knot **V**:

$$T_{0,3}(r, s) = N_{0,3}(r)N_{0,3}(s).$$

Likewise, Fig. 13.5 illustrates the basis function surface obtained from the 'local knot vector' associated with the stencil on the T-mesh.

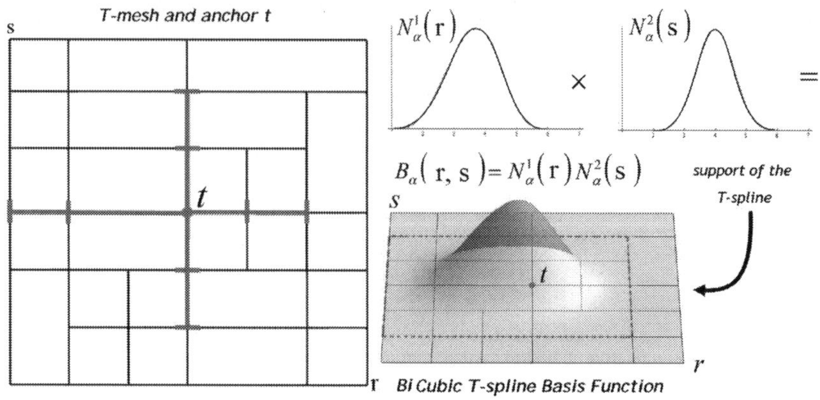

Figure 13.5. T-mesh knot basis functions are based on an Index Space stencil (Brovka *et al.*, 2014)

13.5 Comments

A T-mesh is basically a rectangular grid that allows T-junctions.

A T-spline polynomial is positive in a rectangular local domain 'anchored' to an interior knot or iElement center. Its first and second derivatives all vanish outside the rectangle and on its perimeter.

The T-spline rectangle is split into four by two index space lines passing through its anchor.

A T-mesh requires a †-shaped stencil at every 'anchor' to define the 'local knot vector' and the associated basis functions. A T-mesh usually contains L- or T-junctions.

The perimeter of the rectangular defined by the †-shaped stencil often has segments with reduced continuity.

T-spline elements are regions over which T-spline basis functions are C^∞ continuous.

For odd polynomials of degree p, the T-spline 'local knot vector' has $(p + 2)$ entries with the center being the knot in question.

Tensor product B-splines and NURBS do not allow efficient local refinement, but T-splines do.

Chapter 14

Bezier and Lagrange Extractions

14.1 Introduction

Following the original development of IGA and its iElements, there was still a desire to make IGA more compatible with the vast amount of existing FEA software. Ideally, it would have involved just invoking an optional FEA element subroutine that had the additional IGA features. That was a challenge because most FEA elements are only C^0 continuous between elements for any polynomial of degree p, whereas IGA basis functions automatically default to C^{p-1} continuity at interfaces of iElements (but not for adjacent patches). How can a C^0 continuous set of Bernstein polynomials be extracted from a set of C^{p-1} continuous NURBS and programmed into an optional FEA element subroutine to give C^{p-1} continuity in the enhanced FEA program with improved analysis results due to increased continuity? This question is rather mathematical in nature and reminds one of the expression 'One does not have to know how your stomach works to enjoy your dinner'. The 'how' to enhance existing FEA programs is given in great detail in Borden *et al.* (2011) for Bezier extraction leading to C^0 Bernstein polynomial FEA shape functions and in Schillinger *et al.* (2016) for extracting FEA Lagrange polynomial interpolation functions from the FEA Bernstein interpolation functions.

The results are not only enjoyable but are also spectacular and relatively easy to program. Consider Table 14.1 with the FEA, IGA iElements, and IGA iGe extracts. The right column shows that the iGe extracts meet or exceed FEA technology by every measure.

Table 14.1. Relative benefits of three technologies

Item	FEA	iElements	iGe in FEA
Unstructured mesh	Yes	No	Yes
Structured mesh	Yes	Yes	Yes
Part geometry	Very close fit	Exact	Exact
Nonhomogeneous material	Yes	In patches	Yes
Anisotropic material	Yes	In patches	Yes
Refineability	Great	Okay	Great
C^{p-1} continuity	No	Yes	Yes
Post solution graphics	Excellent	Poor	Excellent

The powerful Bézier decomposition uses repeated knot insertion operations, so the Bézier extraction operator is based on the formulas for new analysis control points formed from the original analysis control points when a knot is inserted into the knot vector. Knot insertion is adding a new knot into the existing knot vector without changing the shape of the curve (or the analysis solution values).

14.2 Knot Insertion

The Bézier decomposition process starts with the concept of inserting a new knot onto the analysis results curve (and/or onto a geometric B-spline curve) and into the analysis knot vector. To paraphrase Professor Shene's very useful web page, the meaning of 'knot insertion' is adding a new knot into the existing knot vector without changing the shape of the geometric curve or the analysis result curve. This new knot may be equal to an existing knot and increasing the multiplicity of that knot by one. Because of the fundamental equality between the number of control (analysis) points, n, the degree of the polynomial, p, and the length of a knot vector, $m = n + p + 1$, after adding a new knot, the value of m is increased by one and, consequently, either the number of control (analysis) points or the degree of the polynomial must be increased by one. Changing the degree of the curve due to adding a knot would undesirably change the shape of the curve (analysis results) globally and that will not be considered. Therefore, inserting a new knot will create a new analysis (control) point to be added. In fact, some existing control points are removed and replaced with new ones by corner cutting.

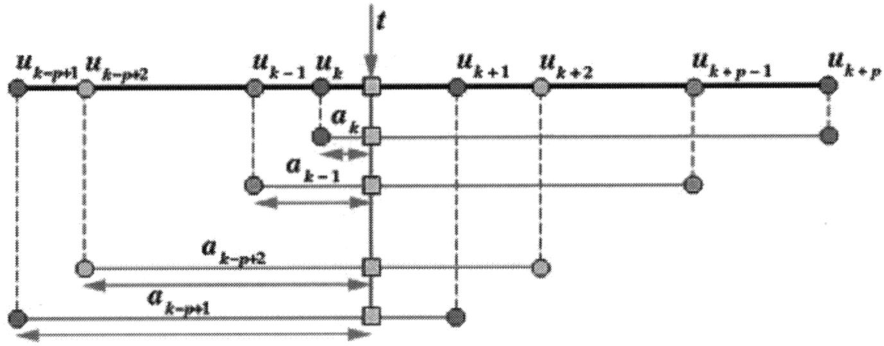

Figure 14.1. Inserting a new knot at $t > u_k$ and moving preceding knots

Although knot insertion may not look very interesting, it is one of the most important algorithms for IGA to FEA enhancement and for geometric enhancements of B-spline curves. The knot insertion calculations are described in the NURBS book (Piegl and Tiller, 1997) and numerous other studies. Clearly, there has to be a new set of control (analysis) points, **Q**, computed from the original data, **P**, and the revised knot vector must have its entries up to and including the inserted knot revised. The Shene website gives some useful images to clarify the two algebraic equations needed to complete a knot insertion at a parametric point, say $u = t$. One is shown as Fig. 14.1.

To paraphrase Shene [2016] again, to insert a new knot at $u = t$, the knot span $[u_k, u_{k+1})$ that bounds t must be located. With knot vector location k available, p new control points $Q_{k+p+1} \ldots Q_k$ are computed with new positions:

$$a_i = \frac{t - u_i}{u_{i+p} - u_i} \quad \text{for } k - p + 1 \leq i \leq k.$$

Finally, the original control polyline between points $\boldsymbol{P}_{k+1}$ and $\boldsymbol{P}_k$ is replaced with the new polyline coordinates (analysis values), $\boldsymbol{Q}_i$, defined by

$$\boldsymbol{Q}_i = (1 - a_i)\boldsymbol{P}_{i-1} - a_i\boldsymbol{P}_i.$$

A number of detailed examples of this process for planar curves are given in *The Nurbs Book* (see problem 51).

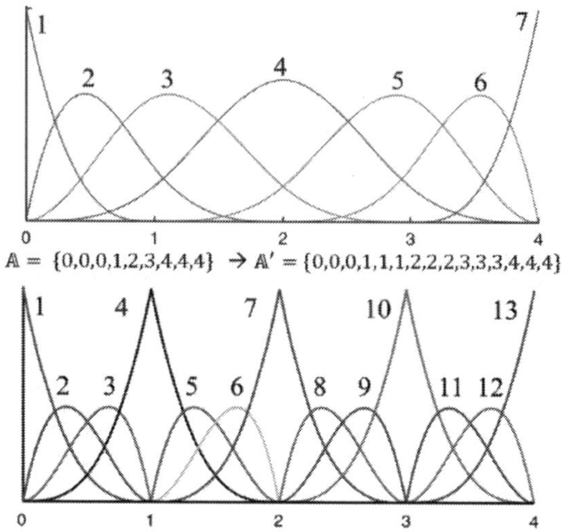

Figure 14.2. Inserting 3pairs of repeated knots decomposes a quadratic B-spline into 4 sets of C^0 quadratic Bernstein polynomials

14.3 Decomposition

As a specific example, let the uniform knot vector be $\mathbb{A}n = \{0, 0, 0, 0, 1, 2, 3, 4, 4, 4, 4\}$. It identifies a single-span cubic B-spline, $p = 3$, with seven APs shown at the top of Fig. 14.2. The interior knots are $\{1,\ 2,\ 3\}$. They are repeated $p - 1 = 2$ times, $\{1,\ 1,\ 2,\ 2,\ 3,\ 3\}$, and inserted into the original knot vector, without changing the analysis results (curve shape), giving $\mathbb{A}n' = \{0, 0, 0, 0, 1, 1, 1, 2, 2, 2, 3, 3, 4, 4, 4, 4\}$. Those four insertions are shown sequentially in Fig. 14.2 below the original cubic. The new basis functions have 13 analysis points and split (decompose) the original single $C^{p-1=2}$ span into four repeating spans with C^0 basis functions. Those four new knot spans have the same identical quadratic basis functions. This process also works with NURBS.

From the knot insertion process, the relation between the 13 Bernstein polynomials and the 7 basis functions can be written as

$$N_1 = 1B_1, \quad N_2 = 1B_2 + \frac{1}{2}B_3 + \frac{1}{4}B_4,$$

$$N_3 = \frac{1}{3}B_3 + \frac{7}{12}B_4 + \frac{2}{3}B_5 + \frac{1}{3}B_6 + \frac{1}{6}B_7,$$

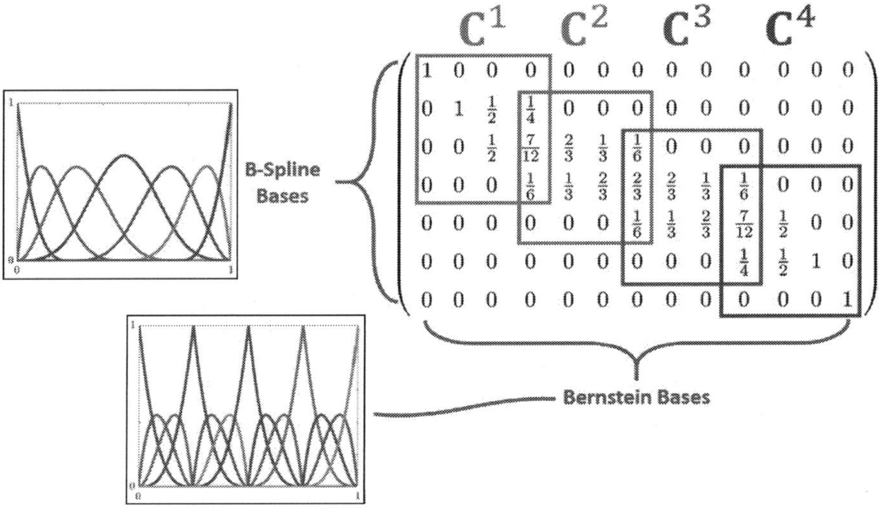

Figure 14.3. Concept of matrices extracting Bernstein polynomials from B-spline bases with an extraction matrix (Vernon, 2020)

$$
\begin{aligned}
N_4 &= \frac{1}{6}B_4 + \frac{1}{3}B_5 + \frac{2}{3}B_6 + \frac{2}{3}B_7 + \frac{1}{3}B_8 + \frac{1}{6}B_9,\\
N_5 &= \frac{1}{6}B_7 + \frac{1}{3}B_8 + \frac{2}{3}B_9 + \frac{7}{12}B_{10} + \frac{1}{2}B_{11},\\
N_6 &= \frac{1}{4}B_{10} + \frac{1}{2}B_{11} + 1B_{12},\\
N_7 &= 1B_{13}.
\end{aligned}
$$

These identities can be inserted into a 7×13 matrix, here denoted by $[\boldsymbol{C}^{\boldsymbol{G}}]$, which contains four 4×4 square matrices along its downward diagonal in Fig. 14.3. The 4×4 square matrices can be precomputed from the NURBS and stored as data for each iGe extract to be included in an FEA code as an element. For a given NURBS, the transpose of the 4×4 can multiply the NURBS basis function to yield the Bernstein polynomials to be used as basis functions for that particular iGe. Those small square matrices are called the Bezier extraction operator for each iGe.

Lagrange extractions are obtained from the Bezier extraction operator by additional manipulations. Lagrange extraction is less efficient that Bezier extraction, but it creates the common interpolation functions with which many FEA users are comfortable.

14.4 Comments

Knot insertion is adding a new knot into the existing knot vector without changing the shape of the curve.

The Bézier square extraction operator matrix for NURBS and T-splines, $\left[\boldsymbol{C}^G\right]$, decomposes their basis functions to the Bernstein polynomials.

The Bézier extraction operator is independent of control points and their basis functions and thus functions the same way for T-splines and NURBS.

The Bezier and Lagrange extraction operators insert extra knots into NURBS or T-spline meshes to decompose the NURBS or T-spline C^{p-1} basis functions into C^0 continuous polynomials over iGe extracts.

Each piecewise C^0 iGe extract is a segment of, and exactly reproduces, the original C^{p-1} analysis results, and the iGe are compatible with FEA data structures.

Bezier and Lagrange extraction results only require additional shape function routines in a finite element code!

An extraction operator, in conjunction with a Bernstein basis function or Lagrange interpolation subroutine, allows the implementation of one version of IGA as an enhanced FEA element-level technology.

When the smooth C^{p-1} 'iElement' basis functions are reduced to C^0 extracted Bernstein polynomials, they will herein be called an 'extract' or an 'iGe'.

To use an 'iGet' only requires a minor extension of the FEA database structure.

Bézier decomposition is a knot insertion operation, so the Bézier extraction operator is based on the formulas for new control points formed from the original control points when a knot is inserted into the knot vector.

Appendix

A1 Selected Analytic IGA Polynomials and Basis Functions

A1.1 *Bernstein polynomials*

B-splines are built with Bernstein polynomials.

$B_{j,p}(u)$ of degree p has $j = 1, 2, \ldots, (p+1)$ terms. In the following, $a = (1 - u)$:

$$
\begin{aligned}
&B_{1,1}(u) = a,\ B_{2,1}(u) = u,\\
&B_{1,2}(u) = a^2,\ B_{2,2}(u) = 2ua, \quad B_{3,2}(u) = u^2,\\
&B_{1,3}(u) = a^3,\ B_{2,3}(u) = 3ua^2, \quad B_{3,3}(u) = 3u^2a, \quad B_{4,3}(u) = u^3,\\
&B_{1,4}(u) = a^4,\ B_{2,4}(u) = 4ua^3, \quad B_{3,4}(u) = 6u^2a^2, \quad B_{4,4}(u) = 4u^3a, \ldots\\
&\qquad B_{4,5}(u) = u^4,\\
&B_{1,5}(u) = a^5,\ B_{2,5}(u) = 5ua^4, \quad B_{3,5}(u) = 10u^2a^3,\ B_{4,5}(u) = 10u^3a^2, \ldots\\
&\qquad B_{5,5}(u) = 5u^4a, \quad B_{5,6}(u) = u^5,\\
&B_{1,6}(u) = a^6,\ B_{2,6}(u) = 6ua^5, \quad B_{3,6}(u) = 15u^2a^4,\ B_{4,6}(u) = 20u^3a^3, \ldots\\
&\qquad B_{5,6}(u) = 15u^4a^2,\ B_{6,6}(u) = 6u^5a, \quad B_{7,6}(u) = u^6,
\end{aligned}
$$

etc., via Pascal's Triangle or the sequence: on [0, 1] as:

$$B_{j,p}(u) = \frac{1}{2}[(1 - u)B_{j,p-1}(u) + (1 + u)B_{j-1,p-1}(u)],$$

where $B_{1,0}(u) = 1$. The derivatives are $\frac{\partial B_{j,p}(u)}{\partial u} = \frac{p}{2}(B_{j-1,p-1}(u) - B_{j,p-1}(u))$.

A1.2 *B-spline curves*

$\sum_k N_k(u) \equiv 1$, and $\sum_k \partial N_k(u)/\partial u \equiv 0$ in each span.

Recursive relation for B-splines of degree p:

$$p = 0: \quad N_{j,0}(u) = \begin{cases} 1 & \text{if } u_j \leq u \leq u_{j+1} \\ 0 & \text{otherwise} \end{cases},$$

$$p = 1, 2, 3, \ldots \quad N_{j,p} = \frac{u - u_j}{u_{j+p} - u_j} N_{j,p-1}(u) + \frac{u_{j+1} - u}{u_{j+p+1} - u_{j+1}} N_{j+1,p-1}(u),$$

$$x(u)/\partial u = [\mathbf{N}(u)/\partial u]\{\mathbf{x}_{CP}\}, \quad \sum_j dN_j(u)/du = 0.$$

Recursive relation for derivatives of B-splines of degree p:

$$N'_{j,p} = \frac{p}{u_{j+p} - u_j} N_{j,p-1}(u) - \frac{p}{u_{j+p+1} - u_{j+1}} N_{j+1,p-1}(u),$$

$$\frac{d^k}{du^k} N'_{j,p} = \frac{p}{u_{j+p} - u_j} \left(\frac{d^{k-1}}{du^{k-1}} N'_{j,p-1}(u) \right) - \frac{p}{u_{j+p+1} - u_{j+1}} \left(\frac{d^{k-1}}{du^{k-1}} N_{j+1,p-1}(u) \right).$$

Both are implemented in Matlab algorithm *symbolic_N_bases.m.*

A1.2.1 *Analytic linear B-splines*

(a) Unit form: Knot Vector $\mathbb{A}u = \{0, 0, 1, 1\}$, Vector Length $= 4$, Degree $= 1$, Number of CPs $= 2$, Number of spans $= 1$ on [0,1) with CP 1-2. Iff CPx $=$ [0 1] L_x, then $|\mathbb{J}_{u:L}| = L_x$.

Count	Basis Functions $N(u)$	Derivatives $dN(u)/du$
1	$[(1-u)]$	$[-1]$
2	$[u]$	$[+1]$
Span	[0, 1)	

(b) Unit form, one center knot, vector $\mathbb{A}u = \{0, 0, 0.5, 1, 1\}$, Length $= 5$, Degree $= 1$, Number of CPs $= 3$, Number of spans $= 2$ on [0, 0.5), [0.5, 1) with CP 1–2, 2–3. Iff CPx $=$ [0 1/2 1]

L_x, then $|\mathbb{J}_{u:L}| = L_x/2$ on each span.

Count	Basis Functions $N(u)$	Derivatives $dN(u)/du$
1, 0	$[(1-2u), \quad 0]$	$[-2, \quad 0]$
2, 3	$[2u, \quad (2-2u)]$	$[2, \quad -2]$
0, 4	$[0, \quad (2u-1)]$	$[0, \quad 2]$
Span	[0, 0.5), [0.5, 1)	

(c) Unit form: knot vector $\mathbb{A}u = \{0, 0, 1/3, 2/3, 1, 1\}$, Length = 6, Degree = 1, Number of CPs = 4, Number of knot spans = 3 on [0, 1/3), [1/3, 2/3), [2/3, 1) with CP 1–2, 2–3, 3–4.

Count	Basis Functions $N(u)$	Derivatives $dN(u)/du$
1, 0, 0	$[(1-3u), \quad 0, \quad 0]$	$[-3, \quad 0, \quad 0]$
2, 3 ,0	$[3u, \quad (2-3u), \quad 0]$	$[3, \quad -3, \quad 0]$
0, 4, 5	$[0, \quad (3u-1), \quad (3-3u)]$	$[0, \quad 3, \quad -3]$
0, 0, 6	$[0, \quad 0, \quad (3u-2)]$	$[0, \quad 0, \quad 3]$
Span	[0, 1/3), [1/3, 2/3), [2/3, 1)	

(d) Integer form: Knot Vector $\mathbb{A}u = \{1, 1, 2, 2\}$, Knot Vector Length = 4, Degree = 1, Number of CPs = 2, Number of knot spans = 1 on [1, 2) with CP 1–2.

Count	Basis Functions $N(u)$	Derivatives $dN(u)/du$
1	$[(2-u)]$	$[-1]$
2	$[(u-1)]$	$[+1]$
Span	[1, 2)	

(e) Integer form, one center knot, knot vector $\mathbb{A}u = \{1, 1, 2, 3, 3\}$, Length = 5, Degree = 1, Number of CPs = 3, Number of spans = 2, on [1 2), [2 3) with CP 1–2, 2–3.

Count	Basis Functions $N(u)$	Derivatives $dN(u)/du$
1, 0	$[(2-u), \quad 0]$	$[-1, \quad 0]$
2, 3	$[(u-1), \quad (3-u)]$	$[1, \quad -1]$
0, 4	$[0, \quad (u-2)]$	$[0, \quad 1]$
Span	[1,2), [2, 3)	

(f) Integer form: two internal knots, vector $\mathbb{A}u = \{1, 1, 2, 3, 4, 4\}$, Length = 6, Degree = 1, Number of CPs = 4, Number of knot spans = 3 on [1, 2), [2, 3), [3 , 4) with CP 1–2, 2–3, 3–4.

Count	Basis Functions $N(u)$	Derivatives $dN(u)/du$
1, 0, 0	$[(2-u),\quad 0,\quad 0]$	$[-1,\quad 0,\quad 0]$
2, 3, 0	$[(u-1),\quad (3-u),\quad 0]$	$[1,\quad -1,\quad 0]$
0, 4, 5	$[0,\quad (u-2),\quad (4-u)]$	$[0,\quad 1,\quad -1]$
0, 0, 6	$[0,\quad 0,\quad (u-3)]$	$[0,\quad 0,\quad 1]$
Span	[1, 2), [2, 3), [3,4)	

A1.2.2 *Quadratic B-splines*

(a) Basic unit form: knot vector $\mathbb{A}u = \{0,0,0,1,1,1\}$, Length = 6, Degree = 2, Number of CPs = 3. Number spans = 1 on [0, 1) with CP 1-3. Iff CPx = [0 1/2 1] L_x, then $|\mathbb{J}_{u:L}| = L_x$.

Count	Basis Functions $N(u)$	Derivatives $dN(u)/du$
1	$[(u-1)\hat{}2]$	$[(2u-2)]$
2	$[2u(1-u)]$	$[(2-4u)]$
3	$[u\hat{}2]$	[2u]
Span	[0, 1)	

(b) One center knot unit form vector $Au = \{0,0,0,0.5,1,1,1\}$, Length = 7, Degree = 2, Number of CPs = 4, Number of spans = 2 on [0, 0.5), [0.5, 1) with CP 1-3, and 2-4. Iff CPx = [0 1/4 3/4 1] L_x, then $|\mathbb{J}_{u:L}| = L_x$.

Count	Basis Functions $N(u)$	Derivatives: $dN(u)/du$
1, 0	$[(2u-1)\hat{}2,\quad 0]$	$[8u-4,\quad 0]$
2, 4	$[-2u(3u-2), (2u-2)(u-1)]$	$[4-12u,\quad 4u-4]$
3, 5	$[2u\hat{}2,\quad -6u\hat{}2+8u-2]$	$[4u,\quad 8-12u]$
0, 6	$[0,\quad (2u-1)\hat{}2]$	$[0,\quad 8u-4]$
Span	[0, 0.5), [0.5, 1)	

Count	$d^2N(u)/du^2$
1, 0	$[8,\quad 0]$
2, 4	$[-12,\quad 4]$
3, 5	$[4,\quad -12]$
0, 6	$[0,\quad 8]$
Span	

(c) Interior third-point unit form, vector $\mathbb{A}u = \{0,0,0,1/3,2/3,1,1,1\}$, Length = 8, Degree = 2, Number of CPs = 5, Number of spans = 3 on [0, 1/3), [1/3, 2/3), [2/3, 1) with CP 1–3, 2–4, 3–5.

Count	Basis Functions $N(u)$
1, 0, 0	$[(3u-1)\hat{}2,\ 0,\ 0]$
2, 4, 0	$[-(3u(9u-4))/2,\ (3u-2)\hat{}2/2,\ 0]$
3, 5, 7	$[(9u\hat{}2)/2,\ -9u\hat{}2+9u-3/2,\ (3u-3)((3u)/2-3/2)]$
0, 6, 8	$[0,\ (3u-1)\hat{}2/2,\ 21u-(27u\hat{}2)/2-15/2]$
0, 0, 9	$[0,\ 0,\ (3u-2)\hat{}2]$
Span	$[0,1/3),\ [1/3,2/3),\ [2/3,1)$

Count	Derivatives: $dN(u)/du$
1, 0, 0	$[18u-6,\ 0,\ 0]$
2, 4, 0	$[6-27u,\ 9u-6,\ 0]$
3, 5, 7	$[9u,\ 9-8u,\ 9u-9]$
0, 6, 8	$[0,\ 9u-3,\ 21-27u]$
0, 0, 9	$[0,\ 0,\ 18u-12]$
Span	

(d) Repeated center knot unit form, vector $\mathbb{A}u = \{0,0,0,1/2,1/2,1,1,1\}$, Length = 8, Degree = 2, Number of CPs = 5, Number of spans = 2 on [0, 1/2), and [1/2, 1), with CP 1–3, 3–5.

Count	Basis Functions $N(u)$	Derivatives: $dN(u)/du$
1, 0	$[(1-2u)\hat{}2,\ 0]$	$[(-4+8u),\ 0]$
2, 4	$[4u(1-2u),\ 4(1-u)\hat{}2]$	$[(4-16u),\ (-8+8u)]$
3, 5	$[4u\hat{}2,\ 4(2u-1)(1-u)]$	$[8u,\ (12-16u)]$
0, 6	$[0,\ (2u-1)\hat{}2]$	$[0,\ (-4+8u)]$
Span	[0, 0.5), [0.5, 1)	

(e) Basic integer form: knot vector $\mathbb{A}u = \{1,1,1,2,2,2\}$, Length = 6, Degree = 2, Number of CPs = 3, Number knot spans = 1 on [0, 1) with CP 1–3.

Count	Basis Functions $N(u)$	Derivatives $dN(u)/du$
1	$[(u-2)\hat{}2]$	$[2u-4]$
2	$[-2(u-1)(u-2)]$	$[6-4u]$
3	$[(u-1)\hat{}2]$	$[2u-2]$
Span	[0, 1)	

(f) Center knot integer form, vector $\mathbb{A}u = \{1,1,1,2,3,3,3\}$, Length = 7, Degree = 2, Number of CPs = 4, Number of spans = 2 on [1, 2), [2, 3) with CP 1–3, 2–4.

Count	Basis Functions $N(u)$
1, 0	$[(u-2)\hat{}2,\ 0]$
2, 4	$[5u-(3u\hat{}2)/2-7/2,\ (u/2-3/2)(u-3)]$
3, 5	$[(u/2-1/2)(u-1),\ 7u-(3u\hat{}2)/2-15/2]$
0, 6	$[0, (u-2)\hat{}2]$
Span	[1, 2), [2, 3)

Count	Derivatives $dN(u)/du$
1, 0	$[2u-4, 0]$
2, 4	$[5-3u, u-3]$
3, 5	$[u-1, 7-3u]$
0, 6	$[0, 2u-4]$
Span	

(g) One center knot unit form vector $Au = \{0, 0, 0, 1, 2, 2, 2\}$, Length = 7, Degree = 2, Number of CPs = 4, Number of spans = 2 on [0, 1), [1,2) with CP 1–3, 2–4.

Count	Basis Functions $N(u)$
1, 0	$[(u-1)\hat{}2,\ 0]$
2, 4	$[-(u(3u-4))/2\ , (u/2-1)(u-2)]$
3, 5	$[u\hat{}2/2,\ 4u-(3u\hat{}2)/2-2]$
0, 6	$[0,\ (u-1)\hat{}2]$
Span	[0, 1), [1, 2)

Count	Derivatives: $dN(u)/du$
1, 0	$[2u-2,\ 0]$
2, 4	$[2-3u,\ u-2]$
3, 5	$[u,\ 4-3u]$
0, 6	$[0,\ 2u-2]$
Span	

(h) Two internal knots integer form: vector $\mathbb{A}u = \{1, 1, 1, 2, 3, 4, 4, 4\}$, Length = 8, Degree = 2, Number of CPs =5, Number of spans = 3 on [1, 2), [2, 3), [3, 4), with CP 1–3, 2–4, 3–5.

Count	Basis Functions $N(u)$
1, 0, 0	$[(u-2)\hat{}2,\ 0,\ 0]$
2, 4, 0	$[5u-(3u\hat{}2)/2-7/2,\ (u/2-3/2)(u-3),\ 0]$
3, 5, 7	$[(u/2-1/2)(u-1),\ -u\hat{}2+5u-11/2,$ $(u/2-2)(u-4)]$

0, 6, 8	$[0, (u/2-1)(u-2), 10u-3u\,\hat{}\,2)/2-16]$
0, 0, 9	$[0, 0, (u-3)\,\hat{}\,2]$
Span	[1, 2), [2, 3), [3, 4)

Count	Derivatives $dN(u)/du$
1, 0, 0	$[2u-4, 0, 0]$
2, 4, 0	$[5-3u, u-3, 0]$
3, 5, 7	$[u-1, 5-2u, u-4]$
0, 6, 8	$[0, u-2, 10-3u]$
0, 0, 9	$[0, 0, 2u-6]$
Span	

A1.2.3 *Cubic B-splines*

(a) Unit form: knot vector $\mathbb{A}u = \{0, 0, 0, 0, 1, 1, 1, 1\}$, Length = 8, Degree = 3, Number of CPs = 4, Number of knot spans = 1 on [0, 1) with CP 1–4.

Count	Basis Functions $N(u)$	Derivatives: $dN(u)/du$
1	$[-(u-1)\,\hat{}\,3]$	$[-3*(u-1)\,\hat{}\,2]$
2	$[3*u*(u-1)\,\hat{}\,2]$	$[9*u\,\hat{}\,2-12*u+3]$
3	$[-3*u\,\hat{}\,2*(u-1)]$	$[-3*u*(3*u-2)]$
4	$[u\,\hat{}\,3]$	$[3*u\,\hat{}\,2]$
Span	[0, 1)	

Count	$d^2N(u)/du^2$
1	$[6-6*u]$
2	$[18*u-12]$
3	$[6-18*u]$
4	$[6*u]$
Span	

(b) One center knot unit form, vector $\mathbb{A}u = \{0, 0, 0, 0, 0.5, 1, 1, 1, 1\}$, Length = 9, Polynomial Degree = 3, Number of CPs = 5, Number of knot spans = 2 on [0, 0.5) with CP 1–4 and [0.5, 1) with CP 2–5.

Count	Basis Functions $N(u)$
1, 0	$[-(2u-1)\,\hat{}\,3,\ \ 0]$
2, 5	$[2u(7u\,\hat{}\,2-9u+3),\ \ -(2u-2)(u-1)\,\hat{}\,2]$

Count	Basis Functions $N(u)$
3, 6	[$-2u^2(4u-3)$, $2(4u-1)(u-1)$^2]
4, 7	[$2u$^3, $-14u$^3 $+ 24u$^2 $- 12u + 2$]
0, 8	[0, $(2u-1)$^3]
Span	[0, 0.5) [0.5, 1)

Count	Derivatives: $dN(u)/du$	$d^2N(u)/du^2$
1, 0	[$-6(2u-1)$^2, 0]	[$24-48u$, 0]
2, 5	[$42u$^2 $- 36u + 6$, $-6(u-1)$^2]	[$84u-36$, $12-12u$]
3, 6	[$-12u(2u-1)$, $24u$^2 $- 36u + 12$]	[$12-48u$, $48u-36$]
4, 7	[$6u$^2, $-42u$^2 $+ 48u - 12$]	[$12u$, $48-84u$]
0, 8	[0, $6(2u-1)$^2]	[0, $48u-24$]
Span	[0, 0.5), [0.5, 1)	

(c) Interior third-point unit form, vector $\mathbb{A}u = \{0,0,0,0,1/3,2/3,1,1,1,1\}$, Length = 10, Degree = 3, Number of CPs = 6, Number of knot spans = 3 on [0, 1/3), [1/3, 2/3), [2/3, 1) with CP 1–4, 2–5, 3–6.

Basis Functions $N(u)$

[$-(3u-1)$^3,	0,	0]
[$(9u(21u$^2, $-18u+4))/4$,	$-(3u-2)$^3/4,	0]
[$-(9u^2(11u-6))/4$,	$(63u^3)/4 - 27u$^2 $+(27u)/2 - 3/2$,	$-(3u-3)((3u)/2$ $-3/2)(u-1)$]
[$9u^3)/2$,	$-(63u^3)/4 + (81u^2)/4$ $-(27u)/4 + 3/4$,	$(9(11u-5)(u-1)$ ^2$)/4$]
[0,	$(3u-1)$^3/4,	$-(189u^3)/4$ $+(405u^2)/4$ $-(279u)/4 + 63/4$]
[0,	0,	$(3u-2)$^3]
Span [0, 1/3)	[1/3, 2/3),	[2/3, 1)

Count	Derivative of Basis Functions $dN(u)/du$
1, 0, 0	[$-9(3u-1)$^2, 0, 0]
2, 5, 0	[$(567u^2)/4 - 81u + 9$, $-(9(3u-2)^2)/4$, 0]
3, 6, 9	[$-(27u(11u-4))/4$, $(189u^2)/4 - 54u + 27/2$, $-(27(u-1)^2)/2$]

Count	Derivative of Basis Functions $dN(u)/du$
4, 7, 10	[(27u^2)/2, (81u)/2 − (189u^2)/4 − 27/4, (297u^2)/4 − (243u)/2 + 189/4]
0, 8, 11	[0, (9(3u − 1)^2)/4, (405u)/2 − (567u^2)/4 − 279/4]
0, 9, 12	[0, 0, 9(3u − 2)^2]
Span	[0, 1/3) [1/3, 2/3), [2/3, 1)

Count	2nd Derivative of Basis Functions, $d^2N(u)/du^2$
1, 0, 0	[54 − 162u, 0, 0]
2, 5, 0	[(567u)/2 − 81, 27 − (81u)/2, 0]
3, 6, 9	[27 − (297u)/2, (189u)/2 − 54, 27 − 27u]
4, 7, 10	[27u, 81/2 − (189u)/2, (297u)/2 − 243/2]
0, 8, 11	[0, (81u)/2 − 27/2, 405/2 − (567u)/2]
0, 9, 12	[0, 0, 162u − 108]
Span	[0, 1/3), [1/3, 2/3), [2/3, 1)

(d) Interior fourth-point unit form, vector $\mathbb{A}u = \{\frac{1}{4}, \frac{2}{4}, \frac{3}{4}1, 1, 1, 1\}$, Length = 11, Degree = 3, Number of CPs = 7, Number of knot spans = 4 on [0, 0.25), [0.25, 0.5), [0.5, 0.75), [0.75, 1) with CP 1–4, 2–5, 3–6, 4–7.

Basis Functions *N(u)*

1	[−(4u − 1)^3,	0, . . .	
1		0,	0]
2	[4u(28u^2 − 18u + 3),	−(2u − 1)^2(4u − 2), . . .	
2		0,	0]
3	[−(8u^2(22u − 9))/3,	(112u^3)/3 − 48u^2	
3		+18u − 3/2, . . . (2u − 3/2)	
		(4u − 3)((4u)/3 − 1),	0]
4	[(32u^3)/3,	−32u^3 + 32u^2 − 8u	
4		+2/3, . . . 32u^3 − 64u^2	−(2u − 2)(4u − 4)
		+40u − 22/3,	((4u)/3 − 4/3)]
5	[0,	(4u − 1)^3/6, . . .	
5		−(112u^3)/3 + 64u^2	(8(22u − 13)
5		34u + 35/6,	(u − 1)^2)/3]
6	[0,	0, . . .	
6		(2u − 1)^2(4u − 2),	−112u^3 + 264u
			^2 − 204u + 52]
7	[0,	0, . . .	
7		0,	(4u − 3)^3]
I	Span [0, 0.25)	Span [0.25, 0.5), . . .	
I		Span (0.5, 0.75),	Span [0.75, 1)

$$\mathbb{A}u = \left\{\frac{1}{4}, \frac{2}{4}, \frac{3}{4}, 1, 1, 1, 1\right\}$$

Count	Derivative of Basis Functions $dN(u)$
1, 0, 0, 0	$[-12(4u-1)^\wedge 2, \quad 0, \quad 0, \quad 0]$
2, 5, 0, 0	$[336u^\wedge 2 - 144u + 12, -12(2u-1)^\wedge 2, \quad 0, \quad 0]$
3, 6, 9, 0	$[-16u(11u-3), \quad 112u^\wedge 2 - 96u + 18, \quad -2(4u-3)^\wedge 2, \quad 0]$
4, 7, 10, 13	$[32u^\wedge 2, \quad -96u^\wedge 2 + 64u - 8, \quad 96u^\wedge 2 - 128u + 40, \quad -32(u-1)^\wedge 2]$
0, 8, 11, 14	$[0, 2(4u-1)^\wedge 2, \quad -112u^\wedge 2 + 128u - 34, \quad 176u^\wedge 2 - 304u + 128]$
0, 0, 12, 15	$[0, \quad 0, \quad 12(2u-1)^\wedge 2, \quad -336u^\wedge 2 + 528u - 204]$
0, 0 , 0, 16	$[0, \quad 0, \quad 0, \quad 12(4u-3)^\wedge 2]$
Span	[0, 0.25), [0.25, 0.5), [0.5, 0.75), [0.75, 1)

Count	2nd Derivative of Basis Functions, $d2N(u)$
1, 0, 0, 0	$[96 - 384u, \quad 0, \quad 0, \quad 0]$
2, 5, 0, 0	$[672u - 144, \quad 48 - 96u, \quad 0, \quad 0]$
3, 6, 9, 0	$[48 - 352u, \quad 224u - 96, \quad 48 - 64u, \quad 0]$
4, 7, 10, 13	$[64u, \quad 64 - 192u, \quad 192u - 128, \quad 64 - 64u]$
0, 8, 11, 14	$[0, \quad 64u - 16, \quad 128 - 224u, \quad 352u - 304]$
0, 0, 12, 15	$[0, \quad 0, \quad 96u - 48, \quad 528 - 672u]$
0, 0, 0, 16	$[0, \quad 0, \quad 0, \quad 384u - 288]$
Span	[0, 0.25), [0.25, 0.5), [0.5, 0.75), [0.75, 1)

A1.2.4 *Analytic quartic B-splines*

(a) Basic unit form: knot vector $\mathbb{A}u = \{0, 0, 0, 0, 0, 1, 1, 1, 1, 1\}$, Length = 10, Degree = 4, Number of CPs = 5, Number of knot spans = 1 on [0, 1) with CP 1–5.

Count	Functions $N(u)$	Derivatives: $dN(u)/du$
1	$[(u-1)^\wedge 4]$	$[4(u-1)^\wedge 3]$
2	$[-4u(u-1)^\wedge 3]$	$[-4(4u-1)(u-1)^\wedge 2]$
3	$[6u^\wedge 2(u-1)^\wedge 2]$	$[12u(2u^\wedge 2 - 3u + 1)]$
4	$[-4u^\wedge 3(u-1)]$	$[-4u^\wedge 2(4u-3)]$
5	$[u^\wedge 4]$	$[4u^\wedge 3]$

Count	$d^2N(u)/du^2$
1	[12(u − 1)^2]
2	[−48u^2 + 72u − 24]
3	[72u^2 − 72u + 12]
4	[−24u(2u − 1)]
5	[12u^2]

(b) Third-point integer form, vector $\mathbb{A}u = \{0, 0, 0, 0, 0, \frac{1}{3}, \frac{2}{3}, 1, 1, 1, 1, 1\}$, Length = 12, Degree = 4, Number of CPs = 7, Number of spans = 3 on [0, 1/3), [1/3, 2/3), [2/3, 1) with CP 1–5, 2–6, 3–7.

Basis Functions $N(u)$

1	[(3u − 1)^4,	0,...
1		0]
2	[−(3u(405u^3 − 504u	(3u − 2)^4/8,...
2	^2 + 216u − 32))/8,	0]
3	[(9u^2(85u^2 − 88u + 24))/8,	−(207u^4)/8 + 63u^3 − 54u^2
3		+18u − 3/2,..., (3u − 3)((3u)/2
3		−3/2)(u − 1)^2]
4	[−(9u^3(13u − 8))/4,	(63u^4)/2 − 63u^3 + (81u^2)/
4		2 − 9u + 3/4,..., −(9(13u − 5)
4		(u − 1)^3)/4]
5	[(9u^4)/2,	−(207u^4)/8 + (81u^3)/2
5		−(81u^2)/4 + (9u)/2 − 3/8,...
5		(9(u − 1)^2(85u^2 − 82u + 21))/8]
6	[0,	(3u − 1)^4/8,...
6		−(1215u^4)/8 + (837u^3)/2
6		−(1701u^2)/4 + (381u)/2 − 255/8]
7	[0,	0,...
7		(3u − 2)^4]
I	Span [0, 1/3),	[1/3, 2/3), ...
I		[2/3, 1)

Derivative of Basis Functions $dN(u)/du$

1	[12(3u − 1)^3,	0,...
1		0]
2	[−(1215u^3)/2 + 567u^2	(3(3u − 2)^3)/2,...
2	−162u + 12,	0]
3	[(9u(85u^2 − 66u + 12))/2,	−(207u^3)/2 + 189u^2
3		−108u + 18,...
3		18(u − 1)^3]
4	[−9u^2(13u − 6),	126u^3 − 189u^2 + 81u − 9,...
4		−9(13u − 7)(u − 1)^2]

5	[18u ^3,	−(207u ^3)/2 + (243u ^2)/2
5		−(81u)/2 + 9/2, ..., (765u ^3)/2
5		−(1701u ^2)/2 + (1215u)/2 − 279/2]
6	[0,	(3(3u − 1) ^3)/2, ...
6		−(1215u ^3)/2 + (2511u ^2)/2
6		−(1701u)/2 + 381/2]
7	[0,	0,...
7		12(3u − 2) ^3]
I	[0, 1/3)	[1/3, 2/3) ...
I		[2/3, 1)

$$\mathbb{A}u = \left\{0, 0, 0, 0, 0, \frac{1}{3}, \frac{2}{3}, 1, 1, 1, 1, 1\right\}$$

I	2nd Derivative of Basis Functions $d^2N(u)/du^2$
1	[108(3u − 1) ^2, 0, 0]
2	[1134u − (3645u ^2)/2 − 162, (27(3u − 2) ^2)/2, 0]
3	[(2295u ^2)/2 − 594u + 54, 378u − (621u ^2)/2 − 108, 54(u − 1) ^2]
4	[−27u(13u − 4), 378u ^2 − 378u + 81, −351u ^2 + 594u − 243]
5	[54u ^2, 243u − (621u ^2)/2 − 81/2, (2295u ^2)/2 −1701u + 1215/2]
6	[0, (27(3u − 1) ^2)/2, 2511u − (3645u ^2)/2 − 1701/2]
7	[0, 0, 108(3u − 2) ^2]
Span	[0, 1/3), [1/3, 2/3), [2/3, 1)

Count	3nd Derivative $d^3N(u)/du^3$
1, 0, 0	[1944u − 648, 0, 0]
2, 6, 0	[1134 − 3645u, 243u − 162, 0]
3, 7, 11	[2295u − 594, 378 − 621u, 108u − 108]
4, 8, 12	[108 − 702u, 756u − 378, 594 − 702u]
5, 9, 13	[108u, 243 − 621u, 2295u − 1701]
0, 10, 14	[0, 243u − 81, 2511 − 3645u]
0, 0, 15	[0, 0, 1944u − 1296]
Span	[0, 1/3), [1/3, 2/3), [2/3, 1)

(c) Basic integer form knot vector, $\mathbb{A}u = \{1, 1, 1, 1, 1, 2, 2, 2, 2, 2\}$, Length = 10, Degree = 4, Number of CPs = 5, Number of knot spans = 1 on [1, 2) with CP 1–5.

Count	Functions $N(u)$	Derivatives: $dN(u)/du$
1	$[(u-2)\hat{}4]$	$[4(u-2)\hat{}3]$
2	$[-4(u-1)(u-2)\hat{}3]$	$[-4(4u-5)(u-2)\hat{}2]$
3	$[6(u-1)\hat{}2(u-2)\hat{}2]$	$[6(2u-2)(u-2)\hat{}2$ $+6(2u-4)(u-1)\hat{}2]$
4	$[-4(u-1)\hat{}3(u-2)]$	$[-4(4u-7)(u-1)\hat{}2]$
5	$[(u-1)\hat{}4]$	$[4(u-1)\hat{}3]$
Span	[1, 2)	

Count	$d^2N(u)/du^2$
1	$[12(u-2)\hat{}2]$
2	$[-48u\hat{}2+168u-144]$
3	$[72u\hat{}2-216u+156]$
4	$[-48u\hat{}2+120u-72]$
5	$[12(u-1)\hat{}2]$

(d) Center knot, integer form, vector $\mathbb{A}u = \{1,1,1,1,1,2,3,3,3,3,3\}$, Length = 11, Degree = 4, Number of CPs = 6, Number of knot spans = 2 on [1, 2), [2, 3) with CP 1–5, 2–6.

Count	Basis Functions *N(u)*
1, 0	$[(u-2)\hat{}4, 0]$
2, 6	$[-(15u\hat{}4)/8+(29u\hat{}3)/2-(165u\hat{}2)/4+(101u)/2$ $-175/8, (u-3)\hat{}4/8]$
3, 7	$[((u-1)\hat{}2(11u\hat{}2-54u+67))/8,$ $-((5u-7)(u-3)\hat{}3)/8]$
4, 8	$[-((5u-13)(u-1)\hat{}3)/8, ((u-3)\hat{}2(11u\hat{}2$ $-34u+27))/8]$
5, 9	$[(u-1)\hat{}4/8, -(15u\hat{}4)/8+(31u\hat{}3)/2$ $-(189u\hat{}2)/4+(127u)/2-255/8]$
0, 10	$[0, (u-2)\hat{}4]$
Span	[1, 2), [2, 3)

$$\mathbb{A}u = \{1,1,1,1,1,2,3,3,3,3,3\}$$

Count	Derivatives: $dN(u)/du$
1, 0	$[4(u-2)\hat{}3, 0]$
2, 6	$[-(15u\hat{}3)/2+(87u\hat{}2)/2-(165u)/2+101/2,$ $(u-3)\hat{}3/2]$
3, 7	$[(11u\hat{}3)/2-(57u\hat{}2)/2+(93u)/2-47/2,$ $-((5u-9)(u-3)\hat{}2)/2]$

Count	Derivatives: $dN(u)/du$
4, 8	$[-((5u-11)(u-1)\hat{}2)/2, (11u\hat{}3)/2-(75u\hat{}2)/2$ $+(165u)/2-117/2]$
5, 9	$[(u-1)\hat{}3/2, -(15u\hat{}3)/2+(93u\hat{}2)/2$ $-(189u)/2+127/2]$
0, 10	$[0, 4(u-2)\hat{}3]$
Span	[1, 2), [2, 3)

Count	Derivatives: $d^2N(u)/du^2$
1, 0	$[12(u-2)\hat{}2, 0]$
2, 6	$[87u-(45u\hat{}2)/2-165/2, (3(u-3)\hat{}2)/2]$
3, 7	$[(33u\hat{}2)/2-57u+93/2, 39u-(15u\hat{}2)/2-99/2]$
4, 8	$[21u-(15u\hat{}2)/2-27/2, (33u\hat{}2)/2-75u+165/2]$
5, 9	$[(3(u-1)\hat{}2)/2, 93u-(45u\hat{}2)/2-189/2]$
0, 10	$[0, 12(u-2)\hat{}2]$
Span	[1, 2), [2, 3)

A1.2.5 *Analytic quintic B-splines*

(a) Basic unit form: knot vector $\mathbb{A}u = \{0,0,0,0,0,0,1,1,1,1,1,1\}$, Length = 12, Degree = 5, Number of CPs = 6, Number of knot spans = 1 on [0, 1) with CP 1–6.

Count	Basis Functions $N(u)$	Derivatives: $dN(u)/du$
1	$[-(u-1)\hat{}5]$	$[-5(u-1)\hat{}4]$
2	$[5u(u-1)\hat{}4]$	$[5(5u-1)(u-1)\hat{}3]$
3	$[-10u\hat{}2(u-1)\hat{}3]$	$[-10u(5u-2)(u-1)\hat{}2]$
4	$[10u\hat{}3(u-1)\hat{}2]$	$[10u\hat{}2(5u\hat{}2-8u+3)]$
5	$[-5u\hat{}4(u-1)]$	$[-5u\hat{}3(5u-4)]$
6	$[u\hat{}5]$	$[5u\hat{}4]$
Span	[0, 1)	

Count	2nd Derivative $d^2N(u)/du^2$
1	$[-20(u-1)\hat{}3]$
2	$[20(5u-2)(u-1)\hat{}2]$
3	$[-200u\hat{}3+360u\hat{}2-180u+20]$
4	$[20u(10u\hat{}2-12u+3)]$
5	$[-20u\hat{}2(5u-3)]$
6	$[20u\hat{}3]$
Span	[0, 1)

Count	3nd Derivative $d^3N(u)/du^3$
1	$[-60(u-1)^\wedge 2]$
2	$[300u^\wedge 2 - 480u + 180]$
3	$[-600u^\wedge 2 + 720u - 180]$
4	$[600u^\wedge 2 - 480u + 60]$
5	$[-60u(5u-2)]$
6	$[60u^\wedge 2]$
Span	

(b) Center knot, unit form, vector $\mathbb{A}u = \{0\ 0\ 0\ 0\ 0\ 0\ 0.5\ 1\ 1\ 1\ 1\ 1\ 1\}$, Length = 13, Degree = 5, Number of CPs = 7, Number of knot spans = 2 on [0, 0.5), [0.5, 1) with CP 1–6, 2–7.

Count	Basis Functions $N(u)$
1, 0	$[-(2u-1)^\wedge 5, 0]$
2, 7	$[2u(31u^\wedge 4 - 75u^\wedge 3 + 70u^\wedge 2 - 30u + 5),$ $-(2u-2)(u-1)^\wedge 4]$
3, 8	$[-2u^\wedge 2(26u^\wedge 3 - 55u^\wedge 2 + 40u - 10),$ $2(6u-1)(u-1)^\wedge 4]$
4, 9	$[2u^\wedge 3(16u^\wedge 2 - 25u + 10),$ $-2(u-1)^\wedge 3(16u^\wedge 2 - 7u + 1)]$
5, 10	$[-2u^\wedge 4(6u-5), 2(u-1)^\wedge 2(26u^\wedge 3$ $-23u^\wedge 2 + 8u - 1)]$
6, 11	$[2u^\wedge 5, -62u^\wedge 5 + 160u^\wedge 4 - 160u^\wedge 3$ $+80u^\wedge 2 - 20u + 2]$
0, 12	$[0, (2u-1)^\wedge 5]$
Span	[0, 0.5), [0.5, 1)

Count	Derivative of Basis Functions $dN(u)/du$
1, 0	$[-10(2u-1)^\wedge 4, 0]$
2, 7	$[310u^\wedge 4 - 600u^\wedge 3 + 420u^\wedge 2 - 120u + 10,$ $-10(u-1)^\wedge 4]$
3, 8	$[-20u(13u^\wedge 3 - 22u^\wedge 2 + 12u - 2),$ $20(3u-1)(u-1)^\wedge 3]$
4, 9	$[20u^\wedge 2(8u^\wedge 2 - 10u + 3), -20(u-1)^\wedge 2$ $(8u^\wedge 2 - 6u + 1)]$
5, 10	$[-20u^\wedge 3(3u-2), 260u^\wedge 4 - 600u^\wedge 3$ $+480u^\wedge 2 - 160u + 20]$

Count	Derivative of Basis Functions $dN(u)/du$
6, 11	$[10u\hat{}4, -310u\hat{}4 + 640u\hat{}3 - 480u\hat{}2 + 160u - 20]$
0, 12	$[0, 10(2u-1)\hat{}4]$
Span	[0, 0.5), [0.5, 1)

$$\mathbb{A}u = \{0\,0\,0\,0\,0\,0\,0.5\,1\,1\,1\,1\,1\,1\}$$

Count	2nd Derivative $d^2N(u)/du^2$
1, 0	$[-80(2u-1)\hat{}3,\ 0]$
2, 7	$[1240u\hat{}3 - 1800u\hat{}2 + 840u - 120,$ $-40(u-1)\hat{}3]$
3, 8	$[-1040u\hat{}3 + 1320u\hat{}2 - 480u + 40,$ $120(2u-1)(u-1)\hat{}2]$
4, 9	$[40u(16u\hat{}2 - 15u + 3),$ $-640u\hat{}3 + 1320u\hat{}2 - 840u + 160]$
5, 10	$[-120u\hat{}2(2u-1),$ $1040u\hat{}3 - 1800u\hat{}2 + 960u - 160]$
6, 11	$[40u\hat{}3, -1240u\hat{}3 + 1920u\hat{}2 - 960u + 160]$
0, 12	$[0, 80(2u-1)\hat{}3]$
Span	[0, 0.5), [0.5, 1)

Count	3nd Derivative $d^3N(u)/du^3$
1, 0	$[-480(2u-1)\hat{}2,\ 0]$
2, 7	$[3720u\hat{}2 - 3600u + 840,\ \ -120(u-1)\hat{}2]$
3, 8	$[-3120u\hat{}2 + 2640u - 480,\ \ 720u\hat{}2 - 1200u + 480]$
4, 9	$[1920u\hat{}2 - 1200u + 120,\ \ -1920u\hat{}2 + 2640u - 840]$
5, 10	$[-240u(3u-1), 3120u\hat{}2 - 3600u + 960]$
6, 11	$[120u\hat{}2,\ \ -3720u\hat{}2 + 3840u - 960]$
0, 12	$[0,\ \ 480(2u-1)\hat{}2]$
Span	[0, 0.5), [0.5, 1)

(c) Basic integer form: vector $\mathbb{A}u = \{1, 1, 1, 1, 1, 1, 2, 2, 2, 2, 2, 2\}$, Length = 12, Degree = 5, Number of CPs = 6, Number of knot spans = 1 on [1, 2) with CP 1–6.

Count	Basis Functions $N(u)$
1	$[-(u-2)$^$5]$
2	$[5(u-1)(u-2)$^$4]$
3	$[-10(u-1)$^$2(u-2)$^$3]$
4	$[10(u-1)$^$3(u-2)$^$2]$
5	$[-5(u-1)$^$4(u-2)]$
6	$[(u-1)$^$5]$
Span	[1, 2)

Count	Derivatives: $dN(u)/du$
1	$[-5(u-2)$^$4]$
2	$[5(5u-6)(u-2)$^$3]$
3	$[-10(u-2)$^$2(5u$^$2-12u+7)]$
4	$[10(u-1)$^$2(5u$^$2-18u+16)]$
5	$[-5(5u-9)(u-1)$^$3]$
6	$[5(u-1)$^$4]$
Span	

Count	2nd Derivative $d^2N(u)/du^2$
1	$[-20(u-2)$^$3]$
2	$[20(5u-7)(u-2)$^$2]$
3	$[-200u$^$3+960u$^$2-1500u+760]$
4	$[200u$^$3-840u$^$2+1140u-500]$
5	$[-20(5u-8)(u-1)$^$2]$
6	$[20(u-1)$^$3]$
Span	[1, 2)

Count	3nd Derivative $d^3N(u)/du^3$
1	$[-60(u-2)$^$2]$
2	$[300u$^$2-1080u+960]$
3	$[-600u$^$2+1920u-1500]$
4	$[600u$^$2-1680u+1140]$
5	$[-300u$^$2+720u-420]$
6	$[60(u-1)$^$2]$
Span	

Algorithm A1.1. Symbolic evaluation of basis functions

```
function [N, dN, d2N, range]=symbolic_N_bases (V, opt)
% Symbolic Evaluation of Cox-Boor Recursion Definition
%     for Burnstein Basis Functions of a B-Spline
%  opt 0 arbitrary vector, opt 1 convert to unit form
% =======================================================
% if opt = 0 utilize an arbitrary knot vector like
% V = [t1, t2, t3, ..., tm] on t1 <= t <= tm, like
% V = [2 2 2 3 4 5 5 5] quadratic with 3 knot spans
%
% if opt = 1 convert the equivalent unit knot vector
% V = [0, ..,0, #, .., #, 1, ..,1] for 0 <= t <= 1
% so [2 2 2 3 4 5 5 5] becomes  [0 0 0 1/3 2/3 1 1 1]
%  before evaluating the associated basis functions
%                    Returned:
% N(t)   = Knot span basis functions,   including zeros
% dN(t)  = dN(t) / dt                 ,   including zeros
% d2N(t) = d2 N(t) / dt2              ,   including zeros
%    These allow for IGA use for PDEs up to order 4
%                      etc.
% P   = number of polynomial coefficients, (p+1)
% m   = length of knot vector
% p   = degree of polynomial, P - 1
% opt = 0 is the default, if not input
if (nargin == 1) ; opt = 0 ; end    ; % arbitrary vector
clc                        ; % clear command window, if used
syms u real              ; % define a real symbolic variable
%            Echo the input knot vector
disp ('Knot Vector') ; disp(V)      ; % echo knot vector
m = length(V);                     % Extract vector length
fprintf ('Knot Vector Length = %i \n', m) % list length
first = V(1)                               ; % first entry
last  = V(m)                            ; % last entry, >0
P = sum( V == first )           ; % Number of coefficients
if ( P < 1 ) ; error ('Knot Vector type invalid'); end;
```

```
p = P - 1                          ; % get polynomial degree
fprintf ('Polynomial Degree  = %i \n', p);% list degree
fprintf ('Number of span CP  = %i \n', P);% CP per span
fprintf ('Number of CP total = %i \n', m-P); % total CP
%          Convert to unit coordinate space
if ( opt > 0 )  ; % Convert to a unit range knot vector
  V = V - first                   ; % first entry is now zero
  V = V / V(m)                     ; % last entry is now one
  disp ('Converted knot vector'); disp (V)   ; % inform
end     ; % if convert to a unit coordinate knot vector

%        Calculate the number of knot spans
spans = 0; repeat = 0              ; % initialize span count
for i = 1:(m-1); % loop over knot vector ---> ---> --->
  if ( V(i) < V(i+1) )               ; % entry not repeated
    spans  = spans + 1             ; % increment span count
  else                            ; % possible interior knot
    repeat = repeat + 1     ; % increment repeated count
  end                                ; % if entry repeats
end ; % for all knot vector entries <--- <--- <--- <---
if ( repeat > 2*p )                  ; % need extra spans
  repeat = repeat - 2*p            ; % remove end repeats?
  fprintf ('Number of repeat knots = %i \n', repeat);
  spans = spans + repeat             ; % update spans count
end                     ; % if internal repeated knts found
fprintf ('Number of knot spans   = %i \n', spans);%list

for i = 1:spans  ; % loop over span ranges ........
  fprintf ('Span: %i [%g, %g) \n', i, V(i+p), V(i+p+1))
  range (i, 1:2) = [V(i+p), V(i+p+1)] ;
end                    ; % for span ranges ...........

%          Allocate symbolic 'N' array
for i = 1:P+spans  ; % Loop over bases   ---> ---> --->
  for j = 1:spans ; % Loop over span columns --> -->
    N(i,j*P) = u ; % Insert dummy symbolic character
  end                          ; % for span insert <-- <--
end  ; % end for basis number i <-- <--- <--- <--- <---
```

```
%          Calculate the basis functions 'N'
d = 1; e = P                              ; % initialize counts
for k = 1:spans   ; % Loop over each span ===> ===> ===>
  for i = 1:(P+spans) ; % Loop over rows --> --> -->
    if ( i==e ) ; N(i,d) = 1    ; % set row i, column d
    else ;        N(i,d) = 0     ; % set row i, column d
    end                          ; % if seed with 0 or 1
  end  ; % for row i <-- <-- <-- <-- <-- <-- <-- <--
  d = d + P ; e = e + 1      ; % increment column and row
end   ; % for all k knot spans <=== <=== <=== <=== <===
r = 0                                     ; % initialize counts
for k = 1:spans   ; % Loop over each span ===> ===> ===>
  for j = 2:P; % Loop over columns---> ---> ---> --->
    for i = 1:P+spans-1; % Rows loop --> --> --> -->
      left = 0 ; right = 0 ; denom = V(i+j-1) - V(i) ;
      if (denom ~= 0)           ; % avoid division by zero
        left = (u - V(i)) / denom ; end      ; % if left
      denom = V(i+j) - V(i+1)             ; % on right side
      if (denom ~= 0)           ; % avoid division by zero
        left = (u - V(i)) / denom ; end      ; % if left
      denom = V(i+j) - V(i+1)             ; % on right side
      if (denom ~= 0)           ; % avoid division by zero
        right = (V(i+j) - u) / denom ; end ; % if right
      % Recursive relations for span (j+r)
      N(i,j+r) = (left  * N(i,    j+r-1))  ...
               + (right * N(i+1, j+r-1))  ; % new bases
    end ; % for loop on i-th row <-- <-- <-- <-- <--
  end ; % for loop on j-th  column <--- <--- <--- <---
  r = r + P                    ; % increment column offset
end   ; % for all k knot spans <=== <=== <=== <=== <===

N(P+spans,:) = []          ; % remove extra workspace rows
% Delete spans workspace columns via vector subscripts
for j=1:spans*P ; I(1,j) = j ; end   ; % fill vector sub
for k = spans:-1:1 ; I(:,k*P) = []; end ; % cull vector
%      Output basis functions and derivatives
N(:,I) = []; N = simplify (N)  ; % simplify the algebra
disp ('Basis Functions N(u)'); disp(N)    ; % list bases
dN = diff (N, u); dN = simplify(dN) ;          % dN(u)/du
disp('Derivative of Basis Functions dN(u)'); disp(dN)
```

```
d2N = diff (dN, u); d2N = simplify(d2N)  ; % d2N(u)/du2
fprintf ('2nd Derivative of Basis Functions,d2N(u) \n')
disp(d2N)                                ; % finished

%  Validate sum(N) == 1, derivative sums ==0
disp ('Sum of span basis functions,    all must = 1'),
disp (simplify (sum (N)  ))                      % == 1 ?
disp ('Sum of span first derivatives,  all must = 0'),
disp (simplify (sum (dN) ))                      % == 0 ?
disp ('Sum of span second derivatives, all must = 0'),
disp (simplify (sum (d2N)))                      % == 0 ?
disp ('ranges '), disp(range)
disp ('Exit symbolic_N_bases')
% end symbolic_N_bases ================================
```

A2 Matrix Procedures for IGA

A2.1 *Vector subscripts*

A programming notation for arrays that is related to the use of vector subscripts is the colon ':' notation. In both Matlab and Fortran, using the colon as an array subscript means it refers to the full range of that subscript. For example, referring to Fig. A.1 and citing $\boldsymbol{u}(:)$ means $\boldsymbol{u}(1)$ through $\boldsymbol{u}(6)$. An array subscript, and a loop, can also be ranged in an incremental fashion from a beginning integer, B, to and ending integer, E, with an integer increment of I. In Matlab, the range is denoted by B:I:E; and in Fortran as B:E:I. For example, $\boldsymbol{u}(1:2:6)$ refers to the odd number of locations: $\boldsymbol{u}(1:2:6) \leftrightarrow [u_1\, u_3\, u_5]$.

Vector subscripts are efficient programming tools in isogeometric analysis (IGA). They are standard in the Fortran language and in the Matlab environment. A *vector subscript* is a named integer vector (one-dimensional array) of values which do not have to be in sequential order and can contain duplicate values. When the named array is placed in the row and/or column locators of any other array, the software treats them as an implied 'for loop' or 'do loop' that uses each integer in the vector subscript, in sequential order, as the range values for that loop over the rows and/or columns of the second array.

```
% extraction of coupled unknowns
% unknowns are D_u and R_k in
%  | S_uu  S_uk | {D_u} = {C_u}
%  | S_ku  S_kk | {D_k}   {C_k + R_k}

Free = [2 3 4]
Fixed = [1 5]
Order = [Free, Fixed]'
n_d = size(size(Order))
n_c = size(size(Fixed))

S   = zeros (n_d,n_d);
R_k = zeros (n_c, 1)
D   = zeros (n_d, 1);
C   = [500, 2000, 1000, 2000, 900]'
D_k = [150, 40]'
D(Fixed) = D_k ;

! coupled square matrix
S=[55, -46,   0,   0,   0 ; ...
  -46, 140, -46,   0,   0 ; ...
    4, -46, 110, -46,   4 ; ...
    0,   0, -46, 142, -46 ; ...
    0,   0,   4, -46,  65 ]

% non-singular partition
S_uu = S(Free, Free)
Det  = det(S_uu)

S_uk = S(Free, Fixed)
C_u  = C(Free)

% solution of unknowns
D_u = inv(S_uu)*(C_u - S_uk*D_k)

% reaction dara
C_k  = C(Fixed)
S_ku = S(Fixed, Free)
S_kk = S(Fixed, Fixed)

% reaction values
R_k  = S_ku*D_u +S_kk*D_k - C_k
```

```
Sample_BC_order
Free  = 2     3     4
Fixed = 1     5
Order = 2 3 4 1 5
n_d = 5
n_c = 2
C =         500  % coupled system
           2000
           1000
           2000
            900
D_k = 150
       40
S = 55    -46      0      0      0
   -46    140    -46      0      0
     4    -46    110    -46      4
     0      0    -46    142    -46
     0      0      4    -46     65
S_uu = 140    -46      0
       -46    110    -46
         0    -46    142
S_uk = -46      0
         4      4
         0    -46
C_u = 2000
      1000
      2000
C_k = 500
      900
D_k = 150
       40

D_u = 81.6802   % solution
      55.1137
      44.8960

S_ku = -46      0      0
         0      4    -46
S_kk = 55      0
        0     65
R_k = 3992.7    % reactions
      -144.8
```

Figure A.1. Matrix partitions using vector subscripts (Free and Fixed)

For example, consider a 4×5 array, $\boldsymbol{A}$, from which four specific non-sequential coefficients, included in angle brackets, are to be

$$\boldsymbol{A} = \begin{bmatrix} 1 & 2 & 3 & 4 & 5 \\ 6 & 7 & 8 & \langle 9 \rangle & \langle 10 \rangle \\ 11 & 12 & 13 & \langle 14 \rangle & \langle 15 \rangle \\ 16 & 17 & 18 & 19 & 20 \end{bmatrix}.$$

The row and column locations of those terms in matrix $\boldsymbol{A}$ can be identified with vector subscripts $\mathbf{Row} = [2\,3], \mathbf{Col} = [4\,5]$. In Fortran and Matlab, those vector subscripts create a new array containing those four terms:

$$\boldsymbol{B} = \boldsymbol{A}(\mathbf{Row}, \mathbf{Col}) = \begin{bmatrix} 9 & 10 \\ 14 & 15 \end{bmatrix}.$$

Now, assume those four terms are to be added (scattered) into the original matrix $\boldsymbol{A}$ in rows 3 and 4 and columns 1 and 3. A new set of vector subscripts is formed: $\mathbf{Row} = [3\,4], \mathbf{Col} = [1\,3]$, and used carry out the addition with a pair of implied loops:

$$\begin{aligned} \boldsymbol{A}(\mathbf{Row},\ \mathbf{Col}) &= \boldsymbol{A}(\mathbf{Row},\ \mathbf{Col}) + \boldsymbol{B}(:,:) \\ &= \begin{bmatrix} 1 & 2 & 3 & 4 & 5 \\ 6 & 7 & 8 & 9 & 10 \\ \langle 20 \rangle & 12 & \langle 23 \rangle & 14 & 15 \\ \langle 30 \rangle & 17 & \langle 33 \rangle & 19 & 20 \end{bmatrix}. \end{aligned}$$

The data needed for scattering the iElement arrays are in the connection lists for each iElement. Those numbers are the system mesh control point (CP). Usually, the system CP numbers are randomly assigned by an automatic mesh generator (and possibly reordered for more efficient storage and solution operations). The list of control points (CPs) define where each local iElement row is to be added into a row in the system equations. For the special case of a single unknown per AP ($n_g = 1$), such a direct equation subscript change defines the scatter destination row as

$$\text{Scatter\_row}(k) = \text{Element\_connection}\ (k), k = 1, \ldots, n_n.$$

For example, to find where the single k-th row term in an iElement source, $\boldsymbol{c}^e$, is to be added into the total system source vector, $\boldsymbol{c}$, the procedure is as follows. From the connection list for the system mesh, extract the row of CP connections for the current iElement. That row contains the list of integer CP numbers that are attached to the iElement. From that list, extract the kth system CP number, say K. Then, increment the system source sum by adding the kth of $\boldsymbol{c}^e$ to the kth row of $\boldsymbol{c}$:

$$\boldsymbol{c}(K) = \boldsymbol{c}(K) + \boldsymbol{c}^e(k).$$

That would add the effect of a single degree of freedom of that iElement into the system matrix. But, all of the iElement coefficients must be added into the system equations. As shown (in A3.1-1), this is accomplished by looping over all of the AP numbers for the current iElement. It would be done with a 'for-loop' in Matlab or a 'do-loop' in Fortran. However, programming a loop is not efficient, especially for parallel computers. Therefore, Fortran 90 and Matlab both offer a more efficient programming option known as a 'vector subscript'. The appearance of a vector subscript is an implied loop calculation to be executed in the most efficient way available in the computer being utilized.

Therefore, rather than write a loop over all the CPs in the iElement connection row, the row of integers is extracted from the input data and employed as a vector subscript. Here, that list is simply given the name of array 'row'. For example, scattering (adding) the iElement column matrix, $\boldsymbol{c}^e$, to the system column matrix, $\boldsymbol{c}$, is written in code as

$$\boldsymbol{c}(\mathbf{row}) = \boldsymbol{c}(\mathbf{row}) + \boldsymbol{c}^e.$$

But, this is actually executed as the implied loop:

```
% Scatter the iElement column matrix into the system column matrix
% Array 'row' contains the system dof numbers connected to
this iElement
 n_dof = size (row)                                % determine number of
                                                      iElement equations
for k = 1:n_dof                                      % loop over all
                                                         iElement dof rows
  c (row (k)) = c (row (k)) + c_e (k)                  % add iElement term
                                                         to system term
end                                                 % for k iElement
                                                       column matrix terms
```

In a similar fashion, the row subscripts and column subscripts are converted using the iElement connection lists. The column numbers are reordered exactly like the row numbers (col $\equiv$ row). In vector subscript notation, the assembly (scattering) of the iElement square matrix into the system square matrix is

$$\boldsymbol{S}(\mathbf{row}, \mathbf{row}) = \boldsymbol{S}(\mathbf{row}, \mathbf{row}) + \boldsymbol{S}^e.$$

Again, this is executed as two implied loops:

```
% Scatter the iElement square matrix into the system square matrix
% Array 'row' contains the system dof numbers connected to this
iElement
n_dof = size (row)                          % determine number of
                                              iElement dof
for m = 1:n_dof                               % loop over all
                                                iElement dof rows
    for n = 1:n_dof                         % loop over all iElement
                                              dof columns
        % add iElement term to the system term
        S (row (m), row (n)) = S (row (m), row (n)) + S_e (m, n)
    end                                     % for n iElement square
                                              matrix column terms
end                                          % for m iElement square
                                               matrix row terms
```

The above-mentioned assembly process is very simple to program in Fortran language and in the Matlab environment by using vector subscripts.

A2.2 *Partitioning the system equations*

In most courses on matrix algebra, the linear systems to be solved usually involve non-singular square matrices. This is because any constraints on the unknowns were satisfied in advance. In IGA applications to be covered here, the constraints are satisfied last, which results in an initially singular square matrix because the reactions due to the constraints appear as unknowns on the column vector side of the equations. In other words, because the essential boundary conditions (EBCs) will be enforced last, there is no a unique solution until that is done. A matrix partition can be used to allow for the solution constraints and to yield a smaller and non-singular system of equations to be solved. The algebraic system can be written in a general matrix form that more clearly defines what must be done to reduce the system to a solvable form by utilizing known essential boundary condition (EBC) values. The system dofs, u, and the full equations could always be rearranged in the following partitioned matrix form:

$$\begin{bmatrix} \mathbf{S_{uu}} & \mathbf{S_{uk}} \\ \mathbf{S_{ku}} & \mathbf{S_{kk}} \end{bmatrix} \begin{Bmatrix} \mathbf{u_u} \\ \mathbf{u_k} \end{Bmatrix} = \begin{Bmatrix} \mathbf{c_u} \\ \mathbf{c_k} + \mathbf{r_k} \end{Bmatrix}, \tag{A.1}$$

where u_u represents the unknown nodal parameters and u_k represents the known essential boundary values of the other parameters. The r_k term represents the fact that there are usually unknown generalized reactions associated with EBCs.

The only unknowns in this matrix system are the vectors $\mathbf{u_u}$ and $\mathbf{r_k}$. The net number of unknowns corresponds to the number of equations, but they must be rearranged before all the remaining unknowns can be computed. Moving the unknown reactions, $\mathbf{r_k}$, to the left side would destroy the very useful symmetry of the square matrix. So, instead, they are obtained as a post-solution calculation.

The sub-matrices $\mathbf{S_{uu}}$ and $\mathbf{S_{kk}}$ are square, whereas $\mathbf{S_{uk}}$ and $\mathbf{S_{ku}}$ are rectangular, in general. In an IGA formulation, all of the coefficients in the $\boldsymbol{S}$ and $\mathbf{c}$ matrices are known. This means that in general after the EBCs ($\mathbf{u_k}$) are prescribed, the remaining unknowns are $\mathbf{u_u}$ and $\mathbf{r_k}$. The unknown dofs are always calculated, but the calculation of the reaction unknowns is optional.

Here, for simplicity, it has been assumed that the equations have been numbered in a manner that places the prescribed parameters (EBCs) at the end of the system equations. The above-mentioned matrix equations can be partitioned as follows:

$$\begin{aligned} \boldsymbol{S_{uu}u_u} + \boldsymbol{S_{uk}u_k} &= \boldsymbol{c_u}, \\ \boldsymbol{S_{ku}u_u} + \boldsymbol{S_{kk}u_k} &= \boldsymbol{c_k} + \mathbf{r_k}, \end{aligned} \tag{A.2}$$

so that the unknown nodal parameters are obtained by inverting the non-singular square matrix $\mathbf{S_{uu}}$ in the top partitioned rows. That is,

$$\boldsymbol{u_u} = \boldsymbol{S_{uu}^{-1}}(\boldsymbol{c_u} - \boldsymbol{S_{uk}u_k}). \tag{A.3}$$

Most books on numerical analysis assume that you have reduced the system to the above-mentioned non-singular form where the effects of the essential conditions, $\boldsymbol{u_k}$, have already been moved to the right-hand side. Many authors use examples with null conditions $\boldsymbol{u_k} = 0$, so the solution is the simplest form, $\boldsymbol{u_u} = \boldsymbol{S_{uu}^{-1}c_u}$. If desired, the values of the necessary reactions, $\boldsymbol{r_k}$, can now be determined from

$$\boldsymbol{r_k} = \boldsymbol{S_{ku}u_u} + \boldsymbol{S_{kk}u_k} - \boldsymbol{c_k}. \tag{A.4}$$

In structural applications using line iElements like bars, beam, trusses, and frames, the reactions on each member are sometimes needed in the design of the member. They are the forces and/or moments that each member transmits at the interface APs. For the individual iElement to be in equilibrium with its now known stiffness, $\boldsymbol{S}^e$, known displacement vector, $\boldsymbol{u}^e$, known iElement source resultant vector, $\boldsymbol{c}^e$, and the unknown external reaction point forces and/or moments, $\boldsymbol{r}^e$, one requires

$$\boldsymbol{S}^e\boldsymbol{u}^e = \boldsymbol{c}^e + \boldsymbol{r}^e \to \boldsymbol{r}^e = \boldsymbol{S}^e\boldsymbol{u}^e - \boldsymbol{c}^e. \tag{A.5}$$

The iElement-level relation is probably easiest to envision for an iElement in the mesh that has no distributed external sources, $\boldsymbol{c}^e \equiv 0$, and then the member reactions are simply $\boldsymbol{r}^e = \boldsymbol{S}^e\boldsymbol{u}^e$ since, like with a linear spring, external forces are required to deform an elastic body.

In nonlinear and time-dependent applications, the reactions can be found from similar calculations. In most applications, the reaction data have physical meanings that are important in their own right or useful in validating the solution. However, this part of the calculations is optional. If one formulates an IGA simulation that satisfies the EBCs in advance, then the second row of the partitioned system $\boldsymbol{S}$ matrix is usually not generated and one cannot recover the reaction data directly from the matrix system.

The above-mentioned matrix partitioning can be accomplished easily with languages that allow vector subscripts. In several examples given later, the Matlab scripts will employ vector subscripts. Figure A.1 illustrates how vector subscripts can extract the above-mentioned partitions in a coupled matrix system and use them to solve for the unknowns and the reactions from the EBC. In that example, the nodal dofs with EBCs occur at the first (1) and last (5) APs, so those two numbers are defined as an integer vector subscript called ***Fixed***. All other dofs are unknown and are placed in the vector subscript ***Free***. The Matlab script is listed on the left side of the figure and the results of the script are on the right side.

A2.3 *Matrix assembly*

The extremely important assembly process (scattering) for combining iElement matrices to form the system matrices is independent of the physical application. The same process is used for fluid flow models, structural analysis, or any other fields of application. The reader does not have to know this material if relying exclusively on such codes. However, if the reader plans to solve small problems by hand, then this information is critical.

The scattering process depends on how many unknowns are associated with each AP in the domain and the iElement connection list that defines which CPs in the mesh are connected to each iElement. It is not necessary that every CP have the same number of unknowns (dofs), but the common case is used here for simplicity. The assembly, the enforcement of boundary conditions (BCs) and constraints, and the solution of the system matrices are easily automated and are hidden from the user of commercial software systems. However, to solve small problems by hand and to really understand the complete process, the reader should master these concepts.

IGA applications all lead to a system of linear algebraic equations to be solved for the set of unknowns. Often, a confusing aspect of this is how the small iElement matrices are assembled (scattered) into the much larger algebraic equations of the system. Unlike most examples included in a linear algebra course, most of the system equations resulting from an IGA study will usually be singular until the EBCs have been imposed on the algebraic system.

The scattering of the iElement matrices is mainly a problem in matrix algebra. The scattering operation could technically be done by using Boolean matrices built from the iElement connection list. However, the process is grossly inefficient from the computational point of view. Here, the critical process of assembling the system matrices is described in terms of the computationally efficient 'direct assembly' using vector subscripts, which is equivalent to the theoretical Boolean matrix process.

To illustrate the concepts of gathering and scattering with vector subscripts, refer to Fig. A.2 where a group of linear 1D iElements is shown along with the list of analysis CPs connected to each iElement. Note that the AP numbers below the iElements are not sequential. There is only one dof per AP, $n_g = 1$, and only one spatial coordinate per AP, $n_s = 1$. When constructing (or post-processing or plotting) the second iElement, its AP connection list is [3 5]. Thus, its physical space location is between AP 3 and 5, so the lower left part of the figure shows that it used that vector subscript to gather the global coordinate x_3 and insert it as the first local knot position in physical

Element	Nodes: 1	2
1	1	3
2	3	5
...	...	
5	6	4
6	4	2

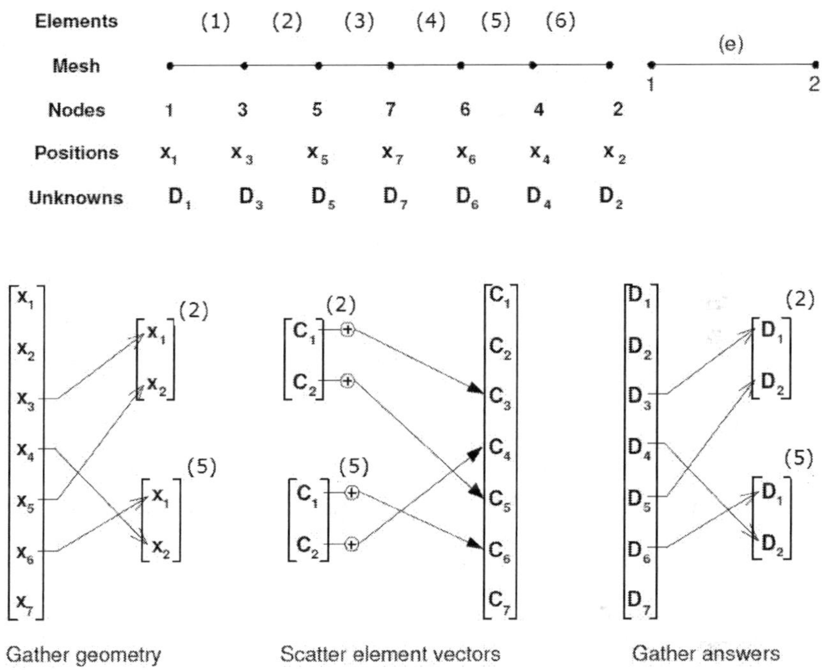

Figure A.2. Gathering and scattering data in a one-dimensional mesh ($n_g = 1$)

space $x_3 \to x_1^{e=1}$. Next, the fifth global x-coordinate, x_5, is gathered and inserted as the second local AP position of that iElement, $x_5 \to x_2^{e=1}$, and so on.

In the center section of the figure, the assembly of heat source resultants of the iElements is illustrated. For iElement number five, its AP connection list is [6 4]. Thus, its first local source resultant, $c_1^{e=5}$, is scattered (added) to the prior system values in the sixth row of the system resultants, $c_6 = c_6 + c_1^{e=5}$, and so on. Note that scatter always involves an addition, but a gather never does.

The right lower portion of that figure illustrates how the computer results at the AP must be gathered in order to compute the gradient

Table A.1. Relating local and system equation numbers

Local AP	dof number	System AP	iElement degree of freedom (dof) numbers: Local	System
I_L	J	$I_S = \text{el\_AP}(I_L)$	$n_g(I_L - 1) + J$	$n_g(I_S - 1) + J$
1	1	el_AP(1)	1	$n_g(\text{el\_AP}(1) - 1) + 1$
1	2	el_AP(1)	2	$n_g(\text{el\_AP}(1) - 1) + 2$
⋮	⋮	⋮	⋮	⋮
1	n_g	el_AP(1)	n_g	$n_g(\text{el\_AP}(1) - 1) + n_g$
2	1	el_AP(2)	$n_g + 1$	$n_g(\text{el\_AP}(2) - 1) + 1$
2	2	el_AP(2)	$n_g + 2$	$n_g(\text{el\_AP}(2) - 1) + 2$
⋮	⋮	⋮	⋮	⋮
k	j	el_AP(k)	$n_g(k - 1) + j$	$n_g(\text{el\_AP(k)} - 1) + j$
⋮	⋮	⋮	⋮	⋮
n_n	1	el_AP(n_n)	$n_g(n_n - 1) + 1$	$n_g(\text{el\_AP}(n_n) - 1) + 1$
⋮	⋮	⋮	⋮	⋮
n_n	n_g	el_AP(n_n)	$n_n * n_g$	$n_g(\text{el\_AP}(n_n) - 1) + n_g$

Notes: n_g = number of dofs per CP,
n_n = number of APs per iElement,
el_AP(k) = control point connection list for iElement number k.

and/or integral of the solution. In addition, the coordinates in the first column are gathered a second time to lost, or plot, where the gradient value is found.

The scattering (assembly) process is the same for all IGA applications and needs to be clearly understood. Sometimes, it is helpful to view the relation between the equation numbers for the iElement's dofs and the equation numbers for the system's dofs in a tabular format. Table A.1 gives such a table when the number of generalized dofs per CP, n_g, is constant. The process is only slightly changed if every analysis CP in the system is allowed to have a different number of dofs.

In practice, all the iElement array coefficients must be added (scattered) to rows and columns of the square matrix, $\boldsymbol{S}$, and source vector, $\boldsymbol{c}$, of the system equilibrium matrices:

$$\boldsymbol{S}\boldsymbol{u} = \boldsymbol{c} + \boldsymbol{c}_{\mathbf{NBC}}.$$

The system column matrix $\boldsymbol{c}$ is built up, from zero, by scattering (adding) the column matrices $\boldsymbol{c}^{\boldsymbol{e}}$ from all of the iElements in the mesh. Likewise, the system square matrix, $\boldsymbol{S}$, is built up, from zero, by scattering into it all of the iElement square matrices, $\boldsymbol{S}^e$. The paired set of numbers can also be used for other purposes. For example, each CP in the mesh and each local AP on an iElement have n_s physical spatial coordinates. So, setting $n_g = n_s$ yields the equations for gathering the coordinates of a single AP (where I is constant) or the physical coordinates of all the local APs on an iElement (for $i = 1 : n_n$). Likewise, some applications have a list of properties assigned to every CP. Then, the above-mentioned equations are easily converted to extract the nodal properties for an iElement.

This numbering system is illustrated in a one-dimensional case in Fig. A.2 where the system has six APs ($n_m = 6$) and each iElement has two APs ($n_n = 2$) each with a single generalized degree of freedom ($n_g = 1$) and where the CPs are numbered randomly. The connection lists in Fig. A.2 are used to gather the one-dimensional spatial coordinates ($n_s = 1$) for each iElement. But, the line iElement in the figure could just as easily have been a planar curve where both coordinates are gathered. For the two AP line iElements, the gathered coordinates are used to calculate some 2×1 source column vector for each iElement, $\boldsymbol{c}^e$. Each of the iElement column vectors is assembled (scattered) into the resultant 6×1 system column, $\boldsymbol{c}$.

A2.4 *Dirichlet conditions*

The actual partitioning of the full matrix is not actually necessary and neither is the use of vector subscripts. Alternate ways to assign a known value to a single degree of freedom, say $\boldsymbol{d_j}$, can be done with numerical manipulations. Since the number of EBCs is often a very small percentage of the number of unknowns, the enforcement of the EBC can be completed by column and row operations that modify both the square matrix and the column vector for each assigned EBC for vector $\boldsymbol{d}$, say $\boldsymbol{d_j} = \boldsymbol{b}$.

The operation first subtracts the assigned value, b times the j-th column of $\boldsymbol{S}$, and then zeros out that column in $\boldsymbol{S}$ along with the j-th row of both $\boldsymbol{S}$ and $\boldsymbol{c}$. At this point, the system is singular (non-unique) because the EBC is still not included in the square matrix. To enforce the EBC, include the EBC as a scalar identity, $d_j = b$, in $\boldsymbol{S}$ and $\boldsymbol{c}$. That is, place a 1 in the matrix diagonal location S_{jj} and in the column vector set $d_j = b$, so the size of the equations remains

$$\begin{array}{c} 1 \\ \vdots \\ j \\ \vdots \\ k \\ \vdots \\ p \end{array} \overset{\begin{array}{ccccccc} 1 & \cdots & j & \cdots & k & \cdots & p \end{array}}{\begin{bmatrix} S_{11} & \cdots & 0 & \cdots & S_{1k} & \cdots & S_{1p} \\ \vdots & \vdots & \vdots & \vdots & \vdots & \vdots & \vdots \\ 0 & \cdots & Big & \cdots & 0 & \cdots & 0 \\ \vdots & \vdots & \vdots & \vdots & \vdots & \vdots & \vdots \\ S_{k1} & \vdots & 0 & \vdots & S_{kk} & \vdots & S_{kp} \\ \vdots & & & & \vdots & & \\ S_{p1} & & & & S_{pk} & & S_{pp} \end{bmatrix}} \begin{Bmatrix} d_1 \\ \vdots \\ d_j \\ \vdots \\ d_k \\ \\ d_p \end{Bmatrix} = \begin{Bmatrix} c_1 - b\,S_{1j} \\ \vdots \\ Big\,b \\ \vdots \\ c_k - b\,S_{kj} \\ \\ c_p - b\,S_{pj} \end{Bmatrix}$$

Figure A.3. Matrix modifications to enforce a Dirichlet boundary condition

the same but the square matrix is rendered non-singular. For the final matrix form of the now non-singular matrix, see Fig. A.3. This approach is stored in the function *enforce_ essential_ BC.m* and/or the related penalty method in the Matlab function *enforce_ MPC_ equations.m.* If the reaction data, r_j, need to be recovered in post-processing, then the original j-th row of $\boldsymbol{S}$ and $\mathbf{c}$ must be saved for later use.

The above-mentioned series of steps is repeated for each and every EBC in the problem. The prior matrix partition approach is correct for any size problem but is more commonly used on very small problems, while the identity manipulation (or the penalty method) is used when there are thousands or millions of equations in the problem.

A2.5 *EBC by a penalty method*

The EBCs can be satisfied, to a selected number of significant figures, by a penalty method that also leaves the original size and equation order unchanged (but which does change the column heights of a sparse storage approach). Let $\boldsymbol{d_j}$ be the dof to be assigned an EBC, say $\boldsymbol{d_j} = \boldsymbol{b}$. A numerical trick to avoid partitions or zeroing out rows and columns is known as the penalty method. For an EBC, it changes only two terms in the governing matrix system. The j-th row of the resultant source vector is changed to

$$c_j = \mathbb{R}\, bS_{j,j}$$

and its diagonal term is changed to

$$S_{j,j} = \mathbb{R}\, S_{j,j},$$

where $\mathbb{R}$ is a really, really big number. Then, the row product of the system matrices becomes

$$s_{j,1}d_1 + \cdots + \mathbb{R}S_{j,j}d_j + S_{j,k}d_k + \cdots + S_{j,n}d_n = \mathbb{R}bS_{j,j}.$$

As the penalty term approaches infinity, $\mathbb{R} \to \infty$, the numerical solution approaches $d_j = b$ to a certain number of significant figures. The actual value of $\mathbb{R}$ to be assigned depends on the hardware word length (which engineering programming languages can display). Values of $\mathbb{R}$ as high as $\mathbb{R} = 10^{23}$ have been used, while the author uses a scaling of the maximum diagonal iElement of the square matrix, say $\mathbb{R} = 10^5 \times max[\boldsymbol{S}]$.

A2.6 *Multiple point constraints (MPCs)*

It is not uncommon for IGA simulations to have MPCs between analysis CP values. Often, the analysis slope has to be imposed. What in FEA would normally be an EBC applied to the slope becomes a constraint involving the displacement values at two APs. The above-mentioned penalty approach is also the easiest way to treat cases where two or more dofs are related by a linear constraint such as $A\,d_j + B\,d_k = C$ and $A\,d_j + B\,d_k + C\,d_m = D$, coupling three dofs, which occur frequently. Let the first dof be the master and the following ones be dependent dofs. Such a constraint equation between multiple rows (points) in the solution vector has long been called a 'Multiple Point Constraint' or MPC. Clearly, the first coefficient is not zero, so the equation is divided through by it to store as many coefficients as there are dofs in the constraint.

The constraint between two dofs is

$$\begin{aligned} A\,d_j + B\,d_k &= C, \\ A/A\,d_j + B/A\,d_k &= C/A, \\ d_j + b\,d_k &= c, \end{aligned}$$

where $A \neq 0$. So, its input equations become $d_j + b\,d_k = c$ where the new coefficients are $b = B/A$, $c = C/A$. Likewise, for a linear constraint between three dofs,

$A\,d_j + B\,d_k + C\,d_m = D$ has an input form of $d_j + b\,d_k + c\,d_m = d$, where $b = B/A, c = C/A$ and $d = D/A$. To input such a constraint equation, you would enter the constrained integer AP number and its integer degree of freedom number for each dof in the constraint, followed by the real values of the coefficients in the constraint.

Note that there is more than one dof in each equation. That is, there are more unknowns than equations. An approximate solution can be obtained from a least squares penalty model (or a singular decomposition algorithm). Either of the example constraints can be written as a null matrix dot product for a Dirichlet boundary condition (BC):

$$[1\ b]\begin{Bmatrix} d_j \\ d_k \end{Bmatrix} - c = 0.$$

For the Dirichlet form, the error is

$$[1\ b]\begin{Bmatrix} d_j \\ d_k \end{Bmatrix} - c \neq 0 = Err.$$

The error squared is

$$Err^T Err = \left(\begin{Bmatrix} d_j \\ d_k \end{Bmatrix}^T [1\ b]^T - c^T\right)\left([1\ b]\begin{Bmatrix} d_j \\ d_k \end{Bmatrix} - c\right),$$

$$Err^2 = [d_j\ d_k]\begin{Bmatrix} 1 \\ b \end{Bmatrix}[1\ b]\begin{Bmatrix} d_j \\ d_k \end{Bmatrix} - [d_j\ d_k]\begin{Bmatrix} 1 \\ b \end{Bmatrix} c$$

$$- c^T[1\ b]\begin{Bmatrix} d_j \\ d_k \end{Bmatrix} + c^T c,$$

$$Err^2 = [d_j\ d_k]\begin{bmatrix} 1 & b \\ b & b^2 \end{bmatrix}\begin{Bmatrix} d_j \\ d_k \end{Bmatrix} - 2[d_j\ d_k]\begin{Bmatrix} 1 \\ b \end{Bmatrix} c + c^T c,$$

$$Err^2 = \frac{1}{2}[d_j\ d_k]\left(\begin{bmatrix} 1 & b \\ b & b^2 \end{bmatrix}\begin{Bmatrix} d_j \\ d_k \end{Bmatrix} - \begin{Bmatrix} c \\ bc \end{Bmatrix}\right) + \frac{1}{2}c^T c.$$

To find the constrained dofs, $\boldsymbol{d^c}$, that minimize the error, solve for

$$\frac{\partial\, Err^2}{\partial\,\{\boldsymbol{d^c}\}} = \{0\},\ \frac{2}{2}\left(\begin{bmatrix} 1 & b \\ b & b^2 \end{bmatrix}\begin{Bmatrix} d_j \\ d_k \end{Bmatrix} - \begin{Bmatrix} c \\ bc \end{Bmatrix}\right) = \{0\}.$$

When multiplied by a really big penalty number, $\mathbb{R}$, these define a pseudo-iElement stiffness and load relation:

$$\boldsymbol{S^c d^c} = \boldsymbol{c^c}.$$

The matrices defined for the first form are

$$\boldsymbol{S^c} = \mathbb{R}\begin{bmatrix} 1 & b \\ b & b^2 \end{bmatrix}, \quad \boldsymbol{c^c} = \mathbb{R}\begin{Bmatrix} c \\ bc \end{Bmatrix}.$$

These matrix identities are treated as pseudo-iElement equations that are assembled into the system equations like actual iElements. Sometimes, you need two dofs to be identical, then $d_j + (-1)\, d_k = 0$. Often, that is called a rigid link and the constraint system is

$$\boldsymbol{S^c} = \mathbb{R}\begin{bmatrix} 1 & -1 \\ -1 & 1 \end{bmatrix}, \quad \boldsymbol{c^c} = \begin{Bmatrix} 0 \\ 0 \end{Bmatrix}.$$

The square matrix is similar to that of a bar or linear spring:

$$\boldsymbol{S^e} = \frac{\mathrm{EA}}{L}\begin{bmatrix} 1 & -1 \\ -1 & 1 \end{bmatrix},$$

where the term EA/L is known as the axial stiffness of a bar. Thus, the constraint equation is thought of as a pseudo-iElement, which acts as a very stiff spring connecting the two equal dofs. This is sometimes called a 'rigid link'.

The elastic modulus of a material, E, is usually a very large number. Without the introduction of $\mathbb{R}$, that almost always would lead to $\boldsymbol{S^e} \gg \boldsymbol{S^c}$ and the numerical constraint would not be accurately satisfied in the numerical solution of the assembled system matrix. The large penalty number scaling is needed such that the constrain equation pseudo-iElement has the proper units and is much larger than a physical iElement, $\boldsymbol{S^c} \gg \boldsymbol{S^e}$. This can be done by multiplying the constraint pseudo-iElement matrices by the huge number $\mathbb{R}$, and by arbitrarily defining $\mathbb{R} = 10^5 \times max\lceil \boldsymbol{S} \rfloor$ it is assured that the constraints have the proper units. Likewise, the constraint equation for the three degree of freedom example is

$$\boldsymbol{S^c} = \mathbb{R}\begin{bmatrix} 1 & b & c \\ b & b^2 & bc \\ c & cb & c^2 \end{bmatrix}, \quad \boldsymbol{c^c} = \mathbb{R}\begin{Bmatrix} d \\ bd \\ cd \end{Bmatrix}.$$

All MPC pseudo-iElement arrays have a connectivity list, and are scattered into the system arrays. The scatter is done last so that the maximum physical diagonal term, $max\lceil \boldsymbol{S} \rfloor$, is known. The penalty method of enforcing the constraints (after complete assembly of all physical iElements) on the system $\boldsymbol{S}$ and $\boldsymbol{c}$ arrays is carried out by the script *enforce_ MPC_ equations.m*.

The most common MPC occurs when a CP in a structure is required to move tangent to an inclined support surface or to have a known slope at one of the interpolation end analysis CPs. For a surface in three dimensions, the components of its unit vector are simply the direction cosines of the line perpendicular to the surface, and those direction cosines can be used in a constraint equation.

There are exact matrix manipulation methods for enforcing MPC relations. The script library can handle full matrix systems having the common type 2 MPC with the matrix manipulator *apply_mpc_type_2.m*. It gives more accurate results than the penalty approach, but for sparse storage systems, it tends to reduce the sparseness more than the penalty approach. This is not a problem when the system square matrix, **S**, is fully populated as in most of the examples herein.

Example A2.1 Given: The two IGA torsional shafts in Fig. A.4 are coupled by two gears. The IGA unknowns are rotations at APs 1 through 4. The shaft and gear rotation at AP 2 is θ_2 and the gear has a radius of 5 inches. The gear engages a smaller gear with a 1-inch radius. The smaller gear and shaft AP 3 have a rotation of θ_3. Determine the constraint equation between the shaft iElement rotations at APs 2 and 3.

Solution: A point on the circumference of the large gear moves through an arc length of $L = \theta_2 R_2$. A point on the circumference of the smaller gear moves through the same arc length $L = \theta_3 R_3$, but with the rotation being in the opposite direction. Therefore, the constraint equation between the two rotations at the APs coupled by those gears is

$$5\theta_2 + 1\theta_3 = 0.$$

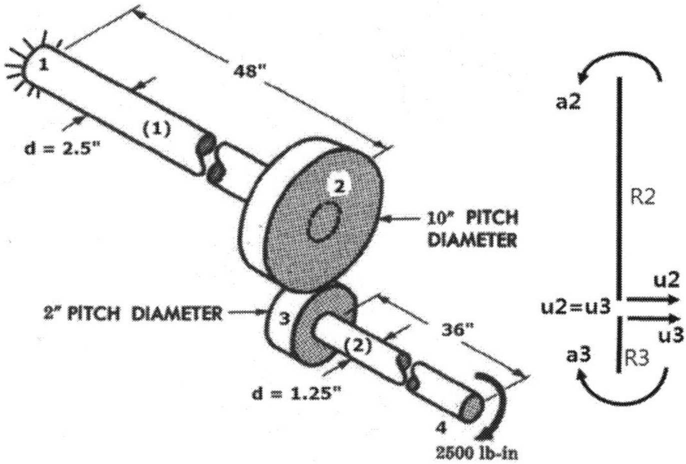

Figure A.4. Torsional rotations of coupled shafts

A2.7 *Equation factorization*

Most of the examples given herein imply that the solution of the equilibrium equations (after enforcing the BCs), $\mathbf{S}\,\mathbf{u} = \mathbf{c}$, will be obtained by inverting the system square matrix and multiplying the inverted matrix times the solution source vector, $\mathbf{c}$, to obtain the system solution vector, $\mathbf{u} = \mathbf{S}^{-1}\mathbf{c}$. While that is theoretically a correct matrix algebra procedure, it is not practical for more than three unknowns. A more efficient solution process is to factor the system square matrix into the product of a lower triangular matrix, L, and a upper triangular matrix, U, as

$$\mathbf{S} \equiv \mathbf{L}\,\mathbf{U},$$

or as a triple matrix product using a diagonal matrix, $\mathbf{S} \equiv \mathbf{L}\,\mathbf{D}\,\mathbf{U}$, where usually $\mathbf{U} = \mathbf{L}^{\mathbf{T}}$.

A factorization solution algorithm greatly reduces the operations count and the required storage. One important issue is that the finite iElement square matrix is sparse and diagonally dominant, but its inverse matrix is fully populated. This contrasts with a factorization process where the matrix factors of S have the same sparsity and can be stored within the original matrix and thus require no additional storage (when the factorizations are programmed efficiently, unlike the Matlab *lu* function).

Example A2.2 Given: Consider the sparse matrix $\boldsymbol{S}$ below. Use the Matlab function *inv* (S) to form the inverse and the function *lu*($\boldsymbol{S}$), the two triangular factors, and compare the storage requirements. Verify that $\boldsymbol{S} \equiv \boldsymbol{L}\,\boldsymbol{U}$.

$$\boldsymbol{S} = \begin{bmatrix} 2 & 1 & 0 & 0 & 0 \\ 1 & 2 & 1 & 0 & 0 \\ 0 & 1 & 2 & 1 & 0 \\ 0 & 0 & 1 & 2 & 1 \\ 0 & 0 & 0 & 1 & 2 \end{bmatrix}.$$

Solution: Its inverse matrix is full populated:

$$inv\,(\mathrm{S}) \rightarrow \boldsymbol{S}^{-1} = \frac{1}{6}\begin{bmatrix} 5 & -4 & 3 & -2 & 1 \\ -4 & 8 & -6 & 4 & -2 \\ 3 & -6 & 9 & -6 & 3 \\ -2 & 4 & -6 & 8 & -4 \\ 1 & -2 & 3 & -4 & 5 \end{bmatrix},$$

but the two factors are sparse (banded), and in some cases can be stored in (overwrite) the original matrix:

$$\boldsymbol{L} = \begin{bmatrix} 1 & 0 & 0 & 0 & 0 \\ 1/2 & 1 & 0 & 0 & 0 \\ 0 & 2/3 & 1 & 0 & 0 \\ 0 & 0 & 3/4 & 1 & 0 \\ 0 & 0 & 0 & 4/5 & 1 \end{bmatrix}, \quad \boldsymbol{U} = \begin{bmatrix} 2 & 1 & 0 & 0 & 0 \\ 0 & 3/2 & 1 & 0 & 0 \\ 0 & 0 & 4/3 & 1 & 0 \\ 0 & 0 & 0 & 5/4 & 1 \\ 0 & 0 & 0 & 0 & 6/5 \end{bmatrix}.$$

Evaluating the Matlab command $\boldsymbol{S} - \boldsymbol{L}\,\boldsymbol{U}$ yields a null matrix as expected.

In practical IGA, the size of the equations is so large that a matrix inversion is not practical and a factorization of the equations or an iterative solution is used. Factorization is especially efficient when the square matrix is symmetric and positive definite. Then, the Crout Factorization, $\boldsymbol{S} \equiv \boldsymbol{L}\,\boldsymbol{D}\,\boldsymbol{U}$, is very popular for large finite iElement systems. The Crout factorization is used herein with the sky line

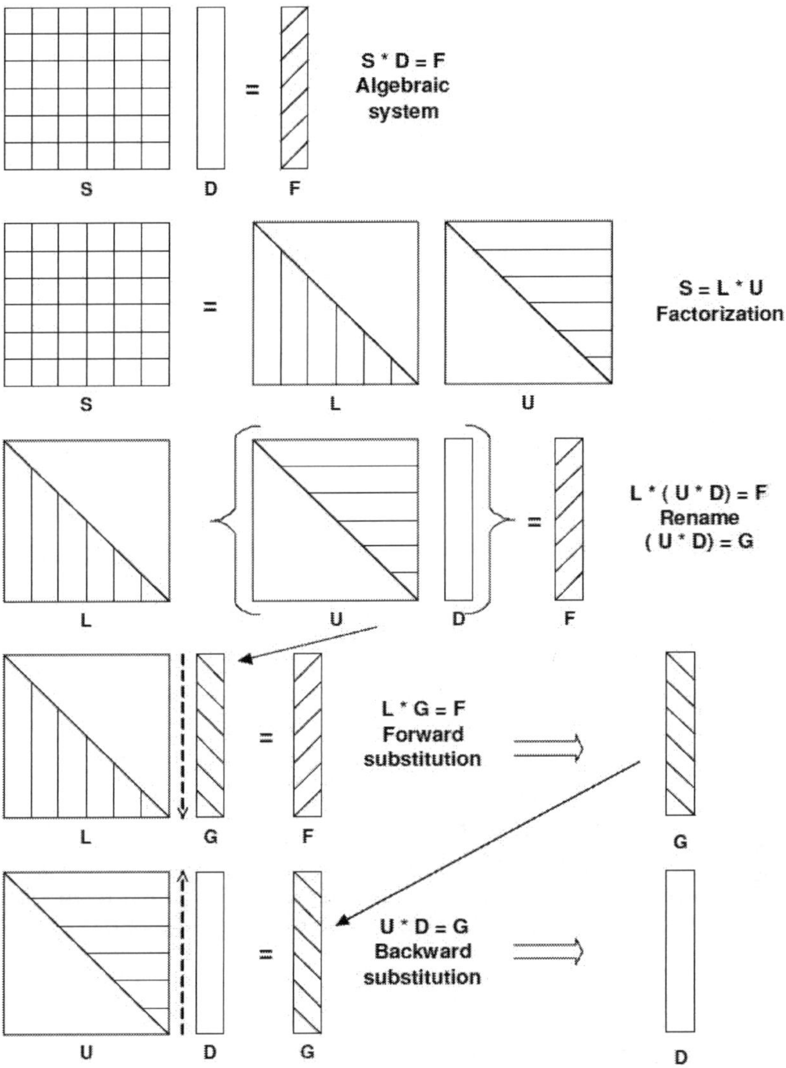

Figure A.5. Utilizing matrix factorization instead of matrix inversion

sparse matrix storage method discussed in the next section. The process for factoring a square matrix into an upper and lower triangle product is sketched in Fig. A.5. After the factorization, a forward substitution loop and a backward substitution loop are run. They each involve only one equation and one unknown when solving for the intermediate dummy unknown and the actual unknown in each row of the matrix system. When the matrix size gets extremely large, an iterative solver, like a pre-conditioned conjugant gradient method, is utilized.

A3 Common Tools of Computer-Aided Geometric Design

A3.1 *Introduction*

Computer-Aided Geometric Design (CAGD) consists of several technologies including Boundary Representations (BRep), Constructive Solid Geometry (CSG), Cells (or Trees), Tessellated Models (STL), Homogeneous Coordinates (HC, or Projection Coordinates), Automatic Mesh Generation (AMG) for FEA, and Interpolation Functions for FEA. Each of these utilizes a different database format and creates different 'exchange file' formats to be utilized by external software systems.

CAGD, and its several terminologies, has evolved into a common tool for modeling solids and surfaces for manufacturing and analysis. Solid modeling allows the designers and analysts to see exactly what has been created. A goal behind IGA procedures is to integrate most aspects of geometric design, mass properties, application analyses, 'associative data' for manufacturing, and documentation tasks. These IGA processes usually begin with CAGD and one or more of its internal technologies. CAGD started with the 'wireframe' display of every edge of a solid (see top of Fig. A.6) but today hidden lines are usually suppressed).

A3.2 *Boundary representations (BRep)*

The (often confusing) wireframe display ability was quickly followed by the boundary representation technology, which is based on Euler's (1758) constant for solid polygons. Euler verified that for certain polyhedra, the number of vertices, V, minus the number of edges, E, and plus the number of faces, F, is a constant number, χ, defined by $V - E + F = \chi$. For a convex polyhedron, $\chi = 2$.

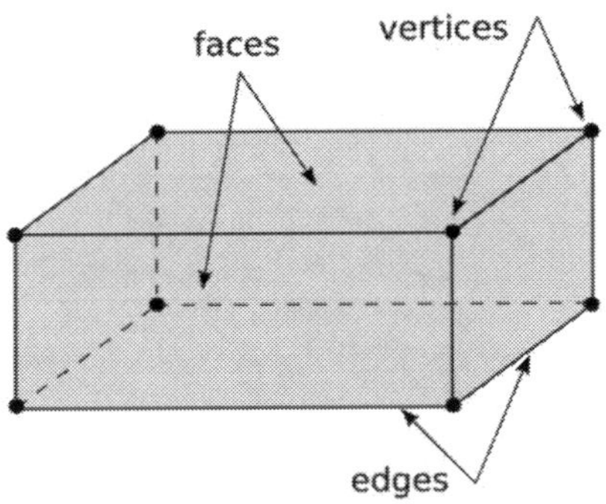

Figure A.6. Boundary representation items counted in Euler's constant

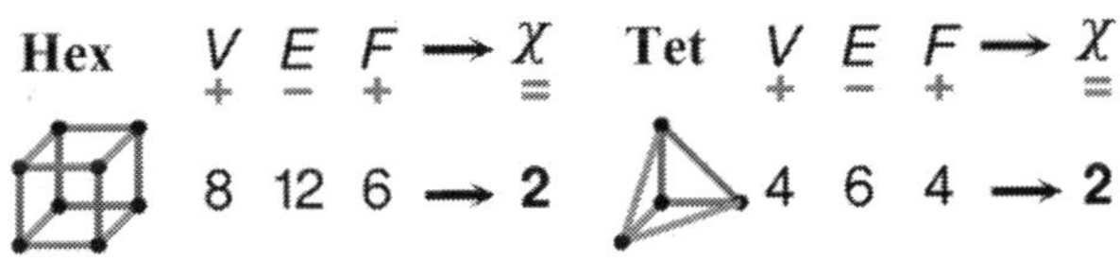

Figure A.7. Euler's Number governs combinations of Hexs or Tets

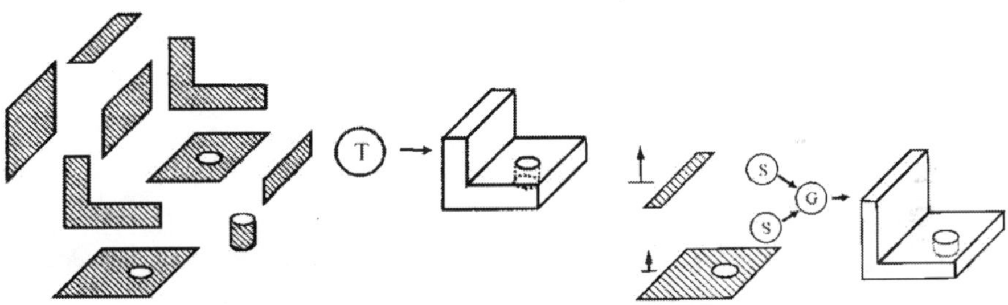

Figure A.8. BRep uses topological rules or sweep (S) and glue (G) to form solids

For early FEA studies in 3D, the shapes in demand were hexahedra and tetrahedra that both satisfied Euler's Number, as shown in Fig. A.7. They continue to satisfy when they are mapped to have curved edges. A solid made from an assemblage of either shape will also be valid (watertight) if the assemblage satisfies the rule. The Boundary Representation (BRep) of any solid surface, seen in Figs. A.6 and A.8, confirms that it is water-tight.

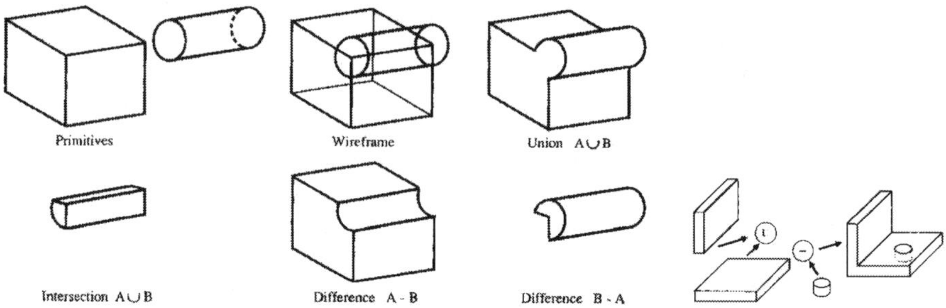

Figure A.9. Primitive solid shapes via Boolean operations form new solids

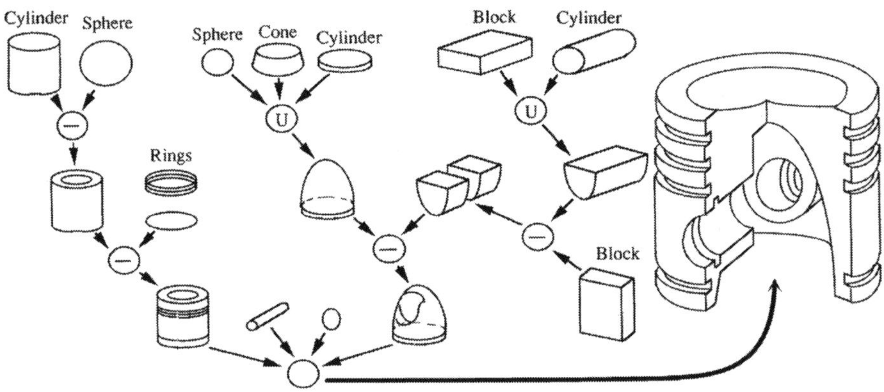

Figure A.10. CSG uses common primitives to build engineered parts

A3.3 *Constructive solid geometry (CSG)*

The constructive solid geometry approach uses a tree-like data stricture where the leaves of the tree are primitive solids. The nodes of the branches denote a Boolean operation applied to branches, A and B to form a new solid, C. The available options are:

1. Intersection, solid C contains all parts common to A and B.
2. Union, solid C contains all points in either A or B.
3. Difference, solid C contains the remainder of A after all points common to A and B are removed from A.

Figures A.9 and A.10 show how CSG can create complex solids from primitive solids.

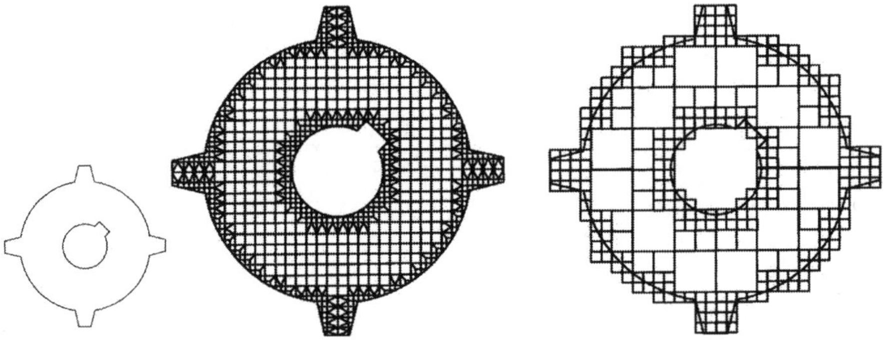

Figure A.11. Uniform (middle) and quadtree mesh generation

A3.4 *Cells and analysis trees*

Both IGA and FEA must eventually have a mesh of patch sub-regions or elements. In IGA, the NURBS-mesh and T-mesh descriptions do not always begin with a CAGD system. For example, some IGA and FEA meshes, and their associated basis functions, are derived from a 'point cloud'. Early attempts to automatically generate a mesh from the perimeter of a part meant dividing the interior with a uniform grid of cells, deleting the outside cells, and splitting the smallest cells intersected by the perimeter, as shown in the center image of Fig. A.11. Then, one had to start with a single square cell bounding the perimeter and search through a loop where the edge length of all cells on the boundary had their edge lengths cut in half before going into the next refinement loop (right image of Fig. A.11). The looping process continues until all cells are full or empty, or until the maximum level number is reached.

The sub-division or refinement process defines a 'quadtree' which makes the next layer of cells a factor of four (quad) smaller. The number of levels needed for an accurate location of the perimeter, and the calculation of an area's geometric properties, is small. For example, for 16 layers of refinement, the area of the final square is 65,536 times smaller than the original area!

The data structures for representing these classes of forms in 1D, 2D, and 3D are called 'bintrees', quadtrees, and octrees, respectively. (Engineers might call the 'graphs' of such data structures upside-down trees where the starting cell is considered to be the ground.) The graphs of each level of the tree usually use a full black circle to

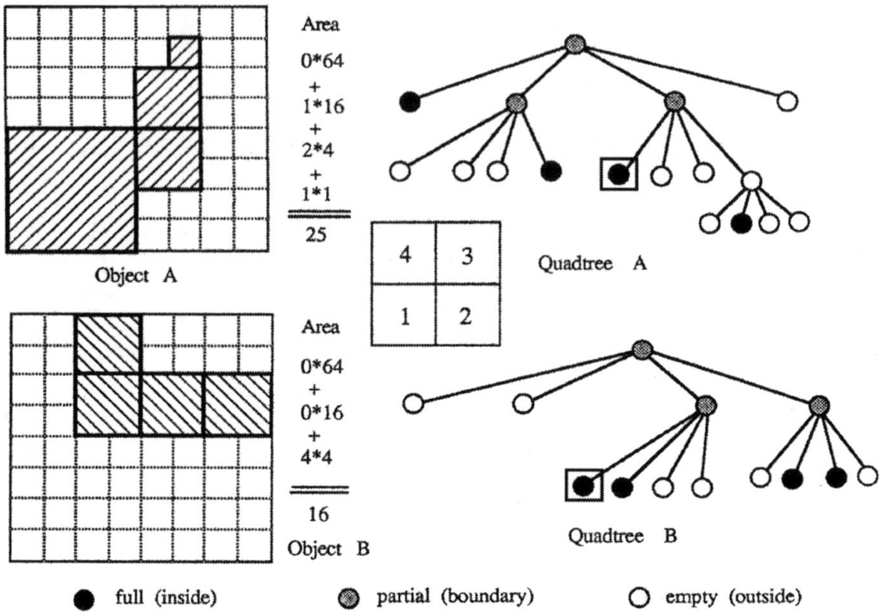

Figure A.12. Two objects with shapes and areas defined by quadtrees

denote a cell that is completely full, an empty circle to represent a completely empty cell, and gray (or intermediate shade) to represent a cell intersected by the boundary and thus only partly filled. In the notation used here, the four smaller cells are numbered counter-clockwise starting with the one in the lower-left corner (LLC).

Hidden from the user, CAGD systems switch back and forth between representations of a part. While the default output will be T-spline, NURBS, etc., quadtrees and octrees are often used, especially for finding the intersection curve of two surfaces, calculating geometric properties, and executing Boolean operations. To illustrate these points, Fig. A.12 shows the quadtree structures of two objects and how the area of each is quickly and easily computed since the cell area at one level is one fourth the area in the previous level. Likewise, since the center of each cell is easily found, their scalar 'first moments' can be summed to give the total first moment, say f_x, and the object's centroid, $\bar{y} = f_x/A$.

Using the quadtrees of the above-mentioned two objects, carrying out Boolean operations with the two objects is easy and fast.

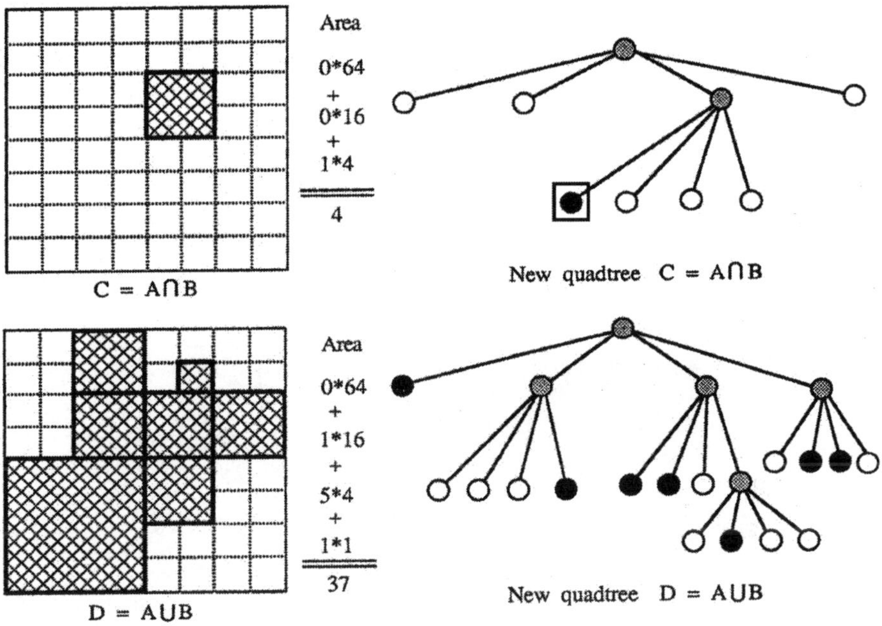

Figure A.13. Quadtrees identify the Boolean intersection (top) and union of two objects along with calculation of their geometric properties

A Boolean set operation simultaneously transverses both given objects' trees while creating a third tree. The third tree contains the results of that operation. For example, if the two grids in Fig. A.12 were overlaid, there is only one fully filled cell where they intersect. An intersection, ∩, determination is a Boolean operation. The intersection of two trees, in the top of Fig. A.13, is found by retaining each cell where the corresponding object cell is full. The new tree created by that comparison is shown in the top of Fig. A.14. The intersection search found that the boxed circles in Figure A.12 are the only two in common, will all other cells at the level being empty. Thus, the new intersection quadtree stopped at that level.

The Boolean union, ∪, or sum operation, creates a third quadtree where it retains a cell if it was full in both the object trees. That union is at the bottom of Fig. A.13.

When an object has an irregular boundary, it will always partially fill the final layer. The smallest of all filled cells could be retained as a

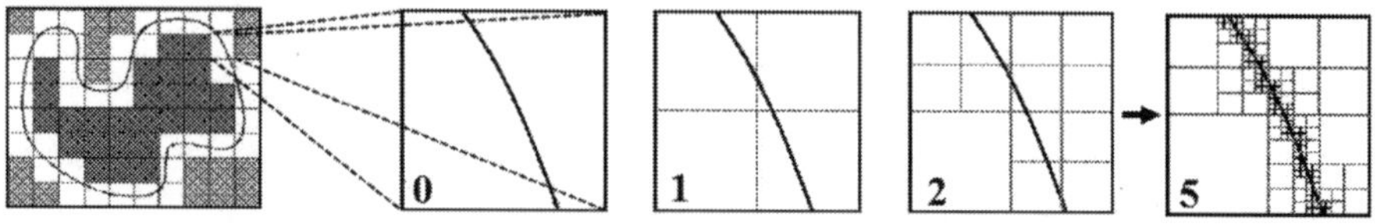

Figure A.14. Quadtree refinements allow very detailed boundary locations

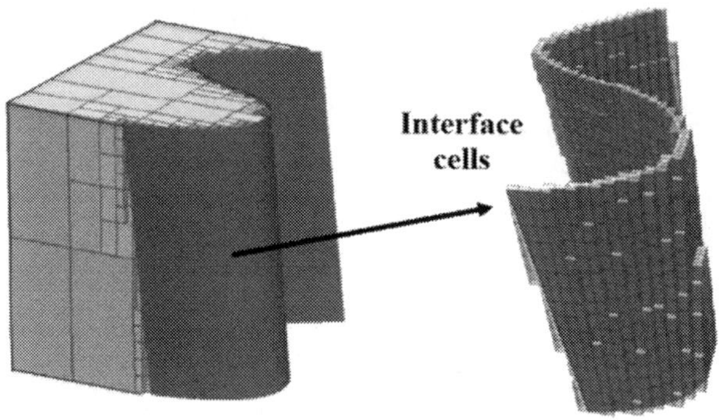

Figure A.15. Octree interface cells on an inclined surface

stair-step approximation of the boundary. The choice might be good for stress analysis, but would not be good for computational fluid dynamics (CFD) studies where important 'boundary layer' physics depend heavily on the roughness of the boundary.

A piecewise cubic, defined by four constants, C^1 boundary curve within a cell could be formed between the two points where the boundary leaves the cell along with the two slopes in the adjacent cells connecting their leaving points. However, assuming that the boundary cells are small enough, a straight line between the leaving points is usually sufficient. The reasonableness of using straight lines rather than continuing the quadtree refinement is illustrated in Fig. A.14.

For biological materials, surfaces or solids cut by curved surfaces (Fig. A.15) can often be represented stepwise by a specified cell size to be smoothed later.

The extension of quadtrees to octrees is easy to visualize. As shown in Fig. A.16, a unit cube is equally divided into eight (oct) identical sized cubes with a numbering system expanded from the

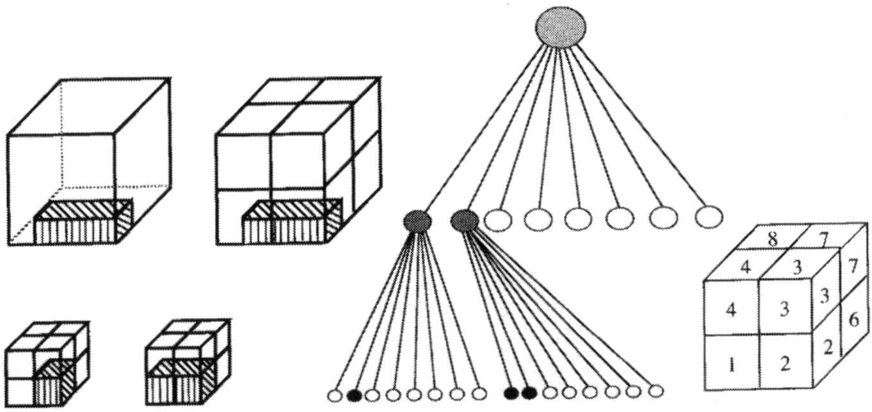

Figure A.16. Octrees extend the factor of two refinements to volumes

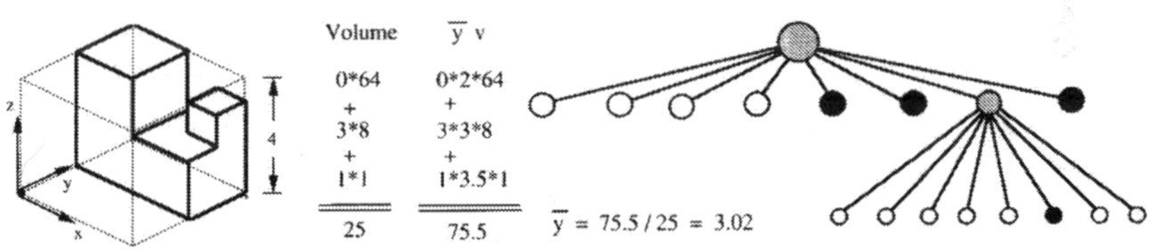

Figure A.17. Volume and mass properties are easily found from an octree

quadtree face ordering. The ordering causes there to be eight limbs in each new sub-cell that is just partially full.

As with quadtrees, an octree structure also allows quick and accurate mass properties to be calculated for other engineering needs such a dynamic analysis or simply determining an object's mass and direction extensions to make shipping crates. Figure A.17 shows how important mass and volume properties can be accurately determined with each new level in the tree. Specifying the number of significant digits for the volume provides one way to cut off the refinement levels.

Quadtree and octrees let the object have both outer boundaries as well as interior voids. When all surfaces are sufficiently refined, the octree can be converted into solid NURBS or solid T-splines for IGA or FEA studies. Most engineering materials are homogeneous with a constant mass density and the mass properties are found from the geometric properties by multiplying them by the mass density.

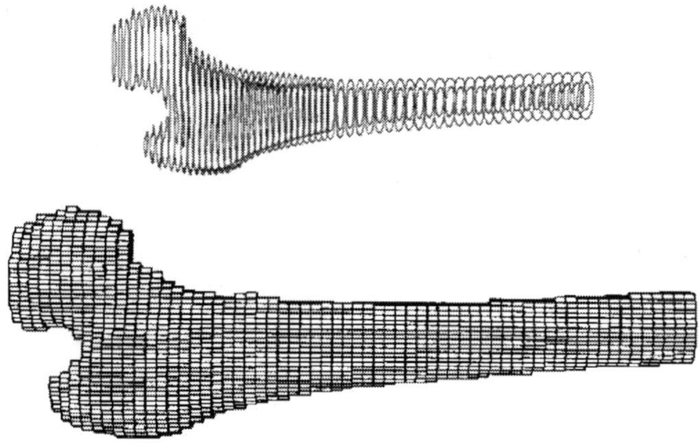

Figure A.18. A medium octree refinement of a human femur bone

This is not true for most biological materials since their mass density and structural properties are very location dependent. Often, models of bones begin as a regular grid of parallel piped volumes, called 'voxels', fitting between slices recorded in a Computer Tomography (CT) machine. Today, such slices are typically less than one mm apart, but that distance is being reduced as the technology advances. Each voxel contains data on its material mass density, which are usually scaled in terms of a 'Hounsfield intensity' value (HU). The density intensity of bone feeds into empirical equations for the modulus of elasticity and Poisson's ratio mechanical properties. An example set of CT slices of a human femur is shown in Fig. A.17. From each slice, the exterior boundary, interior void boundary, and the boundary between compact cortical bone and any adjacent spongy (trabecular) bone can be found.

Those data allow an octree structure to capture each bone class and to define the structural properties in each volumetric cell. Then, each octree cell can have its stiffness matrix formed with an IGA procedure or an FEA procedure. When the structural properties are not smooth, the FEA approach is currently easier. The bottom image of Fig. A.18 shows a medium-level structural octree stair-step approximation of the outer surface of the femur bone. Clearly, further refinement is possible. In this model, every single octree cell is assumed to have a different material and therefore, a different stiffness matrix.

A3.5 *Native and exchange files*

Prior discussions have shown that there are several ways to represent parts or volumes that subsequently need to be subjected to an IGA study or FEA study. There are many software tools that provide ways to supply the data needed to start such studies. It is not surprising that many providers like to store their models in different formats.

Each vendor usually has pet 'native format' files which contain the complete design history of a model. The file format type provides all the model information as created from scratch in that native CAD system. The design parameters and design history in the in-house native format allow the vendor to do fast and easy editing of design variations from the original model. In other words, the design parameters in a native file are easily changed and instantly create a part with a new shape or feature. The native format is the most useful format to obtain, if you have access to the brand name software that created it.

Even within one company, it is not unusual for one department to have design software that uses a file format different from design software in another department. As a user of a huge number of contractors using many different design software platforms, the U.S. government quickly realized that there needed to be a standardized file exchange format that allowed all government contractors to share digital model data. Thus, a government committee with members from the leading design software companies was established. This led to the Initial Graphics Exchange System (IGES). 'Exchange format' files are engineering standards that are designed to allow a CAD model developed on one vendor system to be transferred to another vendor. Fortunately, IGES was just an initial standard. Today, several file formats like Parasolid, and STEP files are used in the design industry, as well as the newer *.bezier files.

Basically, the above-mentioned exchange files existed more or less in two groups, one for users in geometry creation groups and others for FEA study groups. Once Prof. Hughes' team began testing and applying the IGA concept for analysis tasks and those analysis results needed to be locally refined (as common in FEA), it became clear that exchanges between analysis and geometry construction must be a 'two-way street'. Refinement analysis highlighted the need for what became analysis-suitable T-splines and their 3D extension.

The new IGA concept of 'design through analysis' (DTA) to interface IGA with CAGD produced the need for new classes of data and new exchange file formats, with new names. Major progress in DTA exchanges occurred when the software Rhinoceros 3D (generally known as Rhino) released a 2D exchange format, '*.iga' files, and supporting T-spline volume formats, '*.TSM'. Those DTA exchange concepts have been and will continue to be extended. For example, Lai *et al.* (2016) have developed the *.STEP file for volumetric T-spline construction. (Unfortunately, in the author's opinion, another vendor purchased exclusive rights to Rhino exchange technology, providing them with a temporary monopoly on its actual use.) Other DTA exchange file concepts and efficiencies will probably continue to evolve similar to the history where the original *.iges exchange format emphasized the need for today's powerful exchange tools.

Supplementary Material

The supplementary material includes the MATLAB software listed in the text, additional two-dimensional scripts, and supporting text help files.

Online access is automatically assigned if you purchase the ebook online via www.worldscientific.com.

If you have purchased the print copy of this book or the ebook via other sales channels, please follow the instructions below to download the files:

1. Register an account/login @ https://www.worldscientific.com
2. Open a new tab and paste this https://www.worldscientific.com/r/13737-supp
3. Download the files from: https://www.worldscientific.com/worldscibooks/10.1142/13737#t=suppl

For subsequent access, simply log in with the same login details in order to access.

For enquiries, please email: sales@wspc.com.sg.

Bibliography

Adam, C., Bouabdallah, S., Zarroug, M., Maitournam, H.: Improved numerical integration for locking treatment in isogeometric structural elements, Part I: Beams. *Computer Methods in Applied Mechanics and Engineering*, 279, 1–28, 2014.

Adam, C., Hughes, T.J.R., Bouabdallah, S., Zarroug, M., Maitournam, H.: Selective and reduced numerical integrations for NURBS-based isogeometric analysis, *Computer Methods in Applied Mechanics and Engineering*, 284, 732–761, 2015.

Agrawal, V., Gautam, S.S.: IGA: A simplified introduction and implementation details for finite element users. *Journal of The Institution of Engineers (India): Series C*, 100(3), 561–585, 2019.

Ainsworth, M., Oden, J.T.: A posteriori error estimation in finite element analysis. *Computer Methods in Applied Mechanics and Engineering*, 142(1–2), 1 – 88, 1997. https://doi.org/10.1016/S0045-7825(96)01107-3.

Ainsworth, M., Andriamaro, G., Davydov, O.: Bernstein-Bézier finite elements of arbitrary order and optimal assembly procedures. *SIAM Journal on Scientific Computing*, 33(6), 3087–3109, 2011. https://doi.org/10.1137/11082539X.

Ai-zeng, W., Zhao, G.: The analysis of T-spline blending functions linear independence. *Computer-Aided Design and Applications*, 8(5), 735–745, 2011.

Akin, J.E., Berger, M.H.: A finite element algorithm for flow in a gas centrifuge, mathematics of finite elements and applications. In J.R. Whiteman (Ed.), *The Mathematics of Finite Elements and Applications* (Vol. IV, pp. 241–250). Academic Press, London, 1981.

Akin, J.E.: A least squares finite element solution of nonlinear equations. In J.R. Whiteman (Ed.), *The Mathematics of Finite Elements and Applications* (pp. 153–162). Academic Press, London, 1973.

Akin, J.E., Wooten, J.W.: Tokomak plasma equilibria by finite elements. In R.H. Gallagher (Ed.), *Finite Elements in Fluids III* (Chapter 21). pp. 351–361. John Wiley, New York, 1978.

Akin, J.E., Gray, W.H., Zhang, Q.D.: Color displays of isoparametric contours. *Engineering Computations*, 1(1), 36–41, 1984.

Akin, J.E., Tezduyar, T.E., Ungor, M., Mittal, S.: Stabilized formulations and Smagorinsky turbulence model for incompressible flows. In H.A. Mang, F.G. Rammerstorfer, J. Eberhardsteiner (Eds.), *Fifth World Congress on Computational Mechanics*, Vienna, Austria, 2002.

Akin, J.E., Singh, M.: Object-oriented P-adaptive finite element system. *Advances in Engineering Software*, 33(7), 461–468, 2002.

Akin, J.E., Tezduyar, T., Ungor, M., Mittal, S.: Stabilization parameters and Smagorinsky turbulence model. *Journal of Applied Mechanics*, 70(1), 2–9, 2003.

Akin, J.E.: *Finite Element Analysis with Error Estimators.* Elsevier, Amsterdam, 2005. ISBN 0-7506-7622-2.

Akin, J.E.: *Finite Element Analysis, with Numerical and Symbolic Matlab.* World Scientific Publishers, Singapore, 2022. ISBN 9789811250637.

Akkerman I., Bazilevs Y., Kees C.E., Farthing M.W.: Isogeometric analysis of free-surface flow. *Journal of Computational Physics*, 230, 4137–52, 2011. DOI: 10.1016/j.jcp.2010.11.044.

Al Akhras, H., Elguedj, T., Gravouil, A., Rochette, M.: Isogeometric analysis-suitable trivariate NURBS models from standard B-Rep models. *Computer Methods in Applied Mechanics and Engineering*, 307, 256–274, 2016. https://doi.org/10.1016/j.cma.2016.04.028.

American Institute of Steel Construction (AISC), *Manual of Steel Construction*, Chicago, 1989.

Amini, R., Maghsoodi, R., Moghaddam, N.Z.: Simulating free surface problem using isogeometric analysis. *Journal of the Brazilian Society of Mechanical Sciences and Engineering*, 38, 413–421, 2016. https://doi.org/10.1007/s40430-014-0295-z.

Anitescu, C., Jia, Y., Zhang, Y.J., Rabczuk, T.: An isogeometric collocation method using superconvergent points. *Computer Methods in Applied Mechanics and Engineering*, 284, 1073–1097, 2015.

Aung, T.L., Ma, N.: Elastic buckling analysis of a slender column using isogeometric analysis and H-refinement. In *Conference Proceedings The Japan Society of Naval Architects and Ocean Engineers*, Vol. 30, pp. 439–443, The Japan Society of Naval Architects and Ocean Engineers, Tokyo, 2020.

Auricchio, F., Beir, L., Veiga, A.D., Hughes, T.J.R., Reali, A., Sangalli, G.: Isogeometric collocation methods. *Mathematical Models and Methods in Applied Sciences*, 20(11), 2075–2107, 2010. https://doi.org/10.1142/S0218202510004878.

Bazilevs, Y., Hughes, T.J.R.: Weak imposition of Dirichlet boundary conditions in fluid mechanics. *Computers and Fluids*, 36(1), 12–26, 2007. https://doi.org/10.1016/j.compfluid.2005.07.012.

Bazilevs, Y., Calo V.M., Cottrell, J.A., Evans, J.A., Hughes, T.J.R., Lipton, S., Scott, M.A., Sederberg, T.W.: Isogeometric analysis using T-splines. *Computer Methods in Applied Mechanics and Engineering*, 199, 229–263, 2010. https://doi.org/10.1016/j.cma.2009.02.036.

Bazilevs, Y., Takizawa, K., Tezduyar, T.E.: *Basics of the Isogeometric Analysis. Computational Fluid–Structure: Interaction: Methods and Applications.* Wiley, New York, 2013.

Beer, G., Bordas, S. (Eds.), *Isogeometric Methods for Numerical Simulation.* Springer, New York, 2015.

Beer, G., Marussig, B., Duenser, C.: *The Isogeometric Boundary Element Method.* Springer, New York, 2020.

Bekele, Y.W., Kvamsdal, T., Kvarving, A.M., Nordal, S.: Adaptive isogeometric finite element analysis of steady-state groundwater flow. *International Journal for Numerical and Analytical Methods in Geomechanics*, 40(5), 738–765, 2016.

Bekhoucha, F.: Isogeometric analysis for in-plane free vibration of centrifugally stiffened beams including Coriolis effects. *Mechanics Research Communications*, I, 103645, 2021. https://doi.org/10.1016/j.matcom.2015.05.008.

Benson, D.J., Y. Bazilevs, Y., Hsu, M.C., Hughes, T.J.R.: Isogeometric shell analysis: The Reissner-Mindlin shell. *Computer Methods in Applied Mechanics and Engineering*, 199(5–8), 276–289, 2010. https://doi.org/10.1016/j.cma.2009.05.011.

Benson, D.J., de Borst, R., Hughes, T.J.R., Scott, M.A., Verhoosel. C.V.: Isogeometric analysis: A calculus for computational mechanics. In B.H.V. Topping, J.M. Adam, F.J. Pallarés, R. Bru, M.L. Romero, (Eds.), *Developments and Applications in Engineering Computational Technology* (pp. 1–17). Saxe-Coburg Publications, Stirling. 2010.

Benson, D., Bazilevs, Y., Hsu, M., Hughes, T.J.R.: A large deformation, rotation-free, isogeometric shell. *Computer Methods in Applied Mechanics and Engineering*, 200(13–16), 1367–1378, 2011. https://doi.org/10.1016/j.cma.2010.12.003.

Bézier, P.E.: *Numerica Control: Mathematics and Applications*. John Wiley, New York, 1972.

Bézier, P.E.: The first years of CAD/CAM and the UNISURF CAD system. In Les Piegl (Ed.), *Fundamental Developments of Computer-Aided Geometric Modeling* (pp. 13–26). Academic Press, London, 1993.

Blevins, R.D.: *Formulas for Natural Frequency and Mode Shape*. Krieger Pub. Co., Malabar, 2001.

Borden, M.J., Scott, M.A., Evans, J.A., Hughes, T.J.R.: Isogeometric finite element data structures based on Bézier extraction of NURBS. *International Journal for Numerical Methods in Engineering*, 87(1–5), 15–47, 2011.

Bramble, J.H., Schatz, A.H.: Least squares methods for 2m-th order elliptic boundary-value problems. *Mathematics of Computation*, 25(113), 1–32, 1971.

Brebbia, C.A., Dominguez, J.: Boundary element methods for potential problems. *Applied Mathematical Modelling*, 1(7), 372–378, 1977.

Brebbia, C. A.: *The Boundary Element Method for Engineers*. Wiley, New York, 1978. ISBN-10, 0470264381, ISBN-13, 978-0470264386.

Brezzi, F., Fortin, M.: *Mixed and Hybrid Finite Element Methods*, Vol. 15. Springer Science & Business Media, 2012.

Brovka, M., López, J.I., Escobar, J.M., Cascón, J.M., Montenegro, R.: A new method for T-spline parameterization of complex 2D geometries. *Engineering with Computers*, 30, 457–473, 2014a.

Brovka, M., López, J.I., Escobar, J.M., Cascón, J.M., Montenegro, R.: Construction of polynomial spline spaces over quadtree and octree T-meshes. *Procedia Engineering*, 82, 21–33, 2014b.

Brunet, P., Navazo, I.: Geometric modelling using exact octree representation of polyhedral objects. In C.E. Vandoni (Ed.), *Eurographics* 85. Elsevier, Amsterdam, 1985.

Buffa, A., Sangalli, G., Vázquez, R.: Isogeometric methods for computational electromagnetics: B-spline and T-spline discretizations. *Journal of Computational Physics*, 257, 1291–1320, 2014. https://doi.org/10.1016/j.jcp.2013.08.015.

Carlon, L.: Approximation and finite elements in isogeometric problems. PhD Thesis, University of Padua, 2010. https://bugfreeblog.page.link/isogeometric-analysis-presentation.

Carslaw, H.S., Jaeger, J.C.: *Conduction of Heat in Solids* (2nd edn.). Oxford University Press, Oxford, 1959.

Carey, G.F., Shen, Y.: Convergence studies of least-squares finite elements for first-order systems. *Communications in Applied Numerical Methods*, 5(7), 427–434, 1989.

Cazzani, A., Malagù, M., Turco, E.: Isogeometric analysis of plane-curved beams. *Mathematics and Mechanics of Solids*, 21(5), 562–577, 2016. https://doi.org/10.1177/1081286514531265.

Cazzani, A., Stochino, F., Turco, E.: An analytical assessment of finite element and isogeometric analyses of the whole spectrum of Timoshenko beams. *ZAMM*, 96(10), 1220–1244, 2016. https://doi.org/10.1002/zamm.201500280.

Chandrasekhar, K.N.V., Sahithi, N.S.S., Rao, T.M.: Step by step illustrative procedure to perform isogeometric analysis and find the nodal displacements for a two-dimensional plate structure. *i-Manager's Journal on Structural Engineering*, 6(4), 40–52, 2017.

Chandrasekhar, K.N.V., Sahithi, N.S.S.: A step by step procedure to perform isogeometric analysis of beam and bar problems in civil engineering including sizing optimization of a beam. *i-Manager's Journal on Structural Engineering*, 7(1), 13, 2018.

Chen, L., de Borst, R.: Locally refined T-splines. *International Journal for Numerical Methods in Engineering*, 114(6), 637–659, 2018. https://doi.org/10.1002/nme.5759.

Chen, L., Zhang, W., Meng, L., Jiu, L., Feng, S.: An adaptive T-spline finite cell method for structural shape optimization. *Structural and Multidisciplinary Optimization*, 61, 1857–1876, 2020.

Cohen, E., Martin, T., Kirby, R.M., Lyche, T., Riesenfeld, R.F.: Analysis-aware modeling: Understanding quality considerations in modeling for isogeometric analysis. *Computer Methods in Applied Mechanics and Engineering*, 199(5–8), 334–356, 2010. https://doi.org/10.1016/j.cma.2009.09.010.

Cottrell, J.A., Reali, A., Bazilevs, Y., Hughes, T.J.R.: Isogeometric analysis of structural vibrations. *Computer Methods in Applied Mechanics and Engineering*, 195, 5257–5296, 2006. https://doi.org/10.1016/j.cma.2005.09.027.

Cottrell, J.A., Hughes, T.J.R., Reali, A.: Studies of refinement and continuity in isogeometric structural analysis. *Computer Methods in Applied Mechanics and Engineering*, 196(41–44), 4160–4183, 2007. https://doi.org/10.1016/j.cma.2007.04.007.

Cottrell, J.A., Hughes, T.J.R., Bazilevs, Y.: *Isogeometric Analysis: Toward Integration of CAD and FEA*. Wiley, New York, 2009.

Cox, M.: The numerical evaluation of B-splines. *IMA Journal of Applied Mathematics*, 10(2), 134–149, October 1972. https://doi.org/10.1093/imamat/10.2.134.

Crandall, S.H.: *Engineering Analysis*. McGraw-Hill, Columbus, 1956.

Das, S.K., Agrawal, V., Gautam, S.S.: A Comparative assessment of three different refinement strategies in NURBS-based isogeometric analysis. In *66th Congress of ISTAM SM2: Computational Solid Mechanics*, 3rd–5th December. 2021. Andhra Pradesh, India.

Davies, J.: Animated Bézier curves. https://www.jasondavies.com/animated-bezier/.

Davis, P.J., Rabinowitz, P.: *Numerical Integration*. Blaisdell, Waltham, 1967.

de Boor, C.: Package for calculating with B-splines. *SIAM Journal on Numerical Analysis*, 14(3), 441–472, 1977. https://doi.org/10.1137/0714026.

de Boor, C.: *A Practical Guide to Splines*. New York: Springer-Verlag, 1978.

de Borst, R., Chen, L.: The role of Bézier extraction in adaptive isogeometric analysis: Local refinement and hierarchical refinement. *International Journal for Numerical Methods in Engineering*, 113(6), 999–1019, 2018. https://doi.org/10.1002/nme.5696.

de Casteljau, P.: Polar forms for curve and surface modeling as used at Citroen. In L. Piegl (Ed.), *Fundamental Developments of Computer-Aided Geometric Modeling* (pp. 1–12), Academic Press, London, 1993.

de Falco, C., Reali, A., Vazquez, R.: GeoPDEs: A research tool for isogeometric analysis of PDEs. *Advances in Engineering Software*, 42(12), 1020–1034, 2011.

de Lorenzis, L., Evans, J.A., Hughes, T.J.R., Reali, A.: Isogeometric collocation: Neumann boundary conditions and contact. *Computer Methods in Applied Mechanics and Engineering*, 284, 21–54, 2015. https://doi.org/10.1016/j.cma.2014.06.037.

de Luycker, E., Benson, D.J., Belytschko, T., Bazilevs, Y., Hsu, M.C.: X-FEM in isogeometric analysis for linear fracture mechanics. *International Journal for Numerical Methods in Engineering*, 87(6), 541–565, 2011. DOI: 10.1007/978-3-642-17484-1-32.

de Veiga, L.B., Buffa, A., Cho, D., Sangalli, G.: IsoGeometric analysis using T-splines on two-patch geometries. *Computer Methods in Applied Mechanics and Engineering*, 200(21–22), 1787–1803, 2011.

Dedè, L., Borden, M.J., Hughes, T.J.R.: Isogeometric analysis for topology optimization with a phase field model. *Archives of Computational Methods in Engineering*, 19, 427–465, 2012. https://doi.org/10.1007/s11831-012-9075-z.

Delaunay, B.: Sur la sphère vide. *Bulletin de l'Académie des Sciences de l'URSS, Classe des Sciences Mathématiques et Naturelles*, 6, 793–800, 1934.

de Luycker, E., Benson, D.J., Belytschko, T., Bazilevs, Y., Hsu, M.C.: X-FEM in isogeometric analysis for linear fracture mechanics. *International Journal for Numerical Methods in Engineering*, 87, 541–565, 2011.

Do, H.V., Nguyen-Xuan, H.: Computation of limit and shakedown loads for pressure vessel components using isogeometric analysis based on Lagrange extraction. *International Journal of Pressure Vessels and Piping*, 169, 57–70, 2019.

Doha, E.H., Bhrawy, A.H., Saker, M.A.: Integrals of Bernstein polynomials: An application for the solution of high even-order differential equations. *Applied Mathematics Letters*, 24(4), 559–565, 2011.

Dörfel, M., Jüttler, B., Simeon, B.: Adaptive isogeometric analysis by local h-refinement with T-splines. *Computer Methods in Applied Mechanics and Engineering*, 199, 264–275, 2010.

Elguedj, T., Bazilevs, Y., Calo, V., Hughes, T.J.R.: Isogeometric analysis: Cad, finite elements, nurbs, exact geometry and mesh refinement. *Computer Methods in Applied Mechanics and Engineering*, 194, 4135–4195, 2005. https://doi.org/10.1016/j.cma.2004.10.008.

Elguedj, T., Duval, A., Maurin, F., Al-Akhras, H.: Abaqus user element implementation of NURBS based isogeometric analysis. In *6th European Congress on Computational Methods in Applied Sciences and Engineering*, Vienna, Austria, 10–14 September 2012.

Embar, A., Dolbow, J., Harari, I.: Imposing Dirichlet boundary conditions with Nitsche's method and spline-based finite elements. *International Journal for Numerical Methods in Engineering*, 83, 877–898, 2010. DOI: 10.1002/nme.2863.

Ergatoudis, J., Irons, B.M., Zienkiewicz, O.C.: Curved, isoparametric, quadrilateral elements for finite element analysis. *International journal of solids and structures*, 4(1), 31–42, 1968.

Espath, L.F., Braun, A.L., Awruch, A.M.: An introduction to isogeometric analysis applied to solid mechanics. *Mecánica Computacional*, 30(24), 1955–1975, 2011.

Euler, L.: Elementa doctrinae solidorum. *Novi Commentarii Academiae Scientiarum Petropolitanae*, 1(1), 109–140, 1758.

Evans, J.A., Bazilevs, Y., Babuska, I., Hughes, T.J.R.: n-widths, sup-infs, and optimality ratios for the k-version of the isogeometric finite element method. *Computer Methods in Applied Mechanics and Engineering*, 198(21), 1726–1741, 2009. https://doi.org/10.1016/j.cma.2009.01.021.

Finlayson, B.A., Scriven, L.E.: The method of weighted residuals-a review. *Applied Mechanics Reviews*, 19(9), 735–748, 1966.

Finlayson, B.A.: *The Method of Weighted Residuals and Variational Principles*. Society for Industrial and Applied Mathematics, 2013.

Gai, Y., Xing, J., Hu, P.: Efficient MATLAB implementation of NURBS-based IGA and material design using isogeometric topology optimization. *Optimization and Engineering*, 24(3), 1773–1808, 2023.

Gondegaon, S., Voruganti, H.K.: Static structural and modal analysis using isogeometric analysis. *Journal of Theoretical and Applied Mechanics*, 46(4), 36–75, 2016. DOI: 10.1515/jtam-2016–0020

Gotovac, H., Andricevic, R., Gotovac, B.: Multi-resolution adaptive modeling of groundwater flow and transport problems. *Advances in Water Resources*, 30(5), 1105–1126, 2007. https://doi.org/10.1016/j.advwatres.2006.10.007.

Gray, W.H., Akin, J.E.: An Improved method for contouring on isoparametric surfaces. *International Journal for Numerical Methods in Engineering*, 14(3), 451–458, 1979.

Gregory, J.A., Whiteman, J.R.: Local mesh refinement with finite elements for elliptic problems. Brunel University Mathematics Technical Papers Collection, TR/24, 1–16, 1974.

Gregory, J.A., Fishelov, D., Schiff, B., Whiteman, J.R.: Local refinement with finite elements for elliptic problems. *Journal of Computational Physics*, 29(1), 133–140, 1978.

Guo, M., Wang, W., Zhao, G., Du, X., Zhang, R., Yang, J.: T-splines for isogeometric analysis of the large deformation of elastoplastic Kirchhoff–Love shells. *Applied Sciences*, 13(3), 1709, 2023.

Gupta, V., Jameel, A., Verma, S.K., *et al.*: An insight on NURBS based isogeometric analysis, its current status and involvement in mechanical applications. *Archive Computational Methods in Engineering*, 30(2), 1187–1230, 2022. https://doi.org/10.1007/s11831-022-09838-0.

Gupta, V., Verma, S.K., Anand, S., Jameel, A., Anand, Y.: Transient isogeometric heat conduction analysis of stationary fluid in a container. *Proceedings of the Institution of Mechanical Engineers, Part E: Journal of Process Mechanical Engineering*, 09544089221125718, 2022. https://doi.org/10.1177/09544089 221125718.

Hall, S.K., Eaton, M.D., Williams, M.M.R.: The application of isogeometric analysis to the neutron diffusion equation for a pincell problem with an analytic benchmark. *Annals of Nuclear Energy*, 49, 160–169, 2012. https://doi.org/ 10.1016/j.anucene.2012.05.030.

Hartmann, S., Benson, D.J., Lorenz, D.: About isogeometric analysis and the new NURBS based Finite Elements in LSDYNA. In *8th European LS-DYNA User Conference*, Strasbourg, France 2011 May 23, 2016.

Hansbo, P.: Nitsche's method for interface problems in computational mechanics. *GAMM-Mitteilungen*, 28(2), 183–206, 2005. https://doi.org/ 10.1002/gamm.201490018.

Hazel, T.G., Wexler, A.: Variational formulation of the Dirichlet boundary condition. *IEEE Transactions on Microwave Theory and Techniques*, 20(6), 385–390, 1972.

Hekmat, A., Yassin, A.Y.B.M., Yusop, Z.B., Kashefizad, M.: Weak imposition of essential boundary conditions in isogeometric analysis of depth averaged advection dispersion equation. *Journal of Applied Sciences*, 15(2), 184–194, 2015.

Hetényi, M.: *Beams on Elastic Foundation: Theory with Applications in the Fields of Civil and Mechanical Engineering*. University of Michigan Press, Ann Arbor, 1946.

Hilber, H.M., Hughes, T.J.R., Taylor, R.L.: Improved numerical dissipation for time integration algorithms in structural dynamics. *Earthquake Engineering and Structural Dynamics,* 5, 283–292, 1977.

Hughes, T.J.R.: *The Finite Element Method: Linear Static and Dynamic Finite Element Analysis.* Dover Publications, Mineola, 2000.

Hughes, T.J.R., Cottrell, J.A., Bazilevs, Y.: Isogeometric analysis: CAD, finite elements, NURBS, exact geometry and mesh refinement. *Computer Methods in Applied Mechanics and Engineering*, 194(39–41), 4135–4195, 2005. https://doi.org/10.1016/j.cma.2004.10.008.

Hughes, T.J.R., Evans, J.A.: Isogeometric analysis. ICES report 10–18. The Institute for Computational Engineering and Sciences, The University of Texas at Austin, May, 2010.

Hughes, T.J.R., Reali, A., Sangalli, G.: Efficient quadrature for NURBS-based isogeometric analysis. *Computer Methods in Applied Mechanics and Engineering*, 199(5–8), 301–313, 2010. https://doi.org/10.1016/j.cma.2008.12.004.

Hughes, T.J.R., Oden, J.T., Papadrakakis, M. (Eds.): Isogeometric analysis special issue. In *Computer Methods in Applied Mechanics and Engineering*, Vol. 284, No. 1, pp. 1–1182, February, 2015.

Irons, B.M., Razzaque, A.: Experience with the patch test for convergence of the finite element method. In A.R. Aziz (Ed.), *Mathematical Foundation of the Finite Element Method* (pp. 557–587). Academic Press, London, 1972.

Irons, B.M., Ahmad, S.: *Techniques of Finite Elements.* John Wiley, New York, 1980.

Hughes, T.J.R., Baidoo, F., Johnson, M.: *Lecture Notes CSE397: Introduction to Isogeometric Analysis.* Austin: University of Texas, p. 200, Fall 2023.

Inoue, K., Kikuchi, Y., Masuyama, T.: A NURBS finite element method for product shape design. *Journal of Engineering Design*, 16(2), 157–171, 2005. DOI: 10.1080/01405110500033127.

Irzal, F., Remmers, J.J., Verhoosel, C.V., de Borst, R.: Isogeometric finite element analysis of poroelasticity. *International Journal for Numerical and Analytical Methods in Geomechanics*, 37(12), 1891–1907, 2013.

Jardin, S.C.: A triangular finite element with first-derivative continuity applied to fusion MHD applications. *Journal of Computational Physics*, 200(1), 133–152, 2004.

Jaswon, M.: Integral equation methods in potential theory — I. *Proceedings of the Royal Society of London*, A275, 1360, 23–32, 1963 (see Symm, G. for -II).

Jator, S., Zachariah S.: A higher order B-spline collocation method for linear boundary value problems. *Applied Mathematics and Computations*, 191, 100–116, 2007. https://doi.org/10.1016/j.amc.2007.02.027.

Jaxon, N., Qian, X.: Isogeometric analysis on triangulations. *Computer-Aided Design*, 46, 45–57, 2014.

Johnson, R.W.: Higher order B-spline collocation at the Greville Abscissa. *Applied Numerical Mathematics*, 52, 63–75, 2005. https://doi.org/10.1016/j.apnum.2004.04.002.

Joshi, K., Bhatt, A.D.: Building a T-spline-based tri-variate heterogeneous model of human airways: A reverse engineering approach. *Engineering with Computers*, 38, 4085–4097, 2022. https://doi.org/10.1007/s00366-021-01569-3.

Juntunen, M., Stenberg, R.: Nitsche's method for general boundary conditions. *Mathematics of Computation*, 78, 267, 1353–1374, 2009. http://dx.doi.org/10.1090/S0025-5718-08-02183-2.

Kacprzyk, Z., Ostapska-Łuczkowska, K.: Isogeometric analysis as a new FEM formulation-simple problems of steady state thermal analysis. *Procedia Engineering*, 91, 87–92, 2014. https://doi.org/10.1016/j.proeng.2014.12.018.

Kadapa, C., Dettmer, W.G., Perić, D.: NURBS based least-squares finite element methods for fluid and solid mechanics. *International Journal for Numerical Methods in Engineering*, 101(7), 521–539, 2015.

Kamerman, M.: A primer on Bézier curves (Free Book), 2022. https://pomax.github.io/bezierinfo/.

Katili, A.M., Katili, I.: FEA versus IGA in a two-node beam element based on unified and integrated method. *Advances in Applied Mathematics and Mechanics*, 12(6), 1565–1586, 2020.

Katsoulis, T., Wang, X., Kaklis, P.D.: A T-splines-based parametric modeller for computer-aided ship design. *Ocean Engineering*, 191, 106433, 2019. https://doi.org/10.1016/j.oceaneng.2019.106433.

Kostas, K., Ginnis, A., Politis, C., Kaklis, P.: Ship-hull shape optimization with a Tspline based BEM-isogeometric solver. *Computer Methods in Applied Mechanics and Engineering*, 284, 611–622, 2015. https://doi.org/10.1016/j.cma.2014.10.030.

Kumar, M., Kvamsdal T., Johannessen K.A.: Simple a posteriori error estimators in adaptive isogeometric analysis. *Computers and Mathematics with Applications*, 70, 1555–1582, 2015.

Kumar, M., Kvamsdal, T., Johannessen, K.A.: Superconvergent patch recovery and a posteriori error estimation technique in adaptive isogeometric analysis. *Computer Methods in Applied Mechanics and Engineering*, 316, 1086–1156. 2017. https://doi.org/10.1016/j.cma.2016.11.014.

Lai, Y., Liu, L., Zhang, Y.J., Chen, J., Fang, E., Lua, J.: Rhino 3D to Abaqus: A T-spline based isogeometric analysis software framework. In *Advances in Computational Fluid-structure Interaction and Flow Simulation: New Methods and Challenging Computations* (pp. 271–281), 2016. https://doi.org/10.1007/978-3-319-40827-9_21.

Langer, U., Moore, S.E., Neumüller M.: Space–time isogeometric analysis of parabolic evolution problems. *Computer Methods in Applied Mechanics and Engineering,* 306, 342–363, 2016.

LEE, S.J.: Solution of Poisson Equation using Isogeometric Formulation. *Architectural Research*, 13(1), 17–24, 2011.

Lee, S.J., Kim, H.R.: Vibration and buckling of thick plates using isogeometric approach. *Architectural Research*, 15(1), 35–42, 2013.

Li, W.C., Ray, N., Lévy, B.: Automatic and interactive mesh to T-spline conversion. In *4th Eurographics Symposium on Geometry Processing-SGP 2006*, Liu: Sardinia, Italy, June 26–28, 2006.

Lin, H., Cai Y., Gao, S.: Extended T-mesh and data structure for the easy computation of T-spline. *Journal of Information and Computational Science*, 9(3), 583–93, 2012.

Liu, L., Zhang, Y., Hughes, T.J.R., Scott, M.A., Sederberg, T.W.: *Volumetric T-spline Construction Using Boolean Operations.* University of Texas, Austin, ICES REPORT 13–19, July 2013.

Liu, L., Zhang, Y., Hughes, T.J.R., Scott, M.A., Sederberg, T.W.: Volumetric T-spline construction using Boolean operations. *Engineering with Computers*, 30, 425–439, 2014. https://doi.org/10.1007/s00366-013-0346-6.

Liu, Y., Takizawa, K., Otoguro, Y., Kuraishi, T., Tezduyar, T.E.: Flow computation with the space-time isogeometric analysis and higher-order basis functions in time. *Mathematical Models and Methods in Applied Sciences*, 32(12), 2445–2475, 2022.

Liu, W., Huang, Z., Liu, Q., Zeng, J.: An isogeometric analysis approach for solving the Reynolds equation in lubricated piston dynamics. *Tribology International*, 103, 149–166, 2016.

Liu, W.K., Li, S., Park, H.S.: Eighty years of the finite element method: Birth, evolution, and future. *Archives Computational Methods in Engineering*, 29(6), 4431–4453, 2022. https://doi.org/10.1007/s11831-022-09740-9.

Li, X., Zheng, J., Sederberg, T.W., Hughes, T.J.R., Scott, M.A.: On linear independence of T-splines. ICES report 10–40. The Institute for Computational Engineering and Sciences, The University of Texas at Austin, October 2010.

Li, X., Scott, M.A.: Analysis-suitable T-splines: Characterization, refineability, and approximation. *Mathematical Models and Methods in Applied Sciences*, 24(6), 1141–1164, 2014.

Loli, G., Montardini, M., Sangalli, G. and Tani, M.: An efficient solver for space–time isogeometric Galerkin methods for parabolic problems. *Computers and Mathematics with Applications*, 80(11), 2586–2603, 2020.

López, J.I., Brovka, M., Escobar Sánchez, J.M., Cascón Barbero, J.M., Montenegro Armas, R.: Advances on T-spline parameterization based on the Meccano method. In *11th World Congress on Computational Mechanics (WCCM XI)*, Barcelona, 2014.

Lynn, P.P., Arya, S.K.: Finite elements formulated by the weighted discrete least squares method. *International Journal for Numerical Methods in Engineering*, 8(1), 71–90, 1974.

Lynn, P.P., Arya, S.K.: 1973. Use of the least squares criterion in the finite element formulation. *International Journal for Numerical Methods in Engineering*, 6(1), 75–88.

Marchioria, G., Netoa, A.G.: Isogeometric analysis applied to 2D Bernoulli-Euler beam model: Imposition of constraints by Lagrange and penalty methods. http://dx.doi.org/10.1590/1679-78255578.

May, S., Vignollet, J., de Borst, R.: The role of the Bézier extraction operator for T-splines of arbitrary degree: Linear dependencies, partition of unity property, nesting behaviour and local refinement. *International Journal for Numerical Methods in Engineering*, 103(8), 547–581, 2015.

McDonald, B.H., Wexler, A.: Finite-element solution of unbounded field problems. *IEEE Transactions on Microwave Theory and Techniques*, 12, 841–847, 1972.

Mitchell, A.R., Wait, R.: *The Finite Element Method in Partial Differential Equations*. Wiley, New York, 1978. ISBN: 0-471-99405-7.

Mitchell, T., Govindjee, S., Taylor, R.: A method for enforcement of dirichlet boundary conditions in isogeometric analysis. In D. Mueller-Hoeppe, S. Loehnert, S. Reese (Eds.), *Recent Developments and Innovative Applications in Computational Mechanics* (pp. 283–293). Springer, New York, 2011.

Mondal, S.K., Gondegaon, S., Voruganti, H.K.: A novel method to apply Neumann boundary conditions in the isogeometric analysis (IGA) of beam with 1-D formulation. *World Journal of Engineering*, 14(6), 538–544, 2017. https://doi.org/10.1108/WJE-11-2016-0143.

Morgenstern, P, Peterseim, D.: Analysis-suitable adaptive t-mesh refinement with linear complexity, *Computer Aided Geometric Design*, 34, 50–66, 2015.

Morton, K.W.: The convection–diffusion Petrov–Galerkin story. *IMA Journal of Numerical Analysis*, 30(1), 231–240, 2010.

NAFEMS: Benchmark Tests for Thermal Analysis, Test 9. Pub. YR3087, v. 2, Granville, OH, 1986.

Nagy, A.P., Abdalla, M.M., Gürdal, Z.: Isogeometric sizing and shape optimization of beam structures. *Computer Methods in Applied Mechanics and Engineering*, 199, 1216–1230, 2010.

Newmark, N.M.: A method of computation for structural dynamics. *Journal of the Engineering Mechanics Division*, 85(3), 67–94, 1959.

Nitsche, J.: Über ein variationsprinzip zur lösung von dirichlet-problemen bei verwendung von teilräumen, die keinen randbedingungen unterworfen sind. *Abhandlungen aus dem Mathematischen Seminar der Universität Hamburg*, 36(1), 9–15, July 1971. https://doi.org/10.1007/BF02995904.

Nguyen, M.N., Bui, T.Q., Yu, T., Hirose, S.: Isogeometric analysis for unsaturated flow problems. *Computers and Geotechnics*, 62, 257–267, 2014.

Nguyen, T. N., Fredheim, O.J., Mathisen, K.M., Johannessen, K.A.: *Isogeometric Finite Element Analysis Based on Bézier Extraction of NURBS and T-Splines*. In B. Skallerud, H.I. Andersson (Eds.), *MekIT'11 Sixth National Conference on Computational Mechanics*, pp. 239–256. Tapir Academic Press, Trondheim, 23–24 May 2011.

Nguyen, V.P., Rabczuk, T., Bordas, S., Duflot, M.: Meshless methods: A review and computer implementation aspects. *Mathematics and Computers in Simulation*, 79(3), 763–813, 2008.

Nguyen, V.P., P. Kerfriden, M. Brino, S.P.A. Bordas, E. Bonisoli: Nitsche's method for two and three dimensional NURBS patch coupling. *Computational Mechanics*, 53(6), 1163–1182, 2014.

Nguyen, V.P., Anitescu, C., Bordas, S.P.A., Rabczuk, T.: Isogeometric analysis: An overview and computer implementation aspects. *Mathematics and Computers in Simulation*, 117, 89–116, 2015.

Nguyen, L.H., Schillinger, D.: A collocated isogeometric finite element method based on Gauss–Lobatto Lagrange extraction of splines. *Computer Methods in Applied Mechanics and Engineering*, 316, 720–740, 2017.

Nguyen-Thanh, N., Nguyen-Xuan, H., Bordas, S.P.A., Rabczuk, T.: Isogeometric analysis using polynomial splines over hierarchical T-meshes for two-dimensional elastic solids. *Computer Methods in Applied Mechanics and Engineering*, 200(21–22), 1892–1908, 2011. https://doi.org/10.1016/j.cma.2011.01.018.

Nguyen-Xuan, H., Hoang T., Nguyen, V.P.: An isogeometric analysis for elliptic homogenization problems. *Computers and Mathematics with Applications*, 67(9), 1722–1741, 2014.

Niemi, A.H., Collier, N., Dalcin, L., Ghommem, M., Calo, V.M.: Isogeometric shell formulation based on a classical shell model. In *Proceedings of the Eleventh International Conference on Computational Structures Technology*, Paper 221, p. 12, Dubrovnik, Croatia, 4–7 September 2012.

Perduta, A., Putanowicz, R.: Tools and techniques for building models for isogeometric analysis. *Advances in Engineering Software*, 127, 70–81, 2019.

Peterson, J.: How to use knot vectors. In Albert technical memo #5, 1–8, Apple Computers, 1990. https://saccade.com/writing/graphics/KnotVectors.pdf.

Petrov, G.I.: Application of Galerkin's method to the problem of stability of flow of a viscous fluid. (in Russian). *Journal of Applied Mathematics and Mechanics*, 4(3), 3–12, 1940.

Piegl, L., Tiller, W.: *The Tk.* (2nd edn.). Springer-Verlag, New York, 1997.

Qian, J.X.: Isogeometric analysis on triangulations. *Computer-Aided Design*, 46, 45–57, 2014. https://doi.org/10.1016/j.cad.2013.08.017.

Ram-Mohan, L.R.: *Finite Element and Boundary Element Applications in Quantum Mechanics*. Oxford University Press, Oxford, 2002.

Rauen, M., Machado, R.D., Arndt, M.: Isogeometric analysis of free vibration of framed structures: Comparative problems. *Engineering Computations*, 34(2), 377–402, 2017.

Reali, A.: An isogeometric analysis approach for the study of structural vibrations. *Journal of Earthquake Engineering*, 10(1), 1–30, 2006.

Reali, A., Hughes, T.J.R.: An introduction to isogeometric collocation methods. In G. Beerr, S. Bordas (Eds.), *Isogeometric Methods for Numerical Simulation* (pp. 173–204). Springer, New York, 2015.

Reddy, J.N.: *Introduction to the Finite Element Method.* McGraw-Hill, Columbus, 2019.

Richards, D.J., Wexler, A.: Finite-element solutions within curved boundaries. *IEEE Transactions on Microwave Theory and Techniques*, 20(10), 650–657, October 1972. DOI: 10.1109/TMTT.1972.1127839.

Rypl, D., Patzák, B.: Assessment of computation efficiency of numerical quadrature schemes in the isogeometric analysis. *Engineering Mechanics*, 19(4), 249–260, 2012a.

Rypl, D., Patzák, B.: Object oriented implementation of the T-spline based isogeometric analysis. *Advances in Engineering Software*, 50, 137–149, 2012b.

Sangalli, G.: Lecture 3: Definition and theory of T-splines for IGA. In *5th European Women in Mathematics Summer School, Intern.* Centre Theoretical Physics, Trieste, pp. 51, June 2013. https://indico.ictp.it/event/a12191/session/35/contribution/20/material/0/0.pdf.

Schillinger, D.: Isogeometric finite element analysis. In H. Altenbach, A. Öchsner (Eds.), *Encyclopedia of Continuum Mechanics.* Springer, New York, 2020. https://doi.org/10.1007/978-3-662-55771-6_19.

Schillinger, D., Hossain, S.J., Hughes, T.J.R.: Reduced Bézier element quadrature rules for quadratic and cubic splines in isogeometric analysis. *Computer Methods in Applied Mechanics and Engineering*, 277, 1–45, 2014.

Schillinger, D., Dede, L., Scott, M.A., Evans, J.A., Borden, M.J., Rank, E., Hughes, T.J.R.: An isogeometric design-through-analysis methodology based on adaptive hierarchical refinement of NURBS, immersed boundary methods, and T-spline CAD surfaces. *Computer Methods in Applied Mechanics and Engineering*, 249, 116–150, 2012. https://doi.org/10.1016/j.cma.2012.03.017.

Schillinger, D., Ruthala, P.K., Nguyen, L.H.: Lagrange extraction and projection for NURBS basis functions: A direct link between isogeometric and standard nodal finite element formulations. *International Journal for Numerical Methods in Engineering*, 108(6), 515–534, 2016. https://doi.org/10.1002/nme.5216.

Scott, M.A., Hughes, T.J.R., Sederberg, T.W., Sederberg, M.T.: An integrated approach to engineering design and analysis using the Autodesk T-spline plugin for Rhino3d. ICES REPORT 14–33. The Institute for Computational Engineering and Sciences, The University of Texas at Austin, 2014.

Scott, M.A., Li, X., Sederberg, T.W., Hughes, T.J.R.: Local refinement of analysis-suitable T-splines. *Computer Methods in Applied Mechanics and Engineering*, 213, 206–222, 2012. https://doi.org/10.1016/j.cma.2011.11.022.

Scott, M.A., Borden, M.J., Verhoosel, C.V., Sederberg, T.W., Hughes, T.J.R.: Isogeometric finite element data structures based on Bézier extraction of T-splines. *International Journal for Numerical Methods in Engineering*, 88(2), 126–156, 2011. https://doi.org/10.1002/nme.2968.

Scott, M.A., Simpson, R.N., Evans, J.A., Lipton, S., Bordas, S.P., Hughes, T.J.R., Sederberg, T.W.: Isogeometric boundary element analysis using unstructured T-splines. *Computer Methods in Applied Mechanics and Engineering*, 254, 197–221, 2013. https://doi.org/10.1016/j.cma.2012.11.001.

Scott, M.A., Simpson, R.N., Evans, J.A., Lipton, S., Bordas, S.P., Hughes, T.J.R., Sederberg, T.W.: Isogeometric boundary element analysis using unstructured T-splines. *Computer Methods in Applied Mechanics and Engineering*, 254, 197–221, 2013. http://dx.doi.org/10.1016/j.cma.2012.11.001.

Sederberg, T.W., Zhen, J., Bakenov, A., Nasri, A.: T-splines and T-NURCCs. *ACM Transactions on Graphics*, 22(3), 477–84, 2003.

Shahrokhabadi, S., Vahedifard, F., Ghazanfari, E.: Modeling flow regime in shale using isogeometric analysis. In *Shale Energy Engineering 2014: Technical Challenges, Environmental Issues, and Public Policy*, Pittsburgh, 21–23 July 2014.

Shahrokhabadi, S., Vahedifard, F., Bhatia, M.: Head-based isogeometric analysis of transient flow in unsaturated soils. *Computers and Geotechnics*, 84, 183–197, 2017.

Shang, C., Fu, J., Lin, Z., Feng, J., Li, B.: Closed T-spline surface reconstruction from medical image data. *International Journal of Precision Engineering and Manufacturing*, 19, 1659–1671, 2018.

Shene, C.-K.: *Advanced Computer Graphics*. Michigan Technological University, 2016. https://pages.mtu.edu/~shene/COURSES/cs3621/NOTES/spline/Bezier/intro.html.

Shoheib, M.M., Shahrooi, S., Shishehsaz, M., Hamzehei, M.: The application of the isogeometric method based on Bézier extraction for the thermo-plastic analysis of welded steel plate. *Mechanics of Solids*, 58(1), 245–265, 2023. https://doi.org/10.3103/S0025654422600933.

Singh, S.K., Singh, I.V., Mishra, B.K., Bhardwaj, G., Bui T.Q.: A simple, efficient and accurate Bézier extraction based T-spline XIGA for crack simulations. *Theoretical and Applied Fracture Mechanics*, 88, 74–96, 2017.

Simpson, R., Bordas, S., Trevelyan, J., Rabczuk, T.: A two-dimensional isogeometric boundary element method for elastostatic analysis. *Computer Methods in Applied Mechanics and Engineering*, 209–212, 87–100, 2012.

Spetz, A., Tudisco, E., Denzer, R., Dahlblom, O.: Isogeometric analysis of soil plasticity. *Geomaterials*, 7(3), 96–116, 2017.

Spink, D.M.: NURBS toolbox by D. M. Spink, 2010. https://www.mathworks.com/matlabcentral/fileexchange/26390-nurbs-toolbox-by-d-m-spink.

Stahl, A., Kvamsdal, T., Schellewald, C.: Post-processing and visualization techniques for isogeometric analysis results. *Computer Methods in Applied Mechanics and Engineering*, 316, 880–943, 2017.

Stone, M.C., Derose, T.D.: A geometric characterization of parametric cubic curves. *ACM Transactions on Graphics*, 8(3), 147–163, 1989.

Suh, C.H., Radcliffe, C.W.: *Kinematics and Mechanisms Design*. Krieger Publication, Malabar, 1983.

Symm, G.: Integral equation methods in potential theory — II. *Proceedings of the Royal Society of London*, A275, 1360, 33–46, 1963. https://doi.org/10.1098/rspa.1963.0153.

Taher, A.H.S.: Fast and accurate calculations of fourth-order non-self-adjoint Sturm–Liouville eigenvalues for problems in physics and engineering. *International Journal of Applied and Computational Mathematics*, 7(5), 212, 2021.

Takizawa, K., Tezduyar, T.E.: Multiscale space–time fluid-structure interaction techniques. *Computational Mechanics*, 48, 247–267, 2011. https://doi.org/10.1007/s00466-011-0571-z.

Takizawa, K., Tezduyar, T.E.: Space–time computation techniques with continuous representation in time (ST-C), *Computational Mechanics*, 53(1), 91–99, 2014.

Takizawa, K., Henicke, B., Puntel, A., Spielman, T., Tezduyar, T.E.: Space–time computational techniques for the aerodynamics of flapping wings. *Journal of Applied Mechanics*, 79, 010903, 2012. https://doi.org/10.1115/1.4005073.

Taylor, R.L.: Isogeometric analysis of nearly incompressible solids. *International Journal for Numerical Methods in Engineering*, 87(1–5), 273–288, 2011.

Thomas, D.C., Scott, M.A.: Isogeometric analysis based on T-splines. In G. Beer, S. Bordas (Eds.), *Isogeometric Methods for Numerical Simulation* (pp. 205–232). Springer, New York, 2015. https://doi.org/10.1007/978-3-7091-1843-65.

Thomas, D.C., Scott, M.A., Evans, J.A., Tew, K., Evans, E.J.: Bézier projection: A unified approach for local projection and quadrature-free refinement and coarsening of NURBS and T-splines with particular application to isogeometric design and analysis. *Computer Methods in Applied Mechanics and Engineering*, 284, 55–105, 2015.

Tjomsland, M.F.: Isogeometric analysis of higher-order differential equations. Master's Thesis, Department of Mathematical Sciences, Norwegian University of Science and Technology, p. 101, 2013.

Uhm, T.-K., Kim, K.-S., Seo, Y.-D., Youn, S.-K.: A locally refinable T-spline finite element method for CAD/CAE integration. *Structural Engineering and Mechanics*, 30(2), 225–245, 2008. DOI: 10.12989/sem.2008.30.2.225.

Uhm, T.K., Youn, S.K.: T-spline finite element method for the analysis of shell structures. *International Journal for Numerical Methods in Engineering*, 80(4), 507–536, 2009. https://doi.org/10.1002/nme.2648.

van Zwieten, J.S.B., van Zwieten, G.J., Hoitinga, W.: *Nutils 7.0*. Zenodo, 2022.

Verhoosel, C.V., Van Zwieten, G.J., Van Rietbergen, B., de Borst, R.: Image-based goal-oriented adaptive isogeometric analysis with application to the micro-mechanical modeling of trabecular bone. *Computer Methods in Applied Mechanics and Engineering*, 284, 138–64, 2015.

Vernon, G.: Bezier extraction. 2020. https://splinefea.com/2020/01/11/bezier-extraction/.

Wang, W., Zhang, Y., Scott, M.A. Hughes, T.J.R.: Converting an unstructured quadrilateral mesh to a standard T-spline surface. *Computational Mechanics*, 48, 477–498, 2011.

Wang, W., Zhang, Y., Liu, L., Hughes, T.J.R.: Trivariate solid T-spline construction from boundary triangulations with arbitrary genus topology. *Computer-Aided Design*, 45(2), 351–360, 2013. https://doi.org/10.1016/j.cad.2012.10.018.

Wang, W., Zhang, Y., Du, X., Zhao, G.: An efficient data structure for calculation of unstructured T-spline surfaces. *Visual Computing for Industry, Biomedicine, and Art*, 2(1), 1–9, 2019.

Wei, X., Zhang, Y., Liu, L., Hughes, T.J.R.: Truncated T-splines: Fundamentals and methods. *Computer Methods in Applied Mechanics and Engineering*, 316, 349–372, 2017. https://doi.org/10.1016/j.cma.2016.07.020.

Wei, X., Zhang, Y.J., Toshniwal, D., Speleers, H., Li, X., Manni, C., Evans, J.A., Hughes, T.J.R.: Blended B-spline construction on unstructured quadrilateral and hexahedral meshes with optimal convergence rates in isogeometric analysis. *Computer Methods in Applied Mechanics and Engineering*, 341, 609–639, 2018. https://doi.org/10.1016/j.cma.2018.07.013.

Welch, J.A., Kópházi, J., Owens, A.R., Eaton, M.D.: Isogeometric analysis for the multigroup neutron diffusion equation with applications in reactor physics. *Annals of Nuclear Energy*, 101, 465–480, 2017.

Wexler, A.: Computation of electromagnetic fields. *IEEE Transactions on Microwave Theory and Techniques*, 17(8), 416–439, 1969.

Wexler, A., Richards, D.J.: Finite element techniques for the solution of Poisson's equation. In *IEEE GMTT International Microwave Symposium Digest*, pp. 132–133, May 16–19, 1971, Washington.

Wilson, E.L., Farhoomand, I., Bathe, K.J.: Nonlinear dynamic analysis of complex structures. *Earthquake Engineering & Structural Dynamics*, 1(3), 241–252, 1972.

Wu, H., Ye, W., Jiang, W.: Isogeometric finite element analysis of interior acoustic problems. *Applied Acoustics*, 100, 63–73, 2015.

Xiao, W., Liu, Y., Li, R., Wang, W., Zheng, J., Zhao, G.: Reconsideration of T-spline data models and their exchanges using STEP. *Computer-Aided Design*, 79, 36–47, 2016. https://www.sciencedirect.com/science/article/pii/S0010448516300616.

Yerry, M.A., Shephard, M.S.: Automatic mesh generation for three-dimensional solids. *Computers and Structures*, 20(1–3), 31–39, 1985.

Yousefi, S.A., Barikbin, Z., Dehghan, M.: Ritz-Galerkin method with Bernstein polynomial basis for finding the product solution form of heat equation with non-classic boundary conditions. *International Journal of Numerical Methods for Heat & Fluid Flow*, 22(1), 39–48, 2012.

Yu, T., Lai, W., Bui, T.Q.: Three-dimensional elastoplastic solids simulation by an effective IGA based on Bézier extraction of NURBS. *International Journal of Mechanics and Materials in Design*, 15, 175–197, 2019.

Zamani, N.G.: A least squares finite element method applied to b-splines. *Journal of the Franklin Institute*, 311(3), 195–208, 1981a. https://doi.org/10.1016/0016-0032(81)90014-4.

Zamani, N.G.: Least squares finite element approximation of Navier's equation. *Journal of the Franklin Institute*, 311(5), 311–321, 1981b. https://doi.org/10.1016/0016-0032(81)90473-7.

Zhang, G., Alberdi, R., Khandelwal, K.: Analysis of three-dimensional curved beams using isogeometric approach. *Engineering Structures*, 117, 560–574, 2016. http://dx.doi.org/10.1016/j.engstruct.2016.03.035.

Zhang, W.L., Mo, R., Wan, N., Zhang, Q.: Isogeometric analysis of heat transfer in fluids. *Applied Mechanics and Materials*, 105, 2174–2178, 2012.

Zhang, Y., Bazilevs, Y., Goswami, S., Bajaj, C.L., Hughes, T.J.R.: Patient-specific vascular NURBS modeling for isogeometric analysis of blood flow. *Computer methods in applied Mechanics and Engineering*, 196(29–30), 2943–2959, 2007. DOI: 10.1016/j.cma.2007.02.009.

Zhang, Y., Wang, W., Hughes, T.J.R.: Conformal solid T-spline construction from boundary T-spline representations. *Computational Mechanics*, 51, 1051–1059, 2013. https://doi.org/10.1007/s00466-012-0787-6.

Zienkiewicz, O.C.: *The Finite Element Method in Structural and Continuum Mechanics*. McGraw-Hill, Columbus, 1967.

Zienkiewicz, O.C.: *The Finite Element Method in Engineering Science*. McGraw-Hill, Columbus, 1971.

Zienkiewicz, O.C., Taylor R., Too J.: Reduced integration technique in general analysis of plates and shells. *International Journal for Numerical Methods in Engineering*, 3(2), 275–290, 1971.

Zuo, B.Q., Huang, Z.D., Wang, Y.W., Wu, Z.J.: Isogeometric analysis for CSG models. *Computer Methods in Applied Mechanics and Engineering*, 285, 102–124, 2015.

Zou, Z., Hughes, T.J.R., Scott, M.A., Miao, D., Sauer R.A.: Efficient and robust quadratures for isogeometric analysis: Reduced Gauss and Gauss–Greville rules. *Computer Methods in Applied Mechanics and Engineering*, 392, 114722, 2022. https://doi.org/10.1016/j.cma.2022.114722.

Index

O

P

Q

R

S

Made in the USA
Las Vegas, NV
07 January 2025

761b612b-75cf-4ac0-90a3-695dbf3bd97fR01